T0245136

KOMATIITE

Komatiites erupted billions of years ago as pulsating streams of white-hot lava. Their unusual chemical compositions and exceptionally high formation temperatures produced highly fluid lava that crystallized as spectacular layered flows. Investigation of the extreme conditions in which komatiites formed provides important evidence about the thermal and chemical evolution of the planet, and the nature of the Precambrian mantle.

This monograph, written by three experts with long experience in the field, presents a complete account of the characteristics of komatiites including their volcanic structures, textures, mineralogy and chemical compositions. Models for their formation and eruption are evaluated, including discussion of the controversial issue of whether komatiites originated from anhydrous or hydrous magmas. A chapter is also devoted to the valuable nickel and copper ore deposits found in some komatiites.

Komatiite is a key reference for researchers and advanced students interested in petrology, Archean geology, economic geology, and broader questions about the evolution of the Earth's crust and mantle.

NICHOLAS ARNDT was awarded a Ph.D. from the University of Toronto in Canada in 1975. Following academic positions in the United States, Canada and Germany and at the Université de Rennes, France, he moved in 1999 to the Université de Grenoble where he is Professor of Petrology-Geochemistry. Professor Arndt's research interests include petrology and geochemistry of mafic and ultramafic rocks, magmatic ore deposits, and the early-Earth environment. He is Director of the European Science Foundation 'Archean Environment: the Habitat of Early Life' program and an ISI highly cited scientist.

STEVE BARNES is a research scientist at the Commonwealth Scientific and Industrial Research Organisation in Perth, Australia, and MICHAEL LESHER is Professor of Economic Geology and Research Chair in the Mineral Exploration Research Centre and Department of Earth Sciences at Laurentian University in Sudbury, Ontario. Both work on the petrogenesis and physical volcanology of komatiites and associated magmatic Ni–Cu–PGE deposits.

KOMATIITE

NICHOLAS ARNDT

Université Joseph Fourier, Grenoble

With contributions from

C. MICHAEL LESHER

Laurentian University, Ontario

and

STEPHEN J. BARNES

*Division of Exploration and Mining,
CSIRO, Australia*

CAMBRIDGE
UNIVERSITY PRESS

University Printing House, Cambridge CB2 8BS, United Kingdom

One Liberty Plaza, 20th Floor, New York, NY 10006, USA

477 Williamstown Road, Port Melbourne, VIC 3207, Australia

314-321, 3rd Floor, Plot 3, Splendor Forum, Jasola District Centre, New Delhi - 110025, India

79 Anson Road, #06-04/06, Singapore 079906

Cambridge University Press is part of the University of Cambridge.

It furthers the University's mission by disseminating knowledge in the pursuit of education, learning and research at the highest international levels of excellence.

www.cambridge.org
Information on this title: www.cambridge.org/9781108460514

© N. T. Arndt, C. M. Lesher and S. J. Barnes 2008

This publication is in copyright. Subject to statutory exception and to the provisions of relevant collective licensing agreements, no reproduction of any part may take place without the written permission of Cambridge University Press.

First published 2008
First paperback edition 2018

A catalogue record for this publication is available from the British Library

ISBN 978-0-521-87474-8 Hardback
ISBN 978-1-108-46051-4 Paperback

Cambridge University Press has no responsibility for the persistence or accuracy of URLs for external or third-party internet websites referred to in this publication, and does not guarantee that any content on such websites is, or will remain, accurate or appropriate.

To Catherine, Greg and Ben

Contents

Colour plates are located between pages 242 and 243

Preface

The time traveller was becoming despondent. Here she was, 2.7 billion years back in time, and she had spent most of the day flying over a flat, featureless plane. True, some hours back, she had passed over a chain of fiery stratovolcanoes strung out along the margin of the continent, no doubt marking the trace of an active subduction zone. She had marvelled at the spectacle of the thick mat of floating pumice blanketing a vast area of blue-green sea. There was a long mountain range with lofty snow-covered peaks. Imposing, yes, but duller than modern mountains because of their total lack of vegetation. But now all she saw was this dreary plain, feebly lit by a pale yellow sun glinting in the reddish sky. She'd seen it all before, this enormous flow of basaltic lava, so like the flat, monotonous uppermost flows of the Deccan plateau.

Then, far to the north, she saw a small plume of rising vapour. The plume was not large but, as she got closer, she saw in the dull twilight that it was illuminated from below, as if by a powerful flickering searchlight. She reached the plume, circled slowly around it, and realized that she was witnessing, for the very first time, an eruption of komatiite.

The plume shot forth several hundred metres into the air, streaming steadily skyward, like the jet from an enormous fire hose. There was none of the pulsing agitation of the lava fountains she had seen on Hawaii and Iceland. It had to be driven by the density difference between the magma and a thick underlying column of crust rather than being propelled by the expanding gases that drive modern lava fountains. Most remarkable was its colour – not the reddish-orange of basaltic lava but an intense bluish white. Her optical pyrometer gave a reading of 1620 °C! The fountain fed a river of lava that flowed to the north as far as she could see, moving rapidly, jostling and splashing, at close to 40 kilometres per hour. Occasionally, a thin, dark elastic crust formed on the surface, temporarily masking the brilliant whiteness of the lava. But quickly the crust would be destroyed by the turbulence of the flow.

Several kilometres downstream, the lava was noticeably more viscous and the flowage was less agitated and probably laminar. The colour of the lava had shifted to a brilliant yellowish white due to a slight change in temperature. The increase in viscosity must be due to the combined effect of phenocryst growth with a small decrease in temperature, thought the time traveller. She noticed a solid crust was forming at the surface and the lava continued to flow beneath, gradually inflating the roof of the flow so that, a few kilometres farther downstream, the thickness of the flow had doubled.

She started turning over in her mind how she would present this evidence for the eruption of hot, anhydrous, mobile komatiite during her invited talk at the upcoming EGU meeting, three weeks after she got back to the 21st century.

Komatiite is a rare rock type. If we ignore for the moment the glaring anomaly of the Tertiary Gorgona flows, all true komatiites so far discovered are Archean in age. Archean rocks cover only about 10% of the surface of the globe and in these regions komatiite constitutes less than 10% of the volcanic successions or 1% of the entire Archean rock assemblage. Overall, komatiite is no more common than rocks like phonolite, mugearite and other unusual alkaline rocks that excite a few specialists but are of little or no interest to most geologists. Yet, as Bob Thompson pointed out some 24 years ago when he reviewed the first komatiite book (Arndt and Nisbet, 1982a; Thompson, 1983), these rocks have always interested the average geologist, and, when described by a skilled journalist, will capture the attention of the general public. The reasons for this interest are not difficult to find. Komatiite contains spinifex, arguably the most beautiful and intriguing of all igneous textures. Komatiites contain valuable ore deposits, which add economic interest. But above all, komatiite is an extreme magma type with highly unusual physical and chemical characteristics. Komatiites must have formed under conditions quite different from those that yield other types of magma and they provide one of very few tools we have for understanding how the Earth operated 2–3 billion years ago.

Komatiites were far hotter than any other magma that ever erupted on the Earth's surface. Lewis and Williams (1973) developed a series of intriguing arguments to support their conclusion that a komatiite flow in Western Australia erupted as a superheated liquid whose temperature was at least 400 °C higher than that of Hawaiian basalt. We have since realized that komatiite magma forms through partial melting deep in the mantle, at depths far greater than those that yield the modern basaltic magmas. Was this always the case? An idea that pervades the current literature is that the Archean mantle was far hotter than the modern mantle; if true, it is perfectly logical that hot magmas like komatiite were more abundant then than now. All this seems eminently reasonable, yet, for the past decade, a team of petrologists has argued that komatiites do not form deep and hot but shallow, wet and tepid. The long series of papers published by Tim Grove, Stephen Parman, Maarten de Wit and Jesse Dann has done much to stimulate interest in komatiites and, although I agree with very few of their conclusions, I readily acknowledge the contribution they have made to komatiite research. The debate between proponents of wet or dry komatiites has now rumbled on for over a decade and has stimulated a large number of research projects that have greatly improved our knowledge of all aspects of the nature and probable origin of these remarkable magmas.

In this volume I have attempted to provide a complete and comprehensive account of komatiite. I discuss all aspects of the rock type – how and where they occur, their field characteristics, their petrology and geochemistry, and with the addition of two chapters by Mike Lesher and Steve Barnes, their volcanological aspects and their ore deposits. The first part of the book, Chapters 1–7, is mainly descriptive. After a brief chapter dealing with terminology, I illustrate the main features of komatiites by way of summaries of the characteristics of six type examples. Then follow chapters on the field characteristics, mineralogy, geochemistry experimental studies and finally, the physical characteristics of komatiite lavas. My coauthors contributed the lithofacies section in Chapter 1, the Kambalda section in Chapter 3, and part or all of the descriptions of the dimensions of komatiites, of thick dunitic units and of volcanoclastic komatiites in Chapter 4.

The second part of the book opens with the chapters by Lesher and Barnes on ore deposits and volcanology. Then comes a chapter in which I address directly the issue of whether komatiites are wet or dry, and finally there are three chapters that relate current ideas about the origin of komatiite. Chapter 12 deals with the composition of komatiite liquids, Chapter 13 focusses on petrogenesis and finally, in Chapter 14, I discuss the tectonic setting in which they erupt.

Komatiites are beautiful rocks, particularly in thin section, and their textures deserve to be celebrated. I have therefore included a large number of photos of komatiite, particularly in the opening chapters. I have included only a few tables of geochemical data because this information is readily available in on-line databases such as GEOROC http://georoc.mpch-mainz.gwdg.de/georoc/. I tried to keep the book brief and readable, and this has meant that I have avoided going too deeply into discussions of the more arcane aspects of the subject, particularly when these aspects were well treated in published papers. The reference list includes over 600 entries and provides the sources of more detailed information.

During the preparation of the book, which started two decades ago, I received substantial support from the funding agencies first in Germany (DFG) then in France (CNRS) and from the Max-Planck-Institut, the Université de Rennes and most recently the Université de Grenoble. I thank my friends and colleagues in each location for the help and encouragement they provided. Special thanks go to Martine LeFloche, who helped with the figures, and to Rob Hill, Tony Fowler, Claude Herzberg, Tony Naldrett, Allan Wilson, Stephane Guillot, Rich Walker, Phil Thurston and Euan Nisbet, who provided comments on selected chapters, and to Mike Bickle and Bob Nesbitt, who had the patience to read through large portions of the book. Mike Lesher and Steve Barnes thank all the students, postdoctoral researchers and colleagues who worked with them and Steve Beresford and Caroline Perring for reading their chapters. They also acknowledge support from NSERC and CSIRO and mining companies. ... Finally, we all thank Susan Francis, Maureen Storey, Annette Cooper and Jeanette Alfoldi of Cambridge University Press and Devi Rajasundaram of Integra for their patience and encouragement during the long and difficult gestation of the book.

Part I

Background information – description of
the field characteristics, mineralogy and
geochemistry of komatiites

1

A brief history of komatiite studies and a discussion of komatiite nomenclature

1.1 The discovery and early investigations of komatiite

A review of reports written by field geologists in the early twentieth century reveals many references to ultramafic volcanic rocks. These rocks were identified as extrusive because of their volcanic textures and structures, and they seem to have been accepted as a normal component of Archean volcanic successions. Then, between 1920 and 1930, as Bowen and other petrologists conducted the experimental studies that led them to question the very existence of ultramafic magma (Bowen, 1928), and as the influence of these petrologists spread, descriptions of ultramafic volcanics virtually disappeared from the geological literature. There were sporadic reports of serpentinite flows in younger terranes, as reviewed by Bailey and McCallien (1953), and argument persisted about the nature and origin of 'alpine peridotites', but references to ultramafic lavas in Archean areas became rare. In a few Canadian papers, the texture we now know as spinifex was described, and the rocks containing these textures were interpreted as volcanic (Berry, 1940; Prest, 1950; Abraham, 1953). Similar reports appeared of ultramafic lavas in various parts of the (then) Rhodesian craton (Keep, 1929; Wiles, 1957). More commonly, however, the ultramafic rocks were interpreted as intrusive, in most cases as the lower cumulate layers of differentiated sills. This is the case for two type areas of komatiite, Barberton in South Africa and Munro Township in Canada. The ultramafic rocks of the Barberton greenstone belt were originally interpreted as the intrusive component of the 'Jamestown Igneous Complex', described as an association of magnesium-rich intrusive rocks subsequently converted to serpentinites and schists (Hall, 1918; Visser, 1956; de Wit *et al.*, 1987). Satterly (1951b) produced a detailed geological map of Munro Township in Ontario. Satterley's maps are remarkably accurate and the distribution of ultramafic units is shown in detail. Even spinifex texture was identified and carefully mapped. However, instead of interpreting it as volcanic, Satterly attributed the texture to metamorphism that had preferentially affected the margins of ultramafic sills.

The changes that were to come were foreshadowed by Naldrett and Mason (1968), who described in detail skeletal olivine crystals in ultramafic lenses from Dundonald Township, not far from Munro. Naldrett, in his Ph. D. thesis (Naldrett, 1964b), had noticed the similarity between the morphology of these crystals and those of olivine in rapidly cooled experimental mafic liquids, and he argued for the existence of ultramafic liquids. In their paper, Naldrett and Mason (1968) concluded that the ultramafic units were high-level sills or perhaps lava flows that had formed from highly magnesian magmas.

The honour for the discovery of komatiites goes to Morris and Richard Viljoen, twin brothers who, in the course of their detailed mapping and subsequent petrographic and chemical studies of ultramafic units in the Barberton greenstone belt of South Africa, concluded that these units were indeed ultramafic lava flows. In a series of papers published as part of South Africa's contribution to the 1969 Upper Mantle Project (Viljoen and Viljoen, 1969b, c), they presented convincing arguments to support their conclusion. A feature that particularly impressed them was the manner in which the ultramafic units thinned sympathetically with the mafic lavas because this suggested that these units were part of the volcanic succession, and not younger intrusions. Other features such as chilled flow tops, hyaloclastite breccias and what they called 'crystalline quench texture' also pointed to a volcanic origin.

Viljoen and Viljoen (1969c) compared the chemical compositions of the Barberton komatiites with those of other types of ultramafic rocks, particularly alpine peridotites and mantle xenoliths and with other types of mafic lava. In this comparison, the komatiites were found to have lower MgO and higher SiO_2 contents than the ultramafic cumulates or restitic rocks, but were distinguished from the other volcanic rocks by high MgO, low SiO_2 contents and lower contents of TiO_2 and alkali elements. Perhaps the most distinctive chemical feature was the $CaO–Al_2O_3$ ratio, which was significantly higher than in the other rocks. These chemical characteristics were shared by mafic volcanics associated with the ultramafic lavas, and since these mafic rocks also contained crystalline quench textures, the Viljoens argued that they were part of the same magma series. It was for rocks of this ultramafic–mafic suite that the term komatiite was introduced.

In the following five years, komatiites were recognized in many other areas, notably in the Abitibi greenstone belt in Canada, the Yilgarn of Australia and the Belingwe belt in Zimbabwe. Following the studies of Naldrett (1964a) and Naldrett and Mason (1968), Pyke (1970) documented the existence of ultramafic lava flows in Langmuir and Blackstock Townships in the Abitibi greenstone belt, and later worked with Naldrett and Eckstrand in the investigation of the now classic outcrops in Munro Township (Pyke *et al.*, 1973). Eckstrand (1972) discussed the relationship between certain Canadian nickel sulfide deposits and ultramafic volcanic rocks.

McCall and Leishman (1971) published a description of ultramafic and mafic volcanic rocks in Western Australia and, at about the same time, Nesbitt (1971) introduced the term spinifex texture into the international geological literature and discussed how the texture might have formed. Subsequent papers by Williams (1972), Hallberg and Williams (1972) and Barnes *et al.* (1973) demonstrated that ultramafic volcanic rocks were a ubiquitous component of volcanic sequences throughout the Yilgarn belt of Western Australia. Interest in these rocks was heightened by the recognition that ultramafic volcanic rocks hosted the large and rich Ni-sulfide deposits that had been discovered in the mid-1960s in the Kambalda area. These discoveries led to an exploration boom that lasted for a decade and provided the impetus for numerous studies of komatiites and their ore deposits (Ross and Hopkins, 1975; Naldrett and Turner, 1977).

The study of komatiites of the Belingwe greenstone belt by Bickle *et al.* (1975, 1977), and the more detailed report by Nisbet *et al.* (1977) documented the general petrological features of the ultramafic and mafic volcanic rocks in this region, and also provided information about their tectonic environment that had been lacking in other areas. Brooks and Hart (1974) described several examples from Canada, revised the definition of the rock type and, perhaps most importantly, suggested that the rocks might have formed through melting under hydrous conditions. Nesbitt and Sun (1976) and Sun and Nesbitt (1978) undertook the first global investigations of the major- and trace-element geochemistry of komatiites and developed many important ideas concerning the origin of these rocks.

Subsequent reports have confirmed the presence of komatiite in almost all Archean terrains, as shown in Figure 1.1 and Table 1.1. The Proterozoic appears to be devoid of true komatiites, with the possible exceptions of the dunitic hosts of ore deposits at Thompson, Canada (Burnham *et al.*, 2003) and some very poorly preserved occurrences in Guyana (Gruau *et al.*, 1985; Capdevila *et al.*, 1999). Mg-rich basalts, some with spinifex textures, are known in parts of the Circum-Superior Belt in Canada (Baragar and Scoates, 1981), but ultramafic lavas appear to be absent. The only definite example of Phanerozoic komatiite ('dinkum komatiite', to use the expression of Lina Echeverría, who undertook the definitive study of these rocks) are on Gorgona Island off the coast of Colombia. Gansser *et al.* (1979) first reported the existence of these ultramafic lavas, and Echeverría

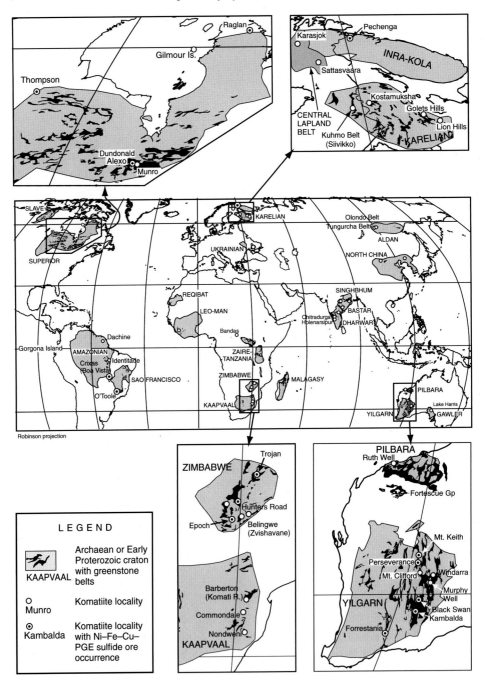

Fig. 1.1. World map showing Archean and older Proterozoic cratons, greenstone belts, principal komatiite localities and major associated nickel deposits. Localities and principal references are listed in Table 1.1.

Table 1.1. *Localities of worldwide komatiite occurrences shown in Figure 1.1*

Country	Locality	Belt	Lat	Long	Reference
Australia	Digger Rocks	Forrestania–Lake Johnston–Ravensthorpe	−32.72	119.81	Perring et al., 1995, 1996.
Australia	Warrawoona Group	Pilbara Craton	117.40	−22.66	Arndt et al., 2001.
Australia	Lake Harris greenstone belt	Lake Harris	−30.78	134.50	Hoatson et al., 2005.
Australia	Kambalda	Norseman–Wiluna	−31.20	121.67	This volume, Chapters 2 and 10
Australia	Mt Keith	Norseman–Wiluna Agnew–Wiluna	−27.24	120.54	This volume, Chapter 10
Australia	Perseverance	Norseman–Wiluna Agnew–Wiluna	−27.81	120.70	Barnes et al., 1988; Barnes 2006.
Australia	Windarra	Norseman–Wiluna East	−28.61	122.24	Marston 1984.
Australia	Black Swan	Norseman–Wiluna Kalgoorlie Terrain	−30.39	121.64	Hill et al., 2004; Dowling et al., 2004; Barnes et al., 2004.
Australia	Scotia	Norseman–Wiluna Kalgoorlie Terrain	−30.21	121.27	Page and Schmulian, 1981.
Australia	Ruth Well	Pilbara West (Karratha–Whim Creek) East Pilbara	−20.85	116.83	Nisbet and Chinner, 1981.
Australia	Fortescue basalts				
Brazil	Boa Vista	Crixas greenstone belt	−14.16	−50.11	Costa et al., 1997; Arndt et al., 1989.
Brazil	O'Toole deposit	Quadrilatero Ferrífero (Rio das Velhas)	−20.90	−46.72	Brenner et al., 1990.
Canada	Alexo	Abitibi greenstone belt	48.65	−80.81	This volume, Chapter 2, Chapter 10.
Canada	Dundonald	Abitibi greenstone belt	48.63	−80.83	Houlé et al., 2007a; This volume, Chapter 3.
Canada	Munro Township	Abitibi greenstone belt	48.59	−80.20	This volume, Chapter 2.
Canada	Dumont Sill	Abitibi greenstone belt	48.87	−79.32	Duke 1986a,b.
Canada	Spinifex Ridge	Abitibi greenstone belt	48.23	−78.22	Gélinas et al., 1977, Dostal and Mueller, 2001.

Table 1.1. *(cont.)*

Country	Locality	Belt	Lat	Long	Reference
Canada	Donaldson	Cape Smith belt, Quebec.	61.67	−73.28	St Onge and Lucas, 1993b; Lesher 2007.
Canada	Katinniq	Cape Smith belt, Quebec.	61.68	−73.67	This volume, Chapter 10.
Canada	Gilmour Island	East Hudson Bay	59.75	−80.05	This volume, Chapter 2.
Canada	Lac Guyer	Superior Province, Quebec	53.42	−77.58	Stamatelopoulou-Seymour and Francis, 1980.
Canada	Thompson	Thompson	55.73	−97.80	This volume, Chapter 10.
Canada	Steep Rock	Wabigoon belt, Ontario	48.80	−91.60	Barnes and Often, 1990; Schaefer and Morton, 1991.
Colombia	Gorgona Island	Gorgona Island	2.95	−78.20	This volume, Chapter 2.
Finland	Sattasvaara	Kittila (Sattasvaara)	67.61	26.68	Hanski et al., 2001.
Finland	Kuhmo Belt (Siivikko)	Kuhmo	64.23	29.09	Sorjonen-Ward et al., 1997; Gruau et al., 1992.
Guyana	Dachine	Gyr-Inini	3.35	−53.30	Capdevila et al., 1999.
Norway	Karasjok	Karasjok	69.10	25.00	Barnes and Often 1990.
Russia	Pechenga	Inari-Kola Craton	69.54	30.53	Hanski 1992; Hanski and Smolkin, 1989; Skufin and Theart 2005.
Russia	Kostamuksha	Kostamuksha-Gimola	64.65	30.75	Puchtel et al., 1997.
Russia	Olondo Belt (Red Hill Massif)	Olondo greenstone belt	56.94	119.33	Puchtel and Zhuralev, 1989.
Russia	Tungurcha Belt	Tungurcha	57.20	120.90	Puchtel et al., 1991.
Russia	Vetreny Belt (Golets Hills)	Vetreny belt	63.78	35.70	Puchtel et al., 2001a,b.
Russia	Lion Hills	Vetreny belt	63.32	37.18	Puchtel and Humayun, 2001.
South Africa	Komati formation	Barberton greenstone belt	−25.95	30.79	This volume, Chapter 2.
South Africa	Commondale	Commondale	−26.90	30.60	Wilson et al., 2003; Wilson, 2003.
Zimbabwe	Zvishavane	Belingwe greenstone belt	−20.33	30.03	Nisbet et al., 1987; Bickle et al., 1993.
Zimbabwe	Trojan	Bindura-Hamva belt	−17.33	31.27	Prendergast, 2003.
Zimbabwe	Epoch	Filabusi	−20.43	29.28	Prendergast, 2003.
Zimbabwe	Hunters Road	Midlands-Chegutu	−19.15	29.78	Prendergast, 2003.

(1980), Cameron and Nisbet (1982) and Aitken and Echeverría (1984) subsequently documented their petrology and chemical compositions, and convincingly demonstrated their origin as eruptive ultramafic magmas.

Viljoen and Viljoen (1982) provide a more detailed account of the early years of komatiite investigation in one of their contributions to the first komatiite book (Arndt and Nisbet, 1982a).

1.2 More recent studies

The cosy world of komatiite studies was upset when, in 1996, Grove, de Wit, Parman and Dann proposed that komatiites were hydrous magmas. In their view, as expressed in a long series of papers (de Wit and Stern, 1980; de Wit *et al.*, 1983, 1987; Grove *et al.*, 1994, 1996; Parman *et al.*, 1996; Grove *et al.*, 1997; Parman *et al.*, 1997; Grove *et al.*, 1999; Parman *et al.*, 2001, 2003; Grove and Parman, 2004; Parman *et al.*, 2004), the komatiites from the type area in the Barberton greenstone belt did not erupt as anhydrous lava flows but were intruded as hydrous magmas at depths of about 6 km in the basaltic lava pile. The idea that komatiites were hydrous was not new, having been mentioned by Brooks and Hart (1974) and Allègre (1982), but Grove and coinvestigators took the model much further and developed a series of petrological and geochemical arguments that persuaded a significant proportion of the geological community of the validity of the hypothesis. The history of the hydrous komatiite hypothesis and its current status are discussed in detail in Chapter 11.

Other important developments include the application of the inductively-coupled plasma mass-spectrometer (ICPMS) method to the analysis of komatiites, which for the first time provided reliable data for a complete range of trace elements (Xie *et al.*, 1993; Xie and Kerrich, 1994). *In-situ* analysis using the ion microprobe and laser-source ICPMS provided important constraints on mineral compositions (Parman *et al.*, 2003), and application of the same techniques on melt inclusions yielded the concentrations of elements that normally were mobile during alteration (McDonough and Ireland, 1993). These developments are discussed in Chapters 4 and 5. New types of isotopic analysis were performed on komatiites, including the Lu–Hf (Gruau *et al.*, 1990a) and Re–Os (Walker *et al.*, 1988) systems and for the first time He isotopes were used (Richard *et al.*, 1996) (Chapter 6). Experimental studies, both at high pressures and under conditions of rapid cooling at low pressure provided valuable constraints on the conditions of formation and crystallization of komatiite (Chapter 7).

Another major development was presented in a remarkable series of papers published by Steve Sparks, Herbert Huppert and coworkers (Huppert and

Sparks, 1985a, b; Turner *et al.*, 1986; Huppert and Sparks, 1989). For the first time fluid dynamic principles were applied in a rigorous manner to the study of komatiites, and this work thoroughly changed our understanding of how these ultramafic lavas erupted and crystallized. The notion that komatiite flowed turbulently and was capable of thermally eroding its substrate stemmed from this work. When combined with detailed studies of field relations of komatiites in Western Australia by R. E. T. Hill and coworkers (Hill *et al.*, 1987, 1995; Perring *et al.*, 1995; Barnes *et al.*, 1999; Hill, 2001) and with C. M. Lesher's studies of Kambalda komatiites (Lesher, 1983, 1989; Lesher *et al.*, 1984), these ideas formed the basis of our understanding of the volcanology of komatiite (Chapter 9). And when combined with work on the Kambalda deposits themselves, they lie at the core of models for the formation of komatiite-hosted ore deposits, as described in Chapter 10.

Komatiites in two particular regions were subjected to rigorous investigation because of their excellent state of preservation. The first of these regions is Gorgona Island, whose extremely fresh komatiites had been known since their initial discovery by Gansser (Gansser 1950; Gansser *et al.* 1979). Subsequent geochemical studies have thoroughly documented the petrological and geochemical characteristics of these rocks (Echeverría and Aitken, 1986; Walker *et al.*, 1991; Kerr *et al.*, 1996; Sylvester *et al.*, 2000; Kerr, 2005). The second region is Zvishavane in the Archean Belingwe belt of Zimbabwe. Euan Nisbet and coworkers mapped the rocks in outcrop and then organized the drilling of the SASKMAR hole, which recovered the core that has provided the best samples available of Archean komatiite. The characteristics of these rocks were first described in the Geology paper of Nisbet *et al.* (1987) and then in many other papers that present the results of numerous petrological and geochemical investigations of these remarkable komatiites (e.g. Bickle *et al.*, 1993; Bickle and Nisbet, 1993; McDonough and Ireland, 1993; Nisbet *et al.*, 1993; Renner *et al.*, 1994; Silva *et al.*, 1997; Danyushevsky *et al.*, 2002).

1.3 Nomenclature

The Penrose conference on komatiites was held in April 1980 in a rather seedy motel in the outskirts of Val d'Or, a gold-mining town in the Quebec portion of the Abitibi greenstone belt. There we reviewed what had been learnt in the decade that followed the initial discovery of the rock type. Examples from South Africa, Canada, Australia, Zimbabwe and Colombia were described and the models for their formation were discussed, but the most important issue of the conference was, perhaps, a new definition of the rock type. It had come to be realized that the distinctive geochemical feature of the

Barberton komatiites – their high $CaO–Al_2O_3$ ratios – was not shared by the majority of ultramafic volcanic lavas that had been discovered in other areas. For a while it had been argued that these komatiites, which had lower, near-chondritic $CaO–Al_2O_3$ ratios, deserved another name. Nesbitt (1971) suggested 'Archean greenstone peridotite' (a name that nobody really liked) and in my long forgotten first paper on komatiites (Arndt, 1976b), I proposed 'magnesian magmatic series' (which was received equally badly). At the Penrose conference, Allen Zindler pointed out that the term komatiite deserved the status of a rock type. His suggestion was that komatiite should be defined simply as a 'volcanic rock of ultramafic composition', the petrological equivalent of basalt or andesite. It followed that the basalts associated with komatiites could be considered part of the 'komatiitic' magma series and should be named 'komatiitic basalts'. The terms 'peridotitic komatiite' and 'basaltic komatiite', coined in the original papers by Viljoen and Viljoen (1969a,c), were to be abandoned. These ideas were explained and developed by Arndt and Brooks (1980) and Arndt and Nisbet (1982b).

Komatiite can therefore be defined as a rock whose field relations or textures provide evidence of a volcanic or subvolcanic origin and whose mineral assemblages or major-element contents indicate an ultramafic composition. Features that can be used to indicate a volcanic setting include a fine-grained, chilled upper margin or, more conclusively, the presence of breccia or hyaloclastite. Fine grain size throughout the unit, spinifex textures, amygdales, complete conformity with surrounding units (lack of intrusive contacts) are consistent with, but do not prove, a volcanic origin.

The limit between komatiite and other less magnesian volcanic rocks was set, rather arbitrarily, at 18% MgO, a value that corresponds to a minimum in the relative abundances of komatiitic rocks in many, though not all, greenstone belts. The term komatiitic basalt is applied to volcanic rocks containing less than 18% MgO that can be linked, using petrological, textural or geochemical arguments, to komatiites.

Like any petrological term, the definition of komatiite is not without problems and ambiguities. Implicit to the definition of komatiite is the notion – difficult to prove – that komatiites crystallize from liquids that contained more than about 18% MgO. (Throughout the book major-element abundances are quoted as weight per cent oxide, normalized to 100% on an anhydrous basis, unless stated otherwise). The selection of 18% MgO as the limit between komatiite and komatiitic basalt has been questioned in several papers: some authors would prefer to see it set higher, at about 25% MgO for komatiites from the Barberton belt according to Smith and Erlank (1982) or lower, at 12–15% MgO for komatiites from Gorgona Island, in the view of Echeverría

12 A brief history of komatiite studies

(1980). In view of this disparity, 18% MgO seems a reasonable compromise, no better and no worse than the 52% (or is it 55%) SiO_2 used to distinguish basalt from andesite. Another complication arises from the existence of other volcanic rocks containing more than 18% MgO that formed either through the accumulation of olivine from less magnesian liquids, or crystallized from magmas with geochemical characteristics quite unlike those of most komatiites. An example of the first type is a phenocryst-charged basaltic liquid (a picrite according to some definitions); an example of the second is meime-chite (Vasil'yev and Zolotukhin, 1975; Arndt et al., 1998b), a rare alkaline ultramafic lava with unusual major- and trace-element composition.

To distinguish komatiite from other types of highly magnesian volcanic rocks, it is useful to include spinifex texture in the definition of the rock type. Spinifex, a texture characterized by the presence of large skeletal or dendritic crystals of olivine or pyroxene, is present in many, but not all komatiite flows (Nesbitt, 1971). As originally suggested by Brooks and Hart (1974) and re-emphasized by Kerr and Arndt (2001), a workable definition of komatiite should include the phrase 'the term komatiite [should] be reserved *solely* for lavas with characteristic spinifex-textured olivines, or lavas that can be related directly, using field or petrological criteria, to lavas with this texture'. With the last part of the definition we can make allowance for the manner in which texture varies within komatiitic units. For example, many komatiite flows comprise an upper spinifex-textured layer and a lower olivine-cumulate layer; and other flows grade along strike from layered spinifex-textured portions to massive olivine–phyric units. With the inclusion of the phrase about spinifex texture, the lower olivine-cumulate portions of layered flows or the olivine–phyric units can also be described as komatiite. On the other hand, meimechites, picrites and other rock types that contain no spinifex are excluded.

To distinguish between komatiite and alkalic ultramafic lavas such as meimechite, a combination of mineralogy and chemical compositions can be employed. Meimechites and associated alkali picrites commonly contain clinopyroxene phenocrysts, which are rare in komatiites, and phlogopite, amphibole and perovskite in their matrices, minerals that are absent in normal komatiites. Le Maitre et al. (1989) used a value of 2% $Na_2O + K_2O$ to distinguish komatiite and meimechite from picrites and a value of 1% TiO_2 to distinguish between komatiite and meimechite. Le Bas (2000) then refined the geochemical definitions using the diagram reproduced as Figure 1.2. This purely geochemical classification was criticized by Kerr and Arndt (2001), who recommended that the presence of spinifex texture be mentioned in the defini-tion of komatiite. A complementary and in many ways more useful approach was developed by Hanski (1979) who used the molar proportions of Al_2O_3 and

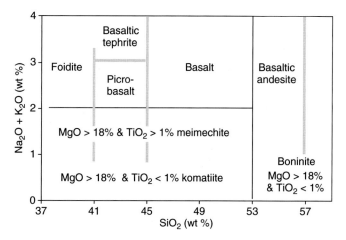

Fig. 1.2. Le Bas's (2000) diagram illustrating the chemical differences between various types of mafic and ultramafic rocks.

TiO_2 to distinguish the various types of highly mafic and ultramafic lava types. This approach is discussed more fully in Chapter 5.

Application of the term komatiite to shallow-level, hypabyssal intrusions has been criticized. One alternative would be to invent a special term, the equivalent of dolerite or diabase, for hypabyssal komatiite, but a more pragmatic solution is simply to speak of komatiitic sills or dykes, rather as we speak of basaltic sills or dykes. The cumulate rocks on the lower parts of thick, differentiated ultra-mafic units pose a special problem. These rocks have relatively coarse textures and commonly are devoid of fine-grained minerals, skeletal morphologies and other textures or structures diagnostic of a volcanic origin. Indeed, in many cases there is no consensus on whether the host unit is extrusive or intrusive. On a lithological basis, these rocks should be described using terms normally applied to intrusive rocks such as dunite, wehrlite and gabbro, or alternatively as olivine cumulate or olivine–clinopyroxene cumulate. Textural variations can be accommodated using normal cumulate terms such as adcumulate or orthocumulate. In practice, therefore, the presence of coarse textures in thick komatiitic units does not constitute an insurmountable problem: intrusive rock names can be used, complemented by an explanation in which it is stated clearly that the rocks in question form part of a volcanic or subvolcanic unit.

1.4 Komatiite lithofacies

The particular conditions that reign during the crystallization of rapidly flowing, highly fluid and extremely hot magmas produce a range of unusual

Table 1.2. *Komatiite lithofacies*

Lithofacies	Characteristics
Non-cumulate	
Volcaniclastic	Predominantly hydroclastic (e.g. flow-top breccias), occasionally epiclastic, rarely pyroclastic (cg heterolithic or fg tuffaceous)
Aphyric	Glassy to aphanitic, rarely vesicular, with fg quench olivine ± pyroxene ± sulfides, < 1% phenocrysts
Olivine phyric	5–25% fg–mg polyhedral or hopper olivine within an aphanitic to glassy matrix comprising fg quench olivine ± pyroxene ± sulfides; rare vesicules
Spinifex	
Random olivine spinifex	10–25% fg–mg randomly-oriented olivine blades in an aphanitic to glassy matrix with fg quench olivine ± pyroxene ± sulfides, occasionally vesicular
Platy olivine spinifex	cg dendritic networks of platy olivine, commonly oriented near parallel to flow margins, interstitial glass with fg quench olivine ± pyroxene ± sulfides
Acicular pyroxene spinifex	cg parallel acicular pyroxene, interstitial glass with fg quench olivine ± pyroxene ± sulfides
Cumulate	
Crescumulate	25–75% mg branching olivine ± interstitial chromite in an aphanitic to glassy matrix with fg quench olivine ± pyroxene ± sulfides, often vesicular
Orthocumulate	50–75% fg–mg polyhedral olivine ± interstitial chromite, 25–75% interstitial glass with fg quench olivine ± pyroxene ± sulfides
Mesocumulate	75–90% mg polyhedral olivine ± interstitial chromite, 10–25% interstitial glass with fg quench olivine ± pyroxene ± sulfides
Adcumulate	> 90% cg polyhedral olivine ± trace chromite, < 10% interstitial glass with fg quench olivine ± pyroxene ± sulfides

fg = fine-grained (< 1 mm), mg = medium-grained (1–2 mm), cg = coarse-grained (> 2 mm)

structures and textures in their crystallization products. In Table 1.2, we employ terms analogous to those used by sedimentologists to define a range of komatiite lithofacies. These then form the basis for a facies analysis of komatiites, developed by Lesher *et al.* (1984), Lesher (1989), Hill and Gole (1990), Hill *et al.* (1995), and Barnes (2006) and discussed in the chapter on volcanology (Chapter 9).

Komatiites exhibit many of the *lithofacies* recognized in basaltic flow fields, including volcaniclastic, massive aphyric, and massive porphyritic facies. However, they also exhibit many lithofacies not found in basalt flow fields, including spinifex-textured, crescumulate (branching), orthocumulate, mesocumulate, and adcumulate facies (Table 1.2). These are the prominent lithological subdivisions of komatiite flows that are readily observable in the field and which are described in detail in the following chapters.

1.5 Conclusion

The definition of komatiite provoked considerable comment at the time it was proposed in the early 1980s, and is still not accepted by all authors (the terms 'peridotitic komatiite' and 'basaltic komatiite' still crop up in various publications, alas). Yet in general the Penrose conference definition of komatiite and of the komatiitic magma series seems to have been accepted and appears to have become the standard way of naming the series of volcanic ultramafic rocks.

Komatiites are remarkable for the large array of distinctive and often spectacular rock types. The diversity is much greater than that seen in typical basalts, and this follows from the wide melting range and high fluidity of komatiite lavas. The broad spectrum of rock types reflects the sensitivity of crystallization mechanisms to subtle variations in cooling rate, nucleation rate and other consequences of volcanic setting.

2
Brief descriptions of six classic komatiite occurrences

2.1 Introduction

In this section I briefly describe six regions that illustrate the main characteristics of komatiites: (1) ~3.5 Ga komatiites from the Barberton greenstone belt in South Africa; (2) 2.7 Ga komatiites from the Abitibi Belt in Canada; (3) ore-bearing 2.7 Ga komatiites from Kambalda in Western Australia; (4) 1.9 Ga komatiitic basalts from the Gilmour Island in Canada; (5) Fe–Ti-rich komatiites from Karasjok, Scandinavia, and (6) Cretaceous (90 Ma) komatiites from Gorgona Island, Colombia. A summary of the main characteristics of komatiites in each of these areas is given in Table 2.2 (page 23). A case could have been made to include various other komatiite localities. The komatiites from Kambalda, for example, were included only after persuasive lobbying from my coauthors, who contributed this section to illustrate the characteristics of a type of thick, highly magnesian komatiitic unit that is abundant throughout the eastern part of the Yilgarn Craton and in many other greenstone belts. In addition, these komatiites contain the type examples of komatiite-hosted Ni-sulfide deposits. Euan Nisbet strongly advocated the inclusion of the 2.7 Ga komatiites from the Belingwe belt in Zimbabwe, which are renowned for their unusually good state of preservation and have been the subject of several excellent studies by his group. However, these rocks are similar in most important respects to the komatiites from the Abitibi belt. Their well-preserved textures and primary mineralogy and their chemical compositions are described in detail in Chapters 3–6.

2.2 Komatiites in the Barberton greenstone belt

These komatiites are described here not just because Barberton is the type locality of komatiite, but mainly because the Barberton belt contains the oldest

Table 2.1. *Representative analyses of komatiites from selected regions*

	Barberton	Abitibi	Norseman–Wiluna	Belingwe	Gilmour	Gorgona
SiO_2	47.7	45.9	47.9	48.7	48.3	46.2
TiO_2	0.36	0.35	0.33	0.36	0.61	0.50
Al_2O_3	4.15	6.49	5.75	7.12	10.3	11.3
Cr_2O_3	0.38	0.38	0.35	0.43		0.31
FeO(tot)	11.2	10.8	10.7	11.0	11.9	10.7
MnO	0.19	0.19	0.18	0.18	0.19	0.19
MgO	28.5	29.2	28.2	24.4	17.6	19.6
NiO	0.20	0.18		0.25		0.14
CaO	6.95	6.25	6.10	6.75	9.75	9.90
Na_2O	0.26	0.22	0.31	0.66	1.16	0.97
K_2O	0.05	0.08	0.16	0.15	0.17	0.06
P_2O_5	0.03	0.03	0.03	0.04	0.00	0.04

Data are normalized to 100% on an anhydrous basis
Principal sources of data: Viljoen and Viljoen (1969b, c 1982); Sun and Nesbitt (1978); Lahaye *et al.* (1995); Lecuyer *et al.* (1994); Smith *et al.* (1980, 1982); Byerly (1999); Chavagnac (2004); Walker *et al.* (1988); Arndt (1986b); Arndt *et al.* (1977, 1997b); Barnes (1983); Gangopadhyay *et al.* (2005); Sproule *et al.* (2002); Bickle *et al.* (1993); Echeverría and Aitken (1986); Aitken and Echeverría (1984); Arndt (unpublished analyses)

well-preserved komatiites and the principal examples of the Al-depleted variety. The Barberton greenstone belt, a small, cusp-shaped succession of volcanic and sedimentary rocks, is invaded on all sides by granitoid plutons (Figure 2.1). In this chapter the emphasis is on the Komati Formation, the lower stratigraphic unit in the main part of the belt. Komatiites make up about 25% of this unit, the remainder being komatiitic and tholeiitic basalts and felsic tuffs. The age of the supracrustal rocks is well constrained by the U–Pb zircon dating of Kröner *et al.* (1996) and Kamo (1994). The oldest komatiites in the region are at least 3550 Ma old, those of the main komatiite-bearing units are 3445 ± 3 to 3472 ± 5 Ma old, and those of the younger Mendon Formation range from 3334 ± 3 to 3258 ± 3 Ma (Byerly *et al.*, 1996).

The Barberton komatiites are exposed in rugged terrain, a series of high ridges cut by deeply incised streams, as illustrated in Figure 2.2. This topography has allowed geologists such as Morris and Richard Viljoen, Maartin de Wit, Don Lowe, Gary Byerly and more recently Jesse Dann to produce excellent geological maps showing the stratigraphy of the volcanic successions and the general organization of different types of flows. At a more detailed scale, however, outcrop is sparse and of relatively poor quality, and only along

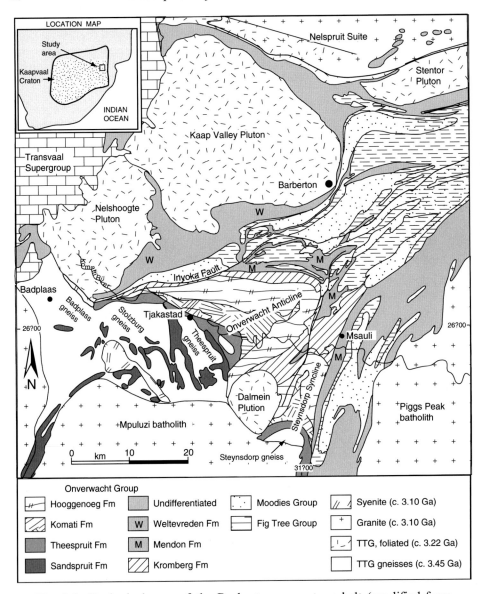

Fig. 2.1. Geological map of the Barberton greenstone belt (modified from Hofmann (2005)).

narrow sections in streambeds can the structures and textures of individual flows be mapped out in detail. It is probably for this reason that Viljoen and Viljoen (1969a, b) were unable to work out the internal structure of spinifex-textured flows during their pioneering study of komatiites: this advance had to await Pyke *et al.*'s (1973) study of the better exposed flows in Munro Township. The more recent paper by Viljoen *et al.* (1983) and fine-scale

Fig. 2.2. Photograph of part of the Komati Formation in the Barberton greenstone belt. The resistant ridges lined with trees consist of thick flows of massive komatiite. These are separated by poorly outcropping series of thin spinifex-textured komatiite and komatiitic and tholeiitic basalts.

mapping of Dann (2000, 2001) have shown that in most respects the Barberton komatiites are very like those in other Archean greenstone belts.

Figure 2.3 compares the characteristics of a thin (~1 m), differentiated spinifex-textured flow from the 'Spinifex Stream' locality in the Barberton belt with those of komatiite flows from several other areas. The textures are illustrated in field photographs in Figure 2.4 and in photomicrographs in Figure 2.5. The thin flow tops, the transition from random to platy spinifex and the abrupt transition to the lower olivine cumulate are all typical of this type of flow. The morphologies and volcanic structures in these flows are described and beautifully illustrated in the papers by Dann (2000, 2001). Particularly interesting are features diagnostic in internal inflation of the lava flows such as tumuli and uplifted portions of the crust.

Spinifex-textured flows make up only a small fraction of the komatiite units in the Komati Formation, between 7 and 10% of the areas mapped by Dann (2000). More abundant are thick sheet flows of massive komatiite, which comprise up to 40% of the volcanic succession. These flows are tens of metres thick, reaching a maximum of 55 m, and are composed of relatively featureless, fine-grained olivine porphyry or cumulate. Descriptions of the mineralogical and textural variations of the massive flows are sparse in later publications, but

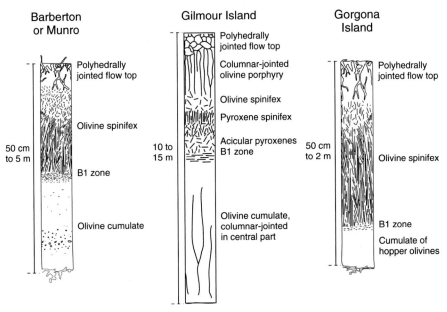

Fig. 2.3. Sketches of spinifex-textured komatiite flows from Archean Barberton or Munro greenstone belts, from Proterozoic Gilmour Island, and from Cretaceous Gorgona Island.

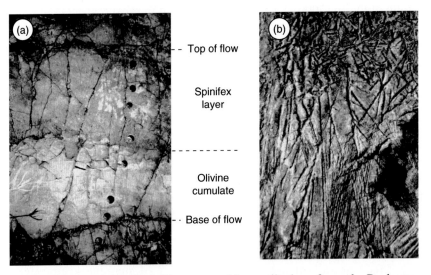

Fig. 2.4. Photographs of spinifex-textured komatiite lava from the Barberton greenstone belt: (a) a single differentiated komatiite flow with spinifex upper portion and olivine cumulate lower portion. The drill holes that deface the flow are about 3 cm in diameter; (b) gradation from random spinifex in the upper part of the photograph to oriented platy spinifex in the lower part.

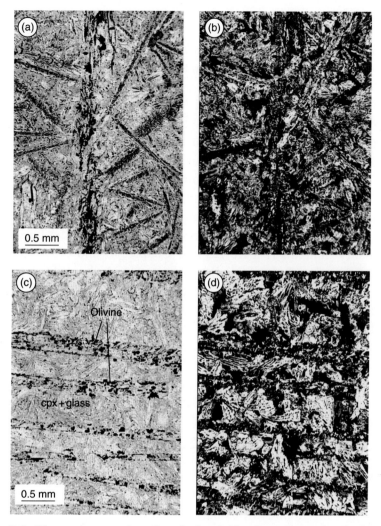

Fig. 2.5. Photomicrographs of typical spinifex lava from the Barberton greenstone belt: (a) random olivine spinifex in plane-polarized light; (b) the same in cross-polarized light; (c) platy olivine spinifex; (d) the same in cross-polarized light. In both samples relict olivine blades are visible but they have been totally replaced by secondary serpentine and opaque minerals. Likewise, a fine-grained assemblage consisting mainly of amphibole and chlorite has replaced the original clinopyroxene and glass in the matrix.

according to Viljoen and Viljoen (1969a), they contain from 40 to 85% olivine (now largely altered to serpentine) and in some cases internal differentiation has produced an olivine-enriched lower portion and an olivine-poor upper portion. Vesicular komatiite, a type of flow in which an upper zone contains

20–25% vesicles, constitutes about 0.1% of the volcanic succession (Dann, 2001).

Komatiitic basalts, identified by their pyroxene spinifex texture, olivine–pyroxene-rich mineralogy and less magnesian chemical compositions, alternate with komatiites and tholeiitic basalts in the upper part of the Komati Formation. A thick (up to 10 m) unit of silicified lapilli tuffs and cherts (the Middle Marker) caps the Formation.

The metamorphic grade throughout the lower stratigraphic units is greenschist facies, reaching amphibolite facies near the margins of granitic intrusions. The rocks are in addition variably affected by hydrothermal alteration and by recent weathering. As a result, the magmatic mineralogy is poorly preserved, even by the loose standards of Archean petrology (see Thompson (1983)). In many samples the entire magmatic assemblage has been converted to secondary minerals (Figure 2.5 and Chapter 4) and only relatively rare examples contain relicts of olivine, pyroxene and chromite. On the other hand, textures commonly are well preserved and the proportions of magmatic minerals can be estimated reliably from the pseudomorphs of primary minerals.

The chemical compositions of Barberton komatiites vary like those of komatiites in other regions (Table 2.1). MgO contents are high in the lower olivine cumulate parts of differentiated flows, reaching a maximum of about 38%, and lower, from about 20 to 30%, in the spinifex zones. (Throughout the book I quote major element abundances as weight per cent oxide, normalized to 100% on an anhydrous basis, unless stated otherwise.) The MgO contents of random spinifex lava, which provides an indication of the composition of the komatiite liquid, range from 26 to 30%. In variation diagrams (Figure 2.6), the concentrations of SiO_2, Al_2O_3, TiO_2 and FeO, when normalized to 100% on an anhydrous basis, plot on or about olivine control lines. In contrast, fluid-mobile elements either fall on linear trends that intercept the MgO axis at values lower than that of olivine (e.g. CaO), or scatter widely (e.g. Na_2O).

One of the criteria used by Viljoen and Viljoen (1969a, b) in their initial definition of komatiite was the $CaO–Al_2O_3$ ratio, which is lower in the Barberton examples than in most other types of ultramafic rocks. Jahn *et al.* (1982), Nesbitt and Sun (1976), Smith *et al.* (1980) and Smith and Erlank (1982) subsequently showed that depletion of Al_2O_3 relative to CaO or to TiO_2 is characteristic of many but not all Barberton komatiites and is unusual in other regions. Figure 2.7 shows that komatiites of the Al-depleted or Barberton type have low Al_2O_3/TiO_2 and relatively high $(Gd/Yb)_N$ (where N indicates that the trace-element ratio is normalized to Hofmann's (1988)

Table 2.2. *Principal features of komatiites from six type areas*

	Barberton Kaapvaal Craton, South Africa	Abitibi Superior Province, Canada	Kambalda Yilgarn craton, Australia	Gilmour Island Circum-Superior Belt, Canada	Karasjok Finland	Gorgona Island Caribbean Plateau, Colombia
Area Location						
Age (Ga)	3.5	2.7	2.7	1.9	2.4	0.09
Dominant komatiite type	Al-depleted	Al-undepleted	Al-undepleted	Al-undepleted	Fe–Ti-rich	Al-enriched
Maximum MgO in liquid	~30%	~30%	~30%	~18–25%	~24%?	~18%–22%
Important features	Type area of komatiite; type locality of Barberton-type Al-depleted komatiite; continuous outcrop provides good stratigraphic control, but komatiites are poorly exposed at a local scale and are relatively poorly preserved due to lower amphibolite facies metamorphism and weathering	Type area of Munro-type Al-undepleted komatiite; contains some excellent outcrops of well-preserved komatiites notably at Munro Township and Alexo; komatiites are associated with thick layered flows and sills, and in places with felsic volcanics; tectonically complicated greenstone belt containing a wide range of lithologies	Type area of komatiite-hosted Ni sulfide deposits and thick olivine-rich flows and intrusions; poor outcrop but good exposure in mines and drill core; amphibolite facies metamorphism; some komatiites and many komatiitic basalts contain xenocrystic zircons and geochemical evidence of contamination with old continental crust	One of very few Proterozoic occurrences of spinifex-textured komatiitic basalts; excellent exposure of fully differentiated komatiite flows and associated tholeiitic flows and intrusions; important source of information about komatiite eruptive processes	Type area of Fe–Ti-rich komatiites; relatively poor outcrop, variable preservation of magmatic textures and structures; abundant fragmental komatiites, probably hyaloclastites	The only example of post-Precambrian komatiite; flows are locally well exposed on beaches and both textures and mineralogy are extremely well preserved; cumulate zones are thin or totally absent; they are composed entirely of skeletal hopper olivines; detailed tectonic setting is not well understood

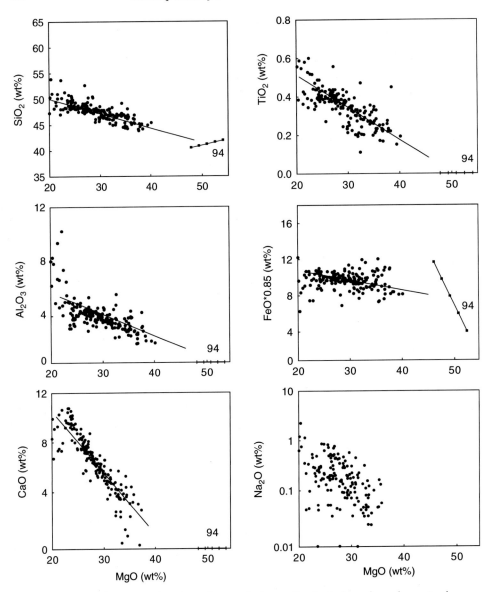

Fig. 2.6. Diagrams showing the variation of selected major elements in Barberton komatiites. Iron is plotted as total FeO*0.85, assuming that 15% of the iron is present as Fe^{3+}. Also shown are the compositions of olivine. The number '94' identifies olivine with the composition Fo_{94}. The best-fit lines through the top four diagrams (MgO vs SiO_2, Al_2O_3, TiO_2 and FeO(tot) * 0.85) correspond to olivine control lines projecting to the composition Fo_{92}. In the plot of MgO vs. CaO the trend does not intersect a magmatic olivine composition and in the MgO vs. Na_2O diagram the data scatter widely. Data are from the GEOROC database http://georoc.mpch-mainz.gwdg.de/georoc/, selected to include only those from the Spinifex Stream section. Principal sources are Nesbitt and Sun (1976), Viljoen *et al.* (1983), Smith *et al.* (1985), Cloete (1999) and Parman *et al.* (2004).

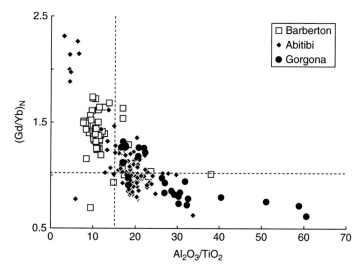

Fig. 2.7. Plot of Al_2O_3/TiO_2 vs $(Gd/Yb)_N$ for the three main types of komatiite. The subscript N indicates that the data are normalized using the primitive mantle values of Hofmann (1988). Data are from the GEOROC database.

estimate of trace-element concentrations in the primitive mantle) compared with most komatiites from other regions. The high Gd/Yb of Barberton komatiites is a measure of the fractionation of the heavy rare-earth elements (HREE): as seen in Figure 2.8, the mantle-normalized abundances of the heavier elements such as Yb are generally lower than those of the lighter elements. This is in marked contrast to komatiites from other regions, in which the more incompatible elements such as the light rare-earth elements (LREE) are relatively depleted.

Figure 2.9 shows that at a given MgO content, the Zr contents of Barberton komatiites are higher on average than those in the two other types of komatiite. Zirconium acts as a relatively incompatible element during high-degree mantle melting and its high concentrations in Barberton komatiites indicate that these magmas were probably produced by lower degrees of partial melting than for other Archean komatiites.

The 3.3 Ga Mendon Formation, which is located higher in the Barberton stratigraphic succession (Lowe and Byerly, 1999a,b), contains three types of komatiite: Al-depleted, Al-undepleted and Al-enriched (Byerly, 1999). Included in the last type are remarkable layered flows with olivine–orthopyroxene spinifex upper portions (Kareem and Byerly, 2003). Komatiites with similar mineralogy and textures are also found in the 3.34 Ga Commondale greenstone belt, which is located about 100 km south of the Barberton belt (Wilson, 2003; Wilson *et al.*, 2003). Both olivine and orthopyroxene have very high Mg/(Mg + Fe) – the forsterite contents of olivine

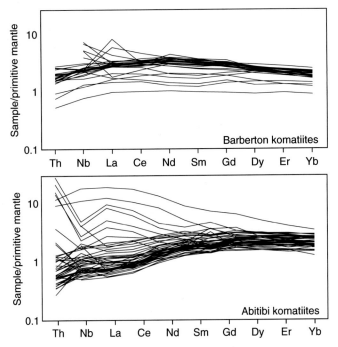

Fig. 2.8. Trace-element concentrations for komatiites from the Barberton and Abitibi belts. The data were first normalized to a constant MgO content of 25% to eliminate the effects of olivine fractionation or accumulation, then normalized using the primitive mantle values of Hofmann (1988). Principal sources of data are: Sun and Nesbitt (1978), Smith *et al.* (1980), Lahaye *et al.* (1995), Lahaye and Arndt (1996), Fan and Kerrich (1997), Cloete (1999), Tomlinson *et al.* (1999), Sproule *et al.* (2002), Parman *et al.* (2004) and Puchtel *et al.* (2004b).

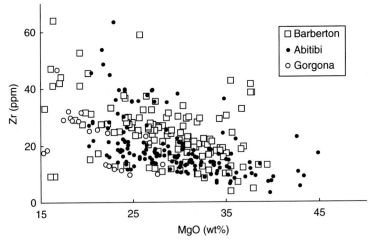

Fig. 2.9. Plot of MgO vs Zr, illustrating the compositions of the three main types of komatiite. Data sources as for Figure 2.8 but also include Aitken and Echeverria (1984), Echeverria and Aitken (1986), Kerr *et al.* (1996) and Arndt *et al.* (1997a).

reach 96.5, amongst the highest recorded in rocks formed by the solidification of magmas. The Commondale komatiites also have extremely low levels of TiO_2 and incompatible trace elements indicating derivation from a highly depleted source.

2.3 Komatiites in the Abitibi greenstone belt

Pyke *et al.*'s (1973) description of the remarkable exposure of komatiite lava flows on what has come to be known as 'Pyke Hill' confirmed Viljoen and Viljoen's (1969a) claim that ultramafic lavas had erupted during the Archean. This outcrop, located in Munro Township in the 2.7 Ga Abitibi greenstone belt (Figure 2.10), is made up of about 40 thin, spinifex-textured komatiite flows or flow units. These flows are exposed in a large glacially polished outcrop (Figure 2.11), and, because of relatively low metamorphic grade (prehnite–pumpellyite facies (Jolly, 1982)) and minimal deformation, the volcanic textures and mineralogy are extremely well preserved. It was during their study of this outcrop that Pyke *et al.* realized that the most prominent textural change in a series of komatiite flows does not mark the

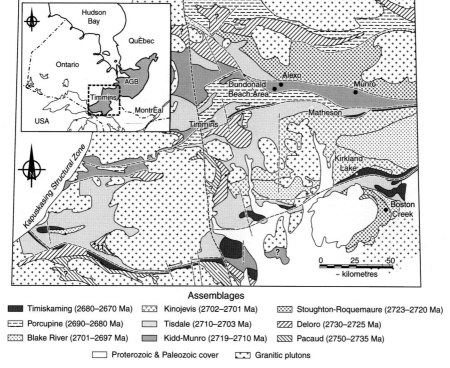

Assemblages

■ Timiskaming (2680–2670 Ma)	Kinojevis (2702–2701 Ma)	Stoughton-Roquemaure (2723–2720 Ma)
Porcupine (2690–2680 Ma)	Tisdale (2710–2703 Ma)	Deloro (2730–2725 Ma)
Blake River (2701–2697 Ma)	Kidd-Munro (2719–2710 Ma)	Pacaud (2750–2735 Ma)
	Proterozoic & Paleozoic cover	Granitic plutons

Fig. 2.10. Map of the Ontario portion of the Abitibi belt showing the geology and some important komatiite localities.

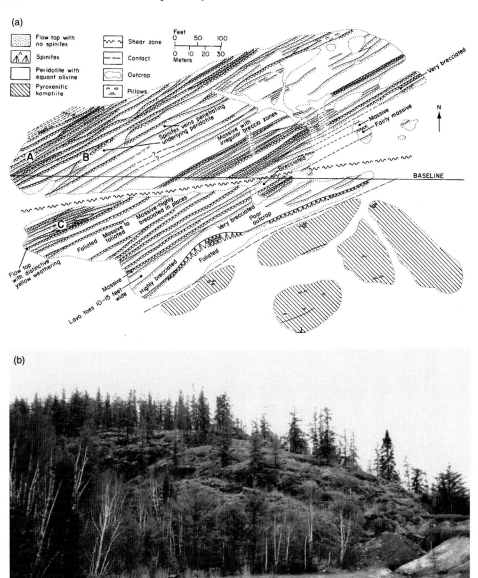

Fig. 2.11. (a) Map of Pyke Hill in Munro Township, from Pyke *et al.* (1973). Some 40 komatiite flows, now tilted so that they are exposed in cross section, are displayed in this superb outcrop. Those in the northwest corner of the outcrop are spinifex-textured but most are massive sheet flows. (b) Photograph of Pyke Hill viewed from the north.

boundary between flow units but is an internal contact between the upper spinifex layer and the lower cumulate layer (Figure 2.12). This discovery has formed the basis of our current interpretation of layering in spinifex-textured komatiite flows.

Other important localities in the Abitibi belt include the Alexo area, where a komatiite and a komatiitic basalt flow have been studied in detail by Barnes (1983), Arndt (1986a) and Lahaye and Arndt (1996); 'spinifex ridge' on the Quebec side of the border, where a series of komatiite flows and volcanoclastic rocks were studied by Lajoie and Gélinas (1978) and more recently by Mueller (2005); and the Boston Creek unit, an unusual Fe-rich, weakly hydrous komatiite studied by Stone *et al.* (1987) and discussed later in this chapter.

The proportion of komatiite in the volcanic successions of the 2.7 Ga Abitibi belt appears on the whole to be less than in the older, 3.5 Ga Barberton belt – perhaps an indication of a secular decline in the abundance of komatiite. Even in areas of the Abitibi belt where komatiite is relatively abundant, they constitute no more than about 10% of the volcanic succession, and large regions contain no ultramafic lavas whatsoever.

The thin (2 m) spinifex-textured flow illustrated in Figures 2.3 and 2.12(a) is just part of a spectrum of flow types. As in the Barberton belt, the majority of komatiites in Munro Township were erupted as thick massive sheet flows, some with thin upper spinifex layers but most devoid of spinifex. A characteristic feature of the thinner sheet flows is polyhedral jointing, the structure illustrated in Figure 2.12(b). Most thick komatiitic units consist of massive olivine porphyry or olivine cumulate.

Another variety of komatiite flow is thick, layered mafic–ultramafic flow characterized by strong internal differentiation, as were first recognized in Munro Township. The two parts of these flows had previously been mapped as separate units, the upper part as a basaltic flow and the lower part as an ultramafic sill. Fred's Flow, the type example of a thick differentiated mafic–ultramafic komatiitic flow (Figure 2.13), has an upper breccia, a layer of olivine spinifex that grades down through pyroxene spinifex into homogeneous gabbro, a thin pyroxene cumulate and a much thicker lower olivine cumulate (Arndt, 1977b). The extent of differentiation in these units, from parental liquid containing about 20% MgO, through gabbro containing only 6% MgO, to olivine cumulates containing 45% MgO, was initially difficult to reconcile with an extrusive context. The explanation only became clear with the development of the internal inflation model for the emplacement of komatiites (Chapter 9) and a better understanding of the formation of spinifex texture (Faure *et al.*, 2006).

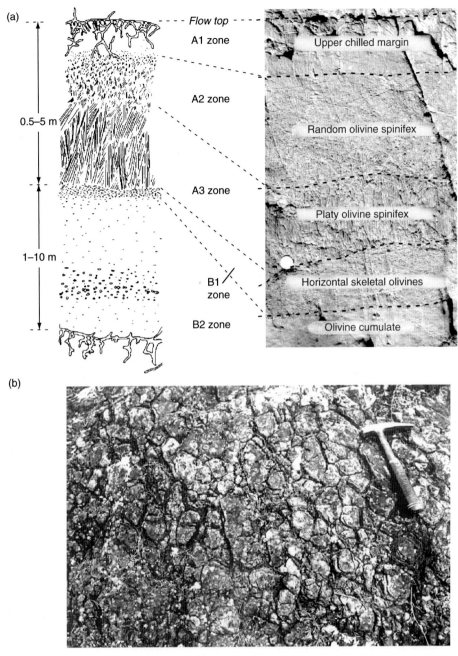

Fig. 2.12. (a) Diagram and photograph of a spinifex-textured differentiated komatiite flow from Pyke Hill, Munro Township. (b) Massive sheet flow of komatiite with polyhedral jointing from the northern part of Munro Township.

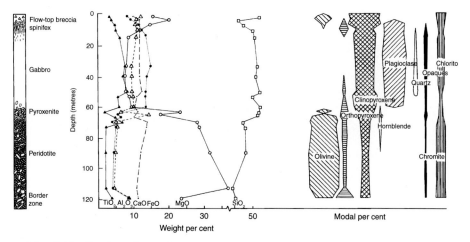

Fig. 2.13. Section through Fred's Flow, a thick, layered komatiitic basalt from Munro Township, showing variations in lithology, major element concentrations and mineralogy (from Arndt (1977b)).

The investigations of the komatiitic flows of the Alexo area were important because they provided indisputable evidence for the existence of extremely magnesian liquids. Figures 2.14(b),(c) show the chilled upper margin of the Alexo komatiite flow, a fine-grained rock containing sparse phenocrysts with the composition Fo_{94}. This olivine grew from a silicate liquid containing about 28% MgO, which is the MgO content of the chilled rock itself.

Pyroxenes are well preserved in some parts of the Alexo komatiite flow and in the underlying layered komatiitic basalt, and analysis of the compositions of composite pigeonite–augite needles in spinifex lavas (Figure 2.15) has helped us to understand the crystallochemistry of these minerals (Fleet and MacRae, 1975; Arndt and Fleet, 1979).

In the 'Dundonald Beach' area, not far from Alexo, a series of thin komatiite units has invaded unconsolidated sediments forming peperites and other features of a near-surface intrusive setting. The komatiites of this area have been described by Houlé *et al.* (2002), Cas *et al.* (1999), Arndt (2004), and Houlé *et al.* (2008a).

Most komatiites in the Abitibi belt have near-chrondritic ratios of Al_2O_3/TiO_2 and $(Gd/Yb)_N$, features that define the Al-undepleted or Munro type of komatiite (Figure 2.7; Sun and Nesbitt (1978), Xie *et al.* (1993)). Exceptions are some relatively rare Al-depleted komatiites such as those from the Newton Township (Cattell and Arndt, 1987; Sproule *et al.*, 2002) and the Boston Creek unit (Stone *et al.*, 1987), which has an unusual Fe-rich composition and a high Al_2O_3/TiO_2. Variation diagrams (Figure 2.16) show the total range of MgO

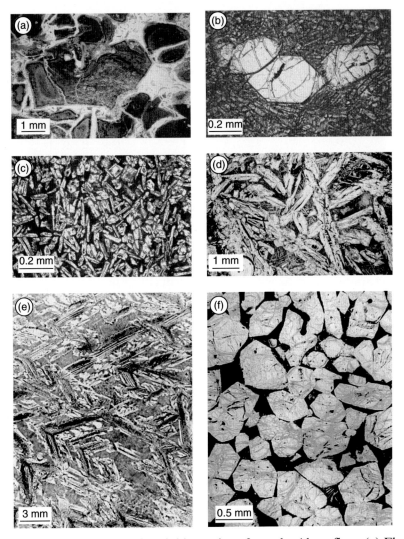

Fig. 2.14. Photomicrographs of thin sections from the Alexo flow. (a) Flow top breccia showing subangular fragments with concave margins typical of hyaloclastite. Most fragments are glassy but some contain small polyhedral olivine phenocrysts (now completely altered to hydrous minerals). (b) Olivine phenocrysts, unaltered except for serpentine in fine fractures, in a matrix of finer skeletal prismatic grains, from the chilled margin. (c) Detail of the fine, skeletal prismatic grains in the matrix of the chilled margin. (d) Random olivine spinifex lava. (e) Chevron spinifex-textured lava, from a section cut parallel to the flow margin. (f) Completely serpentinized polyhedral olivine grains in a dark fine-grained matrix from the lower olivine cumulate portion of the flow.

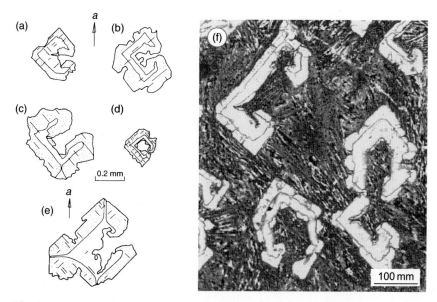

Fig. 2.15. (a)–(e) Sketches of pyroxenes from the Alexo komatiitic basalt flow (from Arndt and Fleet (1979)); (f) photomicrograph of pyroxene spinifex lava. The sample is from the material between large spinifex olivine blades near the base of the spinifex layer of the flow. The large zoned pyroxene grains have cores of pigeonite (dark grey) and margins of augite (pale grey).

contents of komatiites and komatiitic basalts in the belt, from very close to 50% in the adcumulate parts of thick differentiated flows down to 15% MgO. Far less magnesian compositions, as low as 6%, are found in the most evolved parts of thick differentiated flows.

Measurements of the Nd isotopic compositions of Abitibi komatiites show them to have come from a source with long-term depletion of the more incompatible trace elements (Cattell *et al.*, 1984; Dupré *et al.*, 1984; Walker *et al.*, 1988; Lahaye *et al.*, 1995; Lahaye and Arndt, 1996; Dostal and Mueller, 1997; Polat *et al.*, 1999).

2.4 Ore-bearing komatiites at Kambalda

Kambalda holds a special place in the annals of komatiites, being the discovery site of the world's major province of komatiite-hosted ore deposits. It is the type locality for komatiite-associated Ni–Cu–PGE mineralization, as discussed in Chapter 10, and for thick, olivine-rich komatiite flows, as discussed in Chapter 9.

Komatiites are widespread within the extensive greenstone sequence running through south-central Western Australia, commonly referred to in the literature as the Norseman–Wiluna greenstone belt, but now formally termed

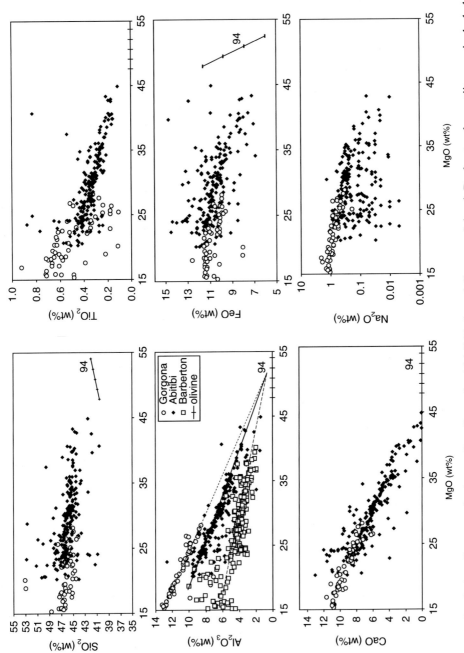

Fig. 2.16. Variation diagrams for komatiites from the Abitibi belt and Gorgona Island. Barberton komatiites are included in the MgO vs. Al$_2$O$_3$ diagram to demonstrate how each series has distinct Al$_2$O$_3$ contents but nonetheless projects to a similar olivine composition. The MgO vs. Na$_2$O diagram illustrates that many Abitibi komatiites have lost most of their sodium. Data from sources listed in the captions of Figures 2.8 and 2.9.

the *Eastern Goldfields Superterrain* (Myers and Swager, 1997). Ni–Cu–PGE deposits occur throughout the region, including the entire length of the 600 km-long by 50 km-wide NNW-trending *Kalgoorlie Terrain* (Figure 2.17). The entire komatiite package has been dated, within the errors of the SHRIMP U–Pb zircon method, between 2600 and 2606 Ma (Nelson, 1997).

The *Kambalda Komatiite Formation* (Gresham and Loftus-Hills, 1981; Cowden and Roberts, 1990) is exposed within the 12 km by 5 km area of the Kambalda Dome, a doubly-plunging anticline (Figure 2.18), and at least two domal structures to the SSE. Exposure is extremely poor, typical of the deeply weathered Yilgarn Craton, but 40 years of exploration and mining in the district have generated an enormous body of detailed data from diamond drill cores and underground workings. The greenstone sequence has undergone at least three phases of deformation, and although folding and thrusting obscure some stratigraphic relationships, most of the sequence appears to be intact.

The ~1 km thick *Kambalda Komatiite Formation* (Figure 2.19) overlies a thick, relatively homogenous sequence of massive, pillowed and locally volcaniclastic tholeiitic basalts (Lunnon basalt), and is overlain by a thick sequence of thick differentiated and thin massive and pillowed komatiitic basalts (Devon Consults and Paringa basalts). The Kambalda Komatiite Formation has been divided into two members: a lower *Silver Lake Member*, which hosts the ores and is composed of thick, olivine-rich flows separated by interflow sediments, and an upper *Tripod Hill Member*, which is barren and consists of thin, massive and spinifex-textured Pyke-Hill-style flow lobes without interflow sediments (Thomson, 1989). The Silver Lake Member is separated in most places from the underlying Lunnon basalt by a regionally continuous unit of sulfidic siliceous sediment.

The Silver Lake Member contains a 'channel-flow facies' made up of thick (up to 100 m) extremely olivine-rich flows and a 'sheet-flow facies' comprising thinner (10–30 m), less olivine-rich flows (Lesher *et al.*, 1984; Lesher, 1989; Cowden and Roberts, 1990). The channel-flow facies represents lava pathways and is characterized by the presence of basal Fe–Ni–Cu sulfide mineralization (Figure 2.17).

The most distinctive features of the ore-bearing lava pathways at Kambalda, and of ore-bearing komatiites worldwide (Lesher, 1989; Lesher and Keays, 2002; Barnes, 2006), are: (a) their great thickness (up to 150 m, although the thickest may have been structurally thickened); (b) the relatively small thicknesses of upper spinifex-textured zones (< 15 m, commonly < 3 m); and (c) the proportionately great thicknesses of the lower olivine cumulate zones.

The komatiites at Kambalda are pervasively metamorphosed to talc–carbonate–chlorite and serpentine–chlorite assemblages and their primary textures are commonly obliterated. In a few areas, however, most notably

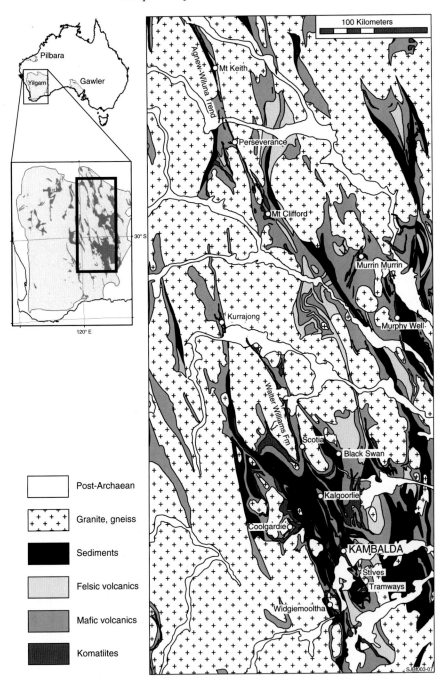

Fig. 2.17. Geology of the Eastern Goldfields Superterrain (formerly the Norseman–Wiluna greenstone belt) showing the distribution of komatiites and the localities mentioned in this volume (modified from Barnes (2006) and Geological Survey of Western Australia mapping).

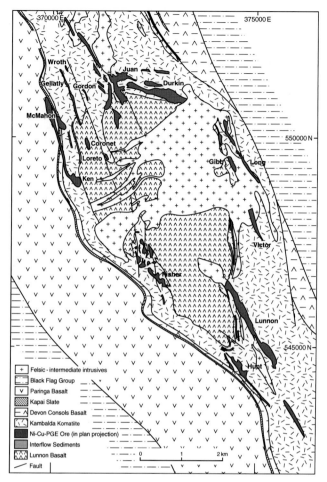

Fig. 2.18. Geological map of Kambalda, modified from Gresham and Loftus Hills (1981).

around the Victor Shoot (Lesher, 1983, 1989; Beresford *et al.*, 2002), volcanic textures and structures are preserved to varying degrees and some igneous minerals have escaped alteration.

Whole-rock MgO contents, Fo contents of olivine (where preserved), and grain sizes of olivine decrease systematically through the upper parts of the cumulate zones (Figure 2.20); whole-rock Cr contents decrease strongly through the lowermost (pyroxene-rich) part of the spinifex-textured zone; and chromite has accumulated in a complementary layer in the uppermost part of the cumulate zones. Vesicular layers occur within many parts of the Silver Lake Member (Beresford *et al.*, 2002). The very-fine-grained interflow sedimentary layers are composed of abundant siliceous albitic chert, carbonaceous siliceous clastic sediments and chlorite-rich sediments, all of

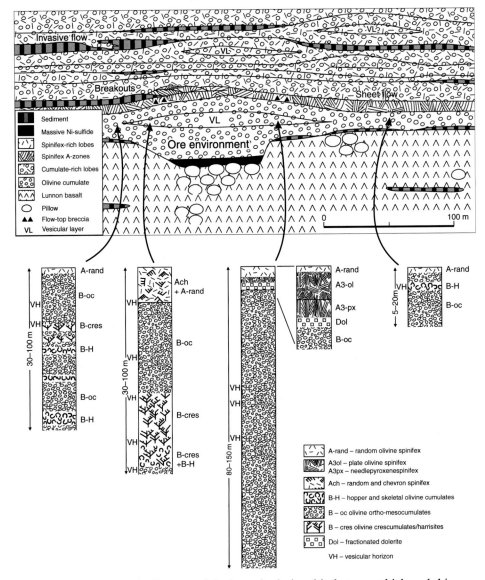

Fig. 2.19. Schematic diagram of the lateral relationship between thick and thin flow units and interflow sediments, and representative textural profiles through various types of flow unit, modified from Beresford *et al.* (2002).

which contain abundant Fe sulfides (Bavinton, 1981). Beresford and Cas (2001) describe rare occurrences of peperite, which formed where komatiite flows invaded unconsolidated wet sediments. McNaughton *et al.* (1988) and Frost and Grove (1989b) describe relatively rare ocellar rocks within the upper flanking sheet flows, which they interpret as melted sediments.

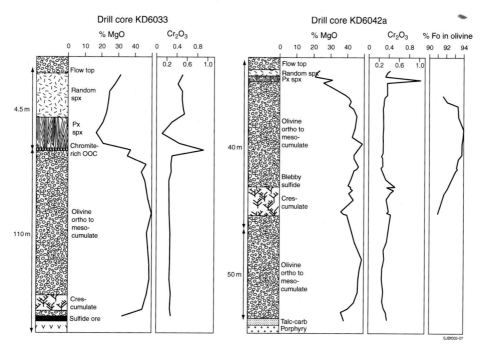

Fig. 2.20. Sections through two flows from the Victor Shoot at Kambalda from Lesher (1989).

Most komatiites at Kambalda belong to the Al-depleted Munro type. Geochemical trends are consistent with crystallization or accumulation of olivine (up to Fo$_{94}$) in a parental magma containing ~31% MgO (Lesher *et al.*, 1981; Redman and Keays, 1985; Lesher and Groves, 1986; Arndt and Jenner, 1986; Lesher, 1989; Lesher and Arndt, 1995). Most of the cumulate komatiites in the lava pathway and sheet-flow facies are depleted in highly incompatible trace elements, consistent with derivation from a depleted mantle source. In contrast, many of the non-cumulate komatiites in the flanking sheet flows are enriched in highly incompatible trace elements and depleted in highly chalcophile elements, consistent with *local* contamination by interflow sediments and sulfide segregation during emplacement (Lesher and Arndt, 1995; Lesher *et al.*, 2001). Komatiitic basalts are moderately to strongly enriched in highly incompatible trace elements and have anomalously low Nb–Th ratios. In addition, they contain inherited 3.4 Ga zircons (Claoué-Long *et al.*, 1988) and exhibit anomalous Sm–Nd and Pb–Pb isotopic signatures, providing unambiguous evidence for contamination by old continental crust (Chauvel *et al.*, 1985; Lesher and Arndt 1995). Finally, they have normal Ni and PGE contents, indicating that they were not saturated in sulfide (Redman and Keays, 1985; Lesher and Arndt, 1995; Lesher *et al.*, 2001).

2.5 Komatiitic basalts on Gilmour Island

The prime interest in these komatiitic basalts comes from their Proterozoic age. Apart from the Gorgona Island komatiites, they are the only known post-Archean example of lavas with true, macroscopic spinifex textures. Equally important, however, is their setting and exposure, which provide major insights into volcanological aspects of komatiitic eruption.

Gilmour is one of the Ottawa Islands, a lonely, uninhabited archipelago some 120 km off the east coast of Hudson Bay (Figure 2.21). These islands are slowly rising out of the sea as the lithosphere rebounds following retreat of the ice caps, and the poorly vegetated, glacially polished surface of the island provides the best exposures of komatiite I have ever seen (Figures 2.21, 2.22; plate I). The rocks are little metamorphosed and relatively well preserved, and

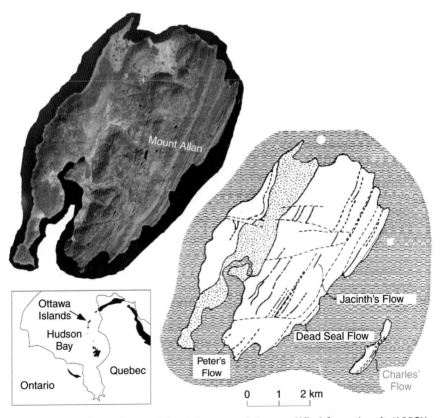

Fig. 2.21. Map of Gilmour Island (bottom right; modified from Arndt (1982)) and an image from Google Earth showing the excellent outcrop of komatiitic flows on the eastern side of the island (top left).

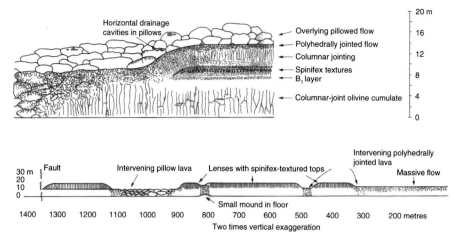

Fig. 2.22. Sketches of a spinifex flow from Gilmour Island from Arndt (1982).

they afford an excellent site to study the textures and morphology of komatiitic lava flows. If only it were easier to get there!

The Ottawa Islands form part of the Circum-Superior Belt (Figure 1.1), a series of volcanic and sedimentary units that girdles the Archean Craton of northern Quebec and extends southwest into central Canada (Baragar and Scoates, 1981). Komatiitic lavas overlie tholeiitic basalts in the Chuckotat Formation but are absent for the largely sedimentary Eskimo Formation (Baragar and Lamontagne, 1980).

The internal structure and textures of a spinifex-textured komatiitic basalt flow from Gilmour Island are illustrated in Figure 2.22. The sketch in the lower part of the diagram shows how thickness and lithology change along a 400 m wide section across the flow. High-standing lenses of spinifex lava, about 14 m thick, alternate with 8–10 m thick segments of massive and pillowed olivine phyric lava. At the rounded extremities of each spinifex lens, the spinifex crystals are oriented perpendicular to the curved upper chilled crust of the flow and this orientation changes progressively to follow the ~40° change in the attitude of the crust. This feature, together with the variations in the thickness and lithology across the section led to the suggestion that the flow had been inflated by internal input of lava (Arndt, 1982). This idea foreshadowed the work of Hon *et al.* (1994) and Self *et al.* (1997) as they developed their important inflation model for the emplacement of basalt flows on Hawaii and along the Columbia River.

Figure 2.3 shows how texture and mineralogy changes from top to base of a Gilmour spinifex flow. At 10–15 m, these flows are thicker overall than the spinifex flows from Barberton and Pyke Hill. There are, to be sure, far thicker units in Archean greenstone belts, as at Kambalda, but in those flows, spinifex forms only a thin layer at the top of a thick section of olivine cumulate. The Gilmour flows are well proportioned and each subunit has substantial thickness. The upper crust of polyhedrally or columnar-jointed olivine porphyry (Figure 2.23) is 1–2 m thick, and it is succeeded by first olivine spinifex, then pyroxene spinifex. The latter grades down into lava composed of fine, randomly oriented, composite pigeonite–augite needles (as in Fred's Flow from Munro Township); then there is an abrupt change to a remarkably thick B_1 layer. In most Gilmour spinifex flows, this layer is more than a metre thick, reaching 150 cm in one flow, far greater than the B_1 layers in Archean komatiites. A strong preferential orientation of the stubby, skeletal-prismatic crystals of the B_1 layer degrades downward through the layer and disappears as the skeletal crystals are displaced by the normal subequant polyhedral crystals of the olivine cumulate.

In parallel with the large variation in mineralogy are changes in chemical composition. In the olivine phyric flow tops, the MgO content is around 15–18%; downwards through the spinifex layers and into the acicular pyroxene rock, MgO decreases from 15 to about 8%; and in the olivine cumulate it rebounds to a maximum of 27%. Olivine in spinifex lavas is totally altered but throughout most of the olivine cumulate it is moderately well preserved and here forsterite contents vary from 91 to 89 mol%. The highest value is that of olivine in equilibrium with liquid with a composition similar to that of the uppermost olivine spinifex lava, and this composition probably represents that of the erupted liquid. In the lower part of the olivine cumulate in several flows, clusters of olivine grains – probable xenocrysts – have more magnesian compositions with forsterite contents up to 92 mol%. This olivine is in equilibrium with a liquid with about 25% MgO. Again with the exception of the Gorgona lavas and the dunites from Thompson (Burnham *et al.*, 2003), this represents the sole evidence for the existence of post-Archean komatiite liquids.

In other respects the Gilmour komatiites are unremarkable. They belong to the Al-undepleted Munro type. Their isotopic compositions were interpreted to indicate derivation of magma from a depleted mantle source, followed by a small amount of contamination with rocks from the continental crust (Arndt *et al.*, 1984).

2.6 Karasjok-type Fe–Ti-rich komatiites and picrites in Fennoscandia

These rocks are included as a type example because of their compositions, which are very different from those of komatiites in the other areas, and

Fig. 2.23. Photomicrographs of thin sections of Gilmour komatiitic basalts.
(a) Olivine porphyritic lava from the polyhedrally jointed upper border zone.
Polyhedral olivine phenocrysts with skeletal overgrowths lie in a fine-grained
clinopyroxene-glass matrix. (b) Fine random olivine spinifex texture from the
upper part of a flow. Note the cluster of skeletal equant grains in the upper
right. (c) Fine random olivine spinifex texture comprising fine olivine plates in
a clinopyroxene-glass matrix. (d) Acicular clinopyroxene rock from below the
olivine spinifex layer of a differentiated flow. (e) The B_1 zone of a differentiated
flow. Knobby skeletal olivine grains in a matrix of relatively coarse skeletal
clinopyroxene grains and a glassy groundmass. (f) Olivine cumulate. Clusters
of coarse olivine grains in a matrix of skeletal clinopyroxene, chromite and
glass.

because of the wide range of rock types with which they are associated. In a broad belt that extends from northern Norway through Finland and into Russia is found a diverse suite of high-magnesium rocks, all of which have high FeO and TiO_2 contents. Variations in their minor- and trace-element contents led Hanski *et al.* (2001) to classify some as komatiite and others as picrite. A Paleoproterozoic age for the sequence is indicated by a Sm–Nd analysis of magmatic pyroxenes and associated whole rocks, which yielded an average age of 2056 ± 25 Ma.

The Fe–Ti-rich komatiites are similar in many ways to the Boston Creek unit in the Abitibi belt (Stone *et al.*, 1987) and the picritic members of the suite are comparable to picrites and meimechites in many Phanerozoic continental flood basalt provinces (Cox, 1988; Arndt *et al.*, 1998b; Thompson *et al.*, 2001). The komatiites have also been compared with porphyritic komatiites and komatiitic volcaniclastic rocks from Dachine in French Guiana (Capdevila *et al.*, 1999) and Steep Rock–Lumby Lake in Canada (Tomlinson *et al.*, 1999), a distribution wide enough, in the view of Barley *et al.* (2000), to justify the use of a new term. These authors coined the term Karasjok-type komatiites after the locality in Norway where they were first described by Barnes and Often (1990).

Spinifex textures are not common in these rocks (Räsänen, 1996) and it might be asked why they should be classified as komatiite. The answer comes in part from the volcanic textures displayed by samples in most outcrops. A peculiarity of these rocks, particularly as they occur in the Sattasvaara area in Finland, is the abundance of fragmental textures. As described by Saverikko (1985, 1987), Barnes and Often (1990) and Hanski *et al.* (2001), many of the units consist of volcanic breccias charged with unsorted, rounded to angular 1–10 cm fragments. These coarsely fragmental rocks form metre-thick bands that alternate with layers of fine-grained laminar tuffs, as well with differentiated mafic–ultramafic sheet flows, pillow lavas and pillow breccias. The magnesian volcanic rocks are associated with cherts, calcareous or argillaceous sediments, indicating eruption in a marine setting. An additional argument is the spinifex texture in the upper part of the Boston Creek unit. (This feature was not enough, however, to convince certain authors (Stone *et al.*, 1987, 2003; Crocket *et al.*, 2005; Stone *et al.*, 2005), who refer to these rocks as ferropicrites – as is so often the case when naming igneous rocks, it is relatively easy to come up with a general definition of a rock type, but very difficult to define its limits.)

Greenschist to amphibolite-facies metamorphism has resulted in the replacement of most magmatic minerals by secondary phases. Volcanic textures are not well preserved but where present they indicate that most of the rocks were

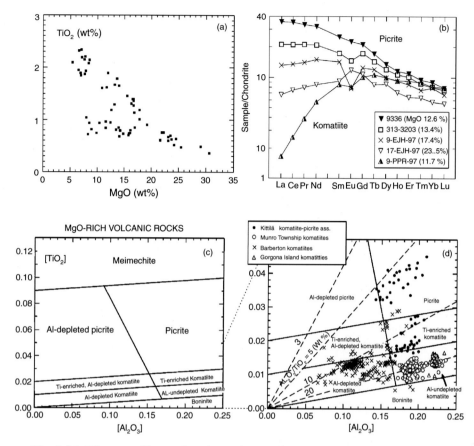

Fig. 2.24. Diagrams illustrating the major- and trace-element compositions of Fe–Ti-rich komatiites from Finnish Lapland. The box in the upper right plot gives sample numbers and MgO contents in weight per cent. (From Hanski *et al.* (2001).) [TiO$_2$] and [Al$_2$O$_3$] are mole proportions, see text for an explanation.

olivine porphyries. Cumulus textures and the preservation of poikilitic clinopyroxene in same samples are taken as evidence that these samples crystallized in the lower parts of layered flows.

The wide range of compositions of the komatiite–picrite suite is illustrated in Figure 2.24. MgO contents range from 5 to 30%, TiO$_2$ contents from 0.5 to 2.5% and FeO(tot) contents from 10 to 14%. In the komatiites, rare-earth element (REE) patterns are hump-shaped, characterized by pronounced depletion of the LREE and more moderate depletion of the HREE, with a maximum at Gd–Dy. The picrites, in contrast, are markedly enriched in LREE, with levels reaching 40 times chondrite.

To illustrate the covariation of Al$_2$O$_3$ and TiO$_2$ contents in the rock suites, Hanski (1992) developed the [Al$_2$O$_3$] vs [TiO$_2$] diagram (Figure 2.24) in which [Al$_2$O$_3$] and [TiO$_2$] are the mole proportions of these components projected from the olivine composition. The proportions are calculated using [Al$_2$O$_3$] = Al$_2$O$_3$/($\frac{2}{3}$ – MgO–FeO) and [TiO$_2$] = TiO$_2$/($\frac{2}{3}$ – MgO–FeO) and all iron is taken as Fe^{2+}. A major advantage of the diagram is that because the compositions are projected from olivine, it eliminates the effects of fractionation or accumulation of this mineral during crystallization of the lava and allows direct comparison of magmas with different MgO contents.

Figure 2.24(a) shows Hanski's (1992) fields for boninite, komatiite and picrite, and Figure 2.24(b) shows the compositions of samples from many of the type areas of mafic–ultramafic lavas. The diagram clearly illustrates the difference between Al-depleted Barberton-type komatiites and Al-undepleted Munro types. It also distinguishes these Archean komatiites, which have low [TiO$_2$], from the Paleoproterozoic komatiites and picrites, which have higher [TiO$_2$].

The Sm–Nd analyses of Hanski *et al.* (2001) yielded positive ε_{Nd} values around +4 for both the depleted komatiites and the enriched picrites, which indicated that the two rock suites came from a source with depleted isotopic composition. Detailed petrogenetic modelling undertaken by Hanski *et al.* (2001) showed that the large spread of trace-element ratios in the two types of lava could not be explained by melting of a single uniform source. To account for the variation they called on 'complex depletion and enrichment processes' that acted shortly before melting augmented by fractionation during dynamic melting.

Stone *et al.* (1987, 1997, 2003), in their detailed descriptions of the Fe-rich komatiite from the Boston Creek locality in the Abitibi belt, Canada, discuss the occurrence and the significance of amphibole in cumulates from the lower part of this unit. They note that this mineral occurs not only in the matrix but also as inclusions in olivine crystals and they infer that amphibole formed relatively early in the crystallization sequence. They then calculate that the original magma contained 1–2% water – in other words, that it is an example of true hydrous komatiite magma. This issue is discussed in more detail in Chapter 11.

2.7 Komatiites and picrites from Gorgona Island

Were it not for Gorgona Island, the komatiite story would be simpler. We could have said that highly ultramafic magmas erupted only in the Archean, when mantle temperatures were far higher than nowadays; that the eruption of komatiites was infrequent during the Proterozoic, their place having been

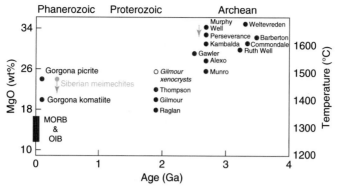

Fig. 2.25. Diagram showing the maximum MgO contents inferred for selected parental komatiite liquids as a function of their age. The procedure used to obtain the MgO values is given in Chapter 12. Also shown are eruption temperatures, calculated using the relation $T(^\circ C) = 1000 + 20 \times MgO$, assuming the magmas were essentially anhydrous. This applies to the komatiites, but not to the meimechite (shown as a grey circle), which probably contained a small amount of H_2O and perhaps CO_2 and erupted at a lower temperature. The Murphy Well sample may also have been hydrous and its temperature may have been lower, as indicated by the arrow.

taken by a few komatiitic basalts; and that they were absent in the Phanerozoic, when mantle temperatures had declined to the extent that only basalts and rare picrites erupted. But in diagrams like Figure 2.25, in which the maximum estimated MgO content and temperature of erupted magmas are plotted against their age, Gorgona stands out conspicuously. In Figure 2.25, magmas produced by high-degree melting of an essentially anhydrous source are shown as black symbols, and meimechites are shown in grey. In this diagram, the Gorgona komatiites and picrites are seen to have MgO contents and eruption temperatures that approach those of Archean lavas and exceed those of any other magma erupted in the past one billion years.

Gorgona is a small tropical island just 30 km off the coast of Colombia, within easy reach of small, high-speed boats from the village of Guapi on the mainland. Palm-fringed and flanked by warm beaches, Gorgona was once a snake-infested prison colony; more recently it has been converted into a snake-infested national park and holiday resort. What attracts geologists to the island are the komatiite lava flows that are found on some of the beaches of the eastern and western coasts, and the remarkable picritic hyaloclastites at the southern tip of the island (Fig. 2.26). The interior of the island is composed mainly of intrusive mafic and ultramafic rocks (Révillon *et al.*, 2000).

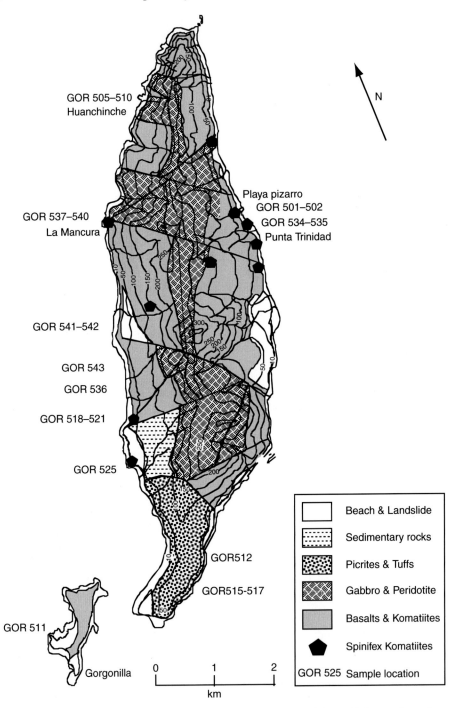

GOR 505–510
Huanchinche

N

Playa pizarro
GOR 501–502
GOR 534–535
Punta Trinidad

GOR 537–540
La Mancura

GOR 541–542

GOR 543

GOR 536

GOR 518–521

GOR 525

GOR512

GOR515-517

GOR 511

Gorgonilla

☐	Beach & Landslide
⬚	Sedimentary rocks
▨	Picrites & Tuffs
▨	Gabbro & Peridotite
▨	Basalts & Komatiites
⬟	Spinifex Komatiites
GOR 525	Sample location

0   1   2
km

Fig. 2.26. Geological map of Gorgona Island (modified from Echeverría and Aitken (1980)).

The komatiites have fine to coarse olivine spinifex textures that are very similar to those of Archean komatiites (Echeverría, 1980, 1982; Aitken and Echeverría, 1984; Kerr, 2005). Highly magnesian olivine-phyric volcaniclastic rocks are restricted to a fault block at the southern tip of the island (Echeverría and Aitken, 1986). These rocks contain the most forsterite-rich olivine of all the Gorgona rocks, and are inferred to have crystallized from the most magnesian liquids, but because they lack spinifex texture they are referred to as picrites and not komatiites. (I realize that this usage is inconsistent with that adopted for the Karasjok Fe-rich komatiites but the term picrite has always been used for the Gorgona rock and it would be confusing to change it). Basalts with both depleted and enriched geochemical characteristics alternate with series of komatiite flows.

Figure 2.3 shows a section through a typical spinifex komatiite from Gorgona. These flows are relatively thin (50 cm to 3 m) and characterized by disproportionately thick spinifex layers and thin, or absent, olivine cumulate layers. The cumulate layers do not contain solid polyhedral crystals as in the lower parts of Precambrian komatiite flows but are made up of beautiful, delicate, highly skeletal hopper olivine grains (Fig. 2.27 and Plate 4). The mineralogy of Gorgona olivine spinifex is subtly different from that of Precambrian komatiite: Ca-poor pyroxene is absent and plagioclase is more abundant. The picritic volcaniclastic rocks, probably hyaloclastites, are composed of solid polyhedral olivine phenocrysts in a matrix of finer clinopyroxene, spinel and glass.

The state of preservation of Gorgona komatiites is excellent, at least as good as in the celebrated ultramafic flows from the Zvishavane region of Zimbabwe (Nisbet *et al.*, 1987), the best-preserved Archean komatiites. In Gorgona samples, most of the olivine and plagioclase crystals retain their magmatic compositions and the clinopyroxene and chrome spinel is pristine. Only the interstitial glass is altered, being devitrified and transformed to an assemblage of fine-grained hydrous minerals. Even in the fragmental picrites, olivine in larger phenocrysts is commonly preserved, although all phases in the matrix are altered.

By using the whole-rock compositions and the most magnesian olivine compositions, the Gorgona komatiites are inferred to have formed from a liquid containing about $18 \pm 3\%$ MgO and the picrites from a liquid with about $22 \pm 3\%$ MgO (Révillon *et al.*, 2000; Herzberg and O'Hara, 2002; Kerr, 2005). These concentrations are lower than in the most magnesian Archean komatiites but far higher than in other post-Archean anhydrous magmas. Gorgona komatiites have relatively high Al_2O_3–TiO_2 ratios, ranging from the chondritic value of around 20 to as high as 50. Highly

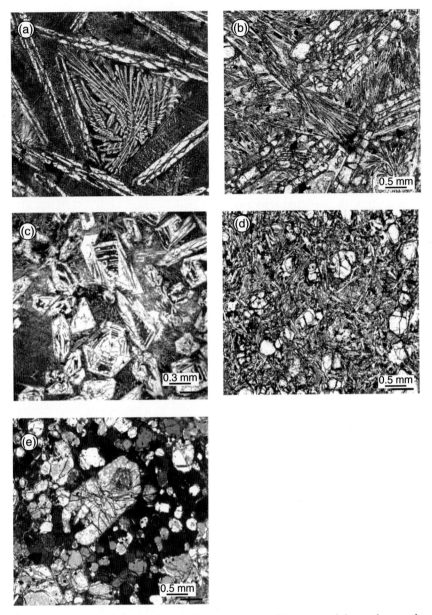

Fig. 2.27. Photomicrographs of thin sections of lavas and intrusive rocks from Gorgona Island. (a) Olivine spinifex-textured komatiite. Olivine plates are randomly oriented in a matrix of feathery clinopyroxene and altered glass. (b) Olivine spinifex layer from a thicker flow. The matrix between the olivine plates has crystallized to a relatively coarse intergrowth of clinopyroxene and plagioclase. (c) Highly skeletal hopper olivine grains in the lower olivine cumulate layer of a komatiite flow. (d) Picrite in a block in hyaloclastite from the southern part of the island: olivine phenocrysts are evenly distributed within a relatively coarse-grained matrix containing finer olivine, augite and plagioclase. (e) Olivine orthocumulate from a peridotitic intrusion in the centre of the island. (Photos from S. Révillon.)

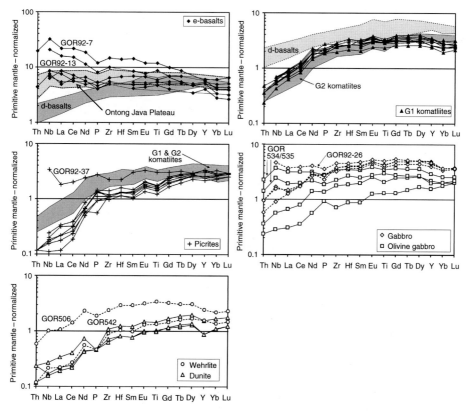

Fig. 2.28. Mantle-normalized trace-element data of various rocks from Gorgona Island (from Révillon *et al.* (2000)).

incompatible elements are strongly depleted in Gorgona komatiites and extremely depleted in the picrites (Figure 2.28). Neodymium isotopic compositions (Figure 2.29) correspond broadly to those of depleted upper mantle but ^{87}Sr–^{86}Sr ratios are elevated, due perhaps to the presence of a seawater-altered component in the source of the komatiites (Kerr *et al.*, 1996; Révillon *et al.*, 2002).

As explained in Chapter 13, the geochemical characteristics of the Gorgona lavas are interpreted to indicate that the parental magmas formed through fractional melting in a mantle plume. The komatiites represent rare examples of advanced fractional melts that reached the surface without mixing with other melts (as seems normally to happen during the formation of common basalts). The komatiites form through a high degree of fractional melting and the picrites through extreme degrees of fractional melting. The so-called E- (for enriched) tholeiites may be fractionated products of low-degree partial melts that formed at an early stage of melting; when they escaped the source

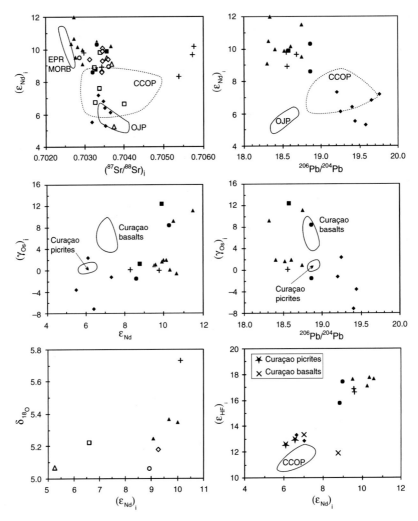

Fig. 2.29. Isotope data from volcanic rocks from Gorgona Island and from other parts of the Caribbean oceanic plateau. EPR MORB – basalts from the East Pacific Rise; CCOP – Caribbean-Colombian oceanic plateau; OJP – Ontong–Java plateau (from Kerr (2005)).

they left a depleted residue that underwent further melting to produce first the komatiites, then the picrites. The D- (for depleted) basalts are mainly the products of fractional crystallization of komatiites and the intrusive mafic–ultramafic rocks in the centre of the island are interpreted as remnants of the magma chambers in which this differentiation took place (Révillon *et al.*, 2000, 2002).

3

Field characteristics, textures and structures

3.1 Introduction

Komatiites exhibit a remarkable range of textures and volcanic structures. The most spectacular ultramafic lava flows have differentiated into olivine spinifex-textured upper parts and olivine cumulate lower parts. Komatiitic basalts display similar textures with pyroxene rather than olivine as the dominant mineral. Pillows, columnar jointing and, in fragmental komatiites, a variety of textures and structures produced during explosive eruption and deposition are also encountered.

The textures and structures of komatiites have been described in many publications. More important early works, which dealt with komatiitic rocks in Southern Africa, Canada and Australia, include Naldrett and Mason (1968), Viljoen and Viljoen (1969a, b), Nesbitt (1971), Williams (1972), Pyke *et al.* (1973), Arndt *et al.* (1977), Nisbet *et al.* (1977) and Imreh (1978). The substance of these papers and the situation that existed after a decade of komatiite research were summarized by Donaldson (1982). Subsequent contributions include: excellent illustrations and descriptions of textures and structures in komatiites in the Crixas greenstone belt, Brazil, by Saboia and Teixeira (1980); detailed descriptions of the Alexo komatiite flow by Barnes (1983) and Arndt (1986a); new work on komatiites in the Quebec portion of the Abitibi belt by Mueller (2005); new descriptions of komatiite flows and lava lakes in the Yilgarn Block, Australia, by Thomson (1989), Hill *et al.* (1987), Barnes *et al.* (1988), Lesher (1989) and Dowling and Hill (1998); accounts of the field relations and textures of komatiites in the Barberton, Nondweni and Commondale greenstone belts (South Africa) by Byerly (1999), Lowe and Byerly (1999a) and Wilson *et al.* (1989, 2003); new and very detailed descriptions of komatiites in the type locality in the Barberton greenstone belt in South Africa by Dann (2000, 2001); and descriptions of the texturally unusual

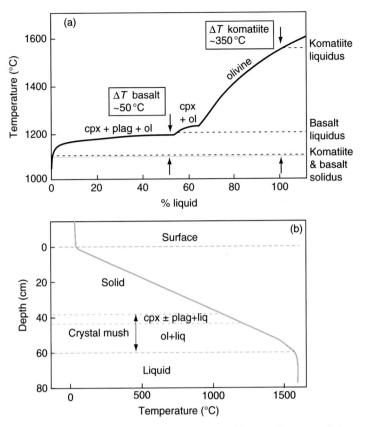

Fig. 3.1. (a) Simplified 1 atm phase relations of komatiite containing about 30% MgO (from Arndt (1976a), Kinzler and Grove (1985), Parman *et al.* (1997)) contrasting the large temperature interval between liquidus and solidus with the much smaller interval in basalt. (b) Sketch of the temperature variation in the upper crust of a komatiite flow. The large liquidus–solidus interval results in a relatively thick layer (~15 cm) of partially molten rock. It is within this interval that spinifex texture grows.

spinifex-textured komatiites on Gorgona Island, Colombia (by Aitken and Echeverria (1984) and Echeverria and Aitken (1986)). Perhaps the most detailed accounts of textures in komatiite flows have emerged from the investigation of the very fresh komatiites sampled by diamond drilling in the Belingwe greenstone belt, Zimbabwe (Nisbet *et al.*, 1987; Renner, 1989) and from the very detailed study of the Pyke Hill komatiites undertaken by Mark Shore during his thesis research (Shore, 1996). It is a great pity that this work was not published.

The diversity of remarkable structures and textures in komatiites results mainly from the unusual physical properties of ultramafic magma. Most obvious are the high temperature and low viscosity of highly magnesian liquids which allowed them to erupt rapidly and flow easily, and in some cases to assimilate wall and floor rocks. The low viscosity also allowed transported olivine phenocrysts, or olivine grains that crystallized during cooling of the lava, to settle towards the base of even the thinnest flow units and in so doing to produce the layered structure that characterizes many komatiite flows. Less evident, but perhaps more significant, is the wide temperature interval between the liquidus (which in most, but not all, cases corresponds to the eruption temperature) and the temperature of appearance of the second silicate mineral. For a komatiite liquid containing 30% MgO, illustrated in Figure 3.1(a), the liquidus is close to 1600 °C but clinopyroxene appears only around 1200 °C. Chromite appears at intermediate temperatures but never in quantities suffi-cient to influence the physical properties of the lava. At 1200 °C, 400 °C below the liquidus, only about 50% of the liquid has crystallized and the proportion of crystals remains low. By comparison, in typical basalt, the second silicate mineral commonly appears within 50 °C of the liquidus and crystallization is complete by the time that the temperature has dropped by 200 °C. The large temperature interval makes it easier for any suspended phenocrysts to settle and contributes to the striking differentiation of many komatiite flows. It also results in the formation of a thick crystal mush zone at the flow margins. Consider, for example, a 1 m thick crust formed at the top of a 30% MgO komatiite flow (Figure 3.1(b)); the lowermost 15 cm of this crust, a quarter of its total thickness, consists of a mesh of olivine crystals in abundant low-viscosity liquid. This property plays an important role in the formation of spinifex textures, as explained in Chapters 8 and 9, as well as in the develop-ment of Ni sulfide ore deposits, as explained in Chapter 10.

3.2 Types of komatiite flows

The best-known and most diagnostic type of komatiite flow is illustrated in Figure 3.2(a). This type of flow, which is superbly exposed in the celebrated 'Pyke Hill' outcrop in Munro Township (Pyke *et al.*, 1973), has a spinifex-textured upper layer and an olivine cumulate lower layer. (The term 'zone', as in the spinifex or A zone and cumulate or B zone, was used, together with 'subunit' and 'portion', in the original descriptions by Pyke *et al.* (1973) of Munro Township komatiite flows, but the two main divisions of komatiite flows are laterally extensive, continuous along strike and originally horizontal. The terms 'unit' or 'layer', as used by Donaldson (1982), are adopted here.) A second type of komatiite flow (Fig. 3.2(c)) is massive, unlayered, olivine phyric and polyhedrally jointed. These flows represent the two end-members of a spectrum of flow types that includes intermediate members containing a poorly developed spinifex layer and a cumulate layer characterized by relatively loose

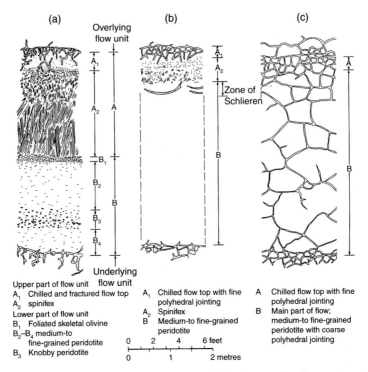

(a)

Overlying flow unit

(b)

(c)

Zone of Schlieren

Underlying flow unit

Upper part of flow unit
A₁ Chilled and fractured flow top
A₂ spinifex
Lower part of flow unit
B₁ Foliated skeletal olivine
B₂–B₄ medium-to
      fine-grained peridotite
B₃ Knobby peridotite

A₁ Chilled flow top with fine
    polyhedral jointing
A₂ Spinifex
B   Medium-to fine-grained
    peridotite

A   Chilled flow top with fine
    polyhedral jointing
B   Main part of flow;
    medium-to fine-grained
    peridotite with coarse
    polyhedral jointing

0   2   4   6 feet

0   1   2 metres

Fig. 3.2. Diagram illustrating three types of thin komatiite flows, from Pyke *et al.* (1973) and Arndt *et al.* (1977): (a) fully differentiated flow with spinifex-textured upper layer and olivine cumulate lower layer; (b) partially differentiated flow with a thin upper spinifex layer and spinifex-textured layer; (c) undifferentiated massive flow cut by polyhedral joints.

packing of olivine grains. The first studies of komatiites focussed in relatively thin flows such as those in Munro Township in Canada and in the type sections in the Barberton greenstone belt in South Africa. Issuing from these studies was a simple three-fold classification of thin komatiite flows into differentiated and non-differentiated units, as illustrated in Figure 3.2. Subsequent detailed work, mainly in the Kambalda region and surrounding parts of the Yilgarn Craton where thick units dominated by olivine cumulate are common (and spinifex-textured units are rare, making up less than 20% of all komatiites), resulted in the more elaborate classification shown in Figure 3.3. The application of this classification is developed in Chapters 9 and 10 where my coauthors discuss the physical volcanology and the ore deposits in komatiites. In this chapter I use the simpler classification, making reference when needed to the thicker olivine-rich units.

Other types of komatiite that do not fit readily into the simple classification include rare pillowed units and various types of fragmental komatiites. These are described later in the chapter.

KOMATIITE SEQUENCE

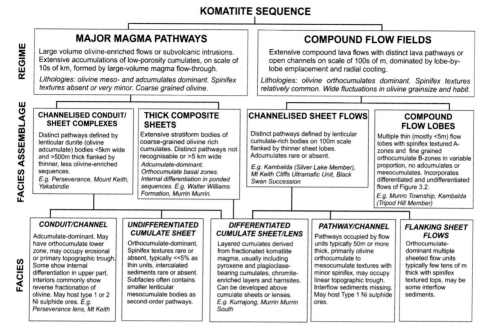

Fig. 3.3. Classification of komatiite flow types from Barnes (2006).

3.3 Dimensions of komatiite flows

The thicknesses of layered komatiite flows vary greatly. At one extreme we have miniature units, less than 50 cm thick, some of which are fully differentiated into an upper spinifex and a lower cumulate layer. Flow lobes within thin flow facies sequences are mostly in the range 1–5 m thick, comparable to lobes in modern compound pahoehoe flows. Units with these dimensions have been described in Munro Township (Pyke *et al.*, 1973; Shore, 1996), and the Barberton greenstone belt (Smith *et al.*, 1980; Viljoen *et al.*, 1983). At the other extreme, we find layered flows more than 100 m thick in the Kambalda area (Gresham and Loftus-Hills, 1981; Lesher *et al.*, 1984), in various other parts of the Yilgarn Craton (Dowling and Hill, 1993; Perring *et al.*, 1995; Hill, 2001) and throughout the Abitibi belt (Jensen and Langford, 1985; Arndt, 1986b). Some of the more extreme pathway facies flows in the East Yilgarn are up to 500 m thick, an example being the stack of vesicular olivine orthocumulates forming the core of the komatiite sequence at Black Swan (Hill *et al.* 2004). The East Yilgarn dunite sheets and lenses are thicker still, but it is unclear in most cases whether they are intrusive or extrusive. Even if we restrict the discussion to the undoubtedly extrusive Kambalda units, we can say that some komatiite flows were remarkably thick. In comparison, the volumetrically vast flood basalt flows of the Columbia River basalts are typically 20–50 m thick (Self *et al.*, 1996).

Many komatiite flow fields contain distinctive successions made up of numerous relatively thin (typically 1–5 m) flows. The individual units in these successions probably all formed during single eruptions and represent portions or segments of complex flows: i.e. they are flow units rather than flows. As was recognized by Pyke *et al.* (1973) and has more recently been discussed by Huppert *et al.* (1984), Huppert and Sparks (1985a, b), Barnes *et al.* (1985, 1988), Dowling and Hill (1993), Perring *et al.* (1995) and Hill (2001), the path followed by erupting lavas was probably complex and tortuous. In parallel with work by Hon *et al.* (1994) in Hawaii and Self *et al.* (1997) on continental flood basalts, it is recognized that large, thick komatiite flows were not emplaced in a single pulse but probably first advanced as a series of thin units that became inflated by the internal intrusion of new lava. Clear evidence of this type of behaviour is exhibited by komatiitic basaltic flows on Gilmour Island (Arndt, 1982), by komatiites in the Barberton greenstone belt (Dann, 2000) and by thick dunitic units in various parts of Western Australia (Hill and Perring, 1996; Hill, 2001). According to the volcanic facies model developed by R. E. T. Hill and coworkers and discussed in detail in Chapter 9, a komatiite flow sequence contains thick central conduits that fed extensive flanking fields of thinner compound flows. The thinner units mapped in areas like Pyke Hill represent lava lobes that issued from the central conduits.

The proportion of spinifex-textured lava to cumulate lava in individual flows is highly variable. In thin flows such as those on Pyke Hill, the cumulate layer has about the same thickness as the spinifex layer, whereas in thicker units like those at Kambalda, the cumulate zone is far thicker (Lesher, 1989). At the other extreme, flows with virtually no olivine cumulate have been described from Gorgona Island (Aitken and Echeverría, 1984) and in some Archean areas (Perring *et al.*, 1995).

Spinifex layers are persistent along strike and commonly can be traced for many tens of metres with little change in thickness, but in some areas of continuous outcrop, the spinifex layers can be shown to diminish in thickness and to disappear along strike. In one example from Munro Township, the thickness of a komatiite flow remains constant as the spinifex layer gradually becomes thinner and less well defined, and then peters out (Figure 3.4). In flows on Gilmour Island, spinifex layers form the upper parts of elongate lenses, one of which is 300 m wide, which separate segments of massive and pillowed lava (Fig. 2.19).

Dann (2000), in his comprehensive study of the Barberton greenstone belt, estimated that only about 25% of the komatiite flows are spinifex-textured, most of the remainder being massive sheet flows. Komatiite comprises 50% of

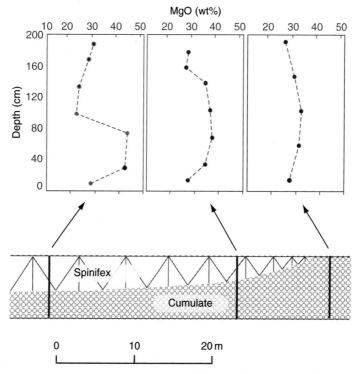

Fig. 3.4. Along-strike variation in the thickness of the spinifex-texture in komatiite flow in Munro Township, from Arndt *et al.* (1977). The MgO profiles illustrate a decrease in the extent of differentiation, from pronounced in the layered spinifex-textured facies on the left, to minimal in the massive facies to the right.

the Lower Komati Formation and 28% of the part of the Barberton greenstone belt that he mapped. Massive units – flows in some cases but probable sills in others – are thicker, reaching at least 55 m in some examples. Some massive flows are laterally continuous for up to 11 km; others are lenticular and pinch and swell, or pinch out entirely along strike. The spinifex flows are thinner, ranging from 0.3 to 8 m. Dann's (2000) third lithological type, vesicular komatiite, comprises less than 0.5% of the sequence. These flows have intermediate thicknesses.

Little is known about the lengths of komatiite flows. Theoretical considerations, as developed by Huppert and Sparks (1985a, b) and Williams *et al.* (2002), suggest that some komatiite flows extend for hundreds of kilometres. Testing this idea is difficult, mainly because of limitations in the extent of outcrop and the amount of deformation that has affected the volcanic sequences.

Terminations of komatiite flows are common in areas of good outcrop, but many or most of these no doubt represent the sides rather than the ends of flow units or lobes. Shore (1996) describes a single flow unit on Pyke Hill in Munro Township, Canada, that is only 70 cm thick but has a fully exposed cross section between terminations 40 m apart. He infers from the symmetrical form of lava tubes at the base of the sequence that the original flow direction was close to perpendicular to the present land surface and he interprets the terminations as sides of a flat lava lobe that must have been much longer in the third dimension.

In most areas of good exposure, a majority of flow units can be traced to the limits of outcrop, often hundreds of metres along strike. Individual komatiite formations in the Abitibi greenstone belt commonly extend many tens of kilometres (Jensen and Langford, 1985; Ayer *et al.*, 2002) and the 2716–2714 Ma komatiites in the Kidd–Munro volcanic assemblage in Ontario and the Malartic group in Quebec are correlative over a distance of at least 400 km (J. A. Ayer, personal communication). Within the assemblage, 85 km of thin discontinuous komatiites have been traced between the Kapuskasing uplift and Kidd Creek, 135 km of thick and fairly continuous komatiites between Kidd Creek and a truncation by the Porcupine–Destor fault zone in eastern Holloway Township, 80 km of thin discontinuous komatiites along the north side of the Porcupine–Destor fault zone in eastern Ontario and western Quebec, and 100 km of thick and fairly continuous komatiites south of the Porcupine–Destor fault zone in the Malartic group to their final truncation by the Porcupine–Destor fault zone in Southeast Abitibi in Quebec (J. A. Ayer, personal communication). Individual flow fields in the Reliance Formation of Zimbabwe extend up to 85 km, but may be correlative for 700 km across the Zimbabwe Craton (Prendergast, 2003).

Individual komatiite formations in the Yilgarn Craton can be correlated up to 85 km, and individual komatiite assemblages with similar age and compositions can be correlated up to 500 km (Hill *et al.*, 1995). The largest known correlatable formation of komatiitic rocks, the Walter Williams dunite sheet, was at least 130 km long, and was comparable in volume to the largest flows found in Phanerozoic flood basalt provinces, but may well be a sill rather than a flow. The well-defined Kambalda Komatiite Formation extends 50 km between Kambalda, St Ives and Tramways (Gresham and Loftus-Hills, 1981) and it may be correlative for at least 180 km between the Scotia, Kambalda–St Ives–Tramways, Widgiemooltha and Pioneer areas of the Kalgoorlie Terrain in the southern part of the Norseman–Wiluna belt (Lesher *et al.*, 1984; Swager *et al.*, 1990). The komatiite stratigraphy in the northern part of the Norseman–Wiluna belt can be correlated over 60 km from Six Mile Well through Mt Keith to Kingston (Hill *et al.*, 1995). If the komatiite units in the northern and southern parts of the Yilgarn block are also

correlative, which seems likely based on their ages and geochemical characteristics, the dimension can be extended up to 500 km (Hill *et al.*, 1995).

Based on these studies, komatiite volcanoes are estimated to have been at
least 85 km, and perhaps as much as 700 km, in diameter. These dimensions are
comparable to the lateral dimensions of the largest modern basaltic shield
volcanoes, but are dwarfed by the dimensions of the largest continental and
oceanic volcanic plateaus. The impressive, 500 km dimensions of the komatiite
assemblages in the Norseman–Wiluna belt of the Yilgarn Craton represent less
than half of the breadth of a Phanerozoic continental flood basalt province
such as Siberia or the Karoo. On the other hand, the Archean sequences are
merely the remnants of larger volcanic provinces that survived the accretion
and deformation attendant on the formation of granite–greenstone terrains
(Chapter 14). There is every reason to believe that komatiites, being low-
viscosity magmas produced by high-degree mantle melting (see Chapters 8
and 13), erupted in large volumes and covered vast areas of the Earth's surface.
Just how vast we will never know.

3.4 Spinifex texture

The term spinifex was introduced into geological literature in the early 1970s in
papers by Lewis (1971) and Nesbitt (1971). The name comes from a type of
grass, *Triodia spinifex*, which is common in Western Australian komatiite
localities. This plant has spiky, randomly oriented leaves or needles that
resemble two-dimensional sections through olivine crystals in certain types
of komatiite (compare Figures 3.5 and 3.6). Prospectors and exploration
geologists in the Yilgarn Block had used the name for years before it was
adopted by Nesbitt (1971). In Canada, flora and fauna provided names such as
'pine needle', 'chicken track', 'herring bone' for similar textures, while in South
Africa 'crystalline quench texture' was used; but these names have now been
superseded by the Australian term.

Several attempts have been made to come up with an acceptable definition
of spinifex texture. Contributions by Lewis (1971), Nesbitt (1971), Williams
(1972), Pyke *et al.*, (1973), Brooks and Hart (1974) and Arndt *et al.* (1977)
were discussed by Donaldson (1982) in his chapter in the first komatiite book.
In the same volume, Arndt and Nesbitt (1982) presented a working definition
based on responses to a questionnaire sent to geologists working with komatiites at that time. A slightly reworded version of this definition, which seems to
have been broadly accepted, is:

Spinifex is a texture characterized by large, skeletal, platy blades of olivine or acicular
needles of pyroxene, found in the upper parts of komatiitic flows, or, less commonly,

Spinifex | Arts Project

Spinifex
Computing

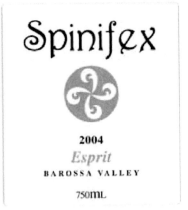

Fig. 3.5. Spinifex in various forms. A tussock of typical spinifex grass is shown in the lower photograph and various ways in which the term has been used are illustrated in the upper part of the diagram.

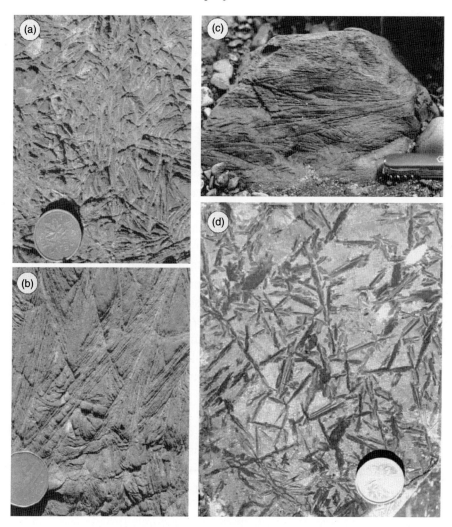

Fig. 3.6. Photographs of spinifex textures in outcrop and in hand samples: (a) random olivine spinifex texture in the upper part of a thin flow on Pyke Hill, Munro Township; (b) platy spinifex from deeper in the flow; (c) platy olivine spinifex in a boulder on the beach of Gorgona Island; (d) large olivine plates in another flow on Pyke Hill.

at the margins of sills and dikes. The texture is believed to form during relatively rapid, *in-situ* crystallization of ultramafic or highly mafic liquids.

In platy olivine spinifex texture, olivine has a plate or lattice habit and forms complex grains made up of many individual plates arranged roughly parallel to one another. The 'books' of parallel grains are oriented approximately perpendicular to flow or intrusive margins. Composite olivine plates may be as long as 1 m but are only 0.5–2 mm thick. Interstitial material is fine skeletal, dendritic or spherulitic pyroxene, cruciform or dendritic chromite, and devitrified altered glass. Plagioclase is present in the matrix of some samples. Random olivine spinifex texture contains smaller, less elongate, randomly oriented olivine plates.

Pyroxene spinifex texture contains augite, pigeonite or in rare cases orthopyroxene as complex skeletal needles that either are arranged in sheaths perpendicular to flow margins or are randomly oriented. The pyroxene needles typically are 1–5 cm long but only 0.5 mm wide and lie in a matrix of much finer augite needles and devitrified glass, or augite, plagioclase and quartz.

Primary phases usually are replaced by secondary minerals such as serpentine, chlorite, tremolite, talc, epidote, albite and metamorphic olivine.

Examples of spinifex[1] textures are illustrated in Figures 3.6 and 3.7.

In contrast to the response to the proposed definition of komatiite (Chapter 2), which provoked considerable comment at the time it was proposed in the early 1980s, little discussion has appeared in the literature about the definition of spinifex. In general the term is applied in a manner consistent with the definition, but some problems remain. 'Spinifex' continues to be used for textures produced during the growth of metamorphic olivines and other bladed or spiky minerals in metamorphosed ultramafic rocks. As has been pointed out by several authors e.g. Oliver *et al.* (1972), Collerson *et al.* (1976), Donaldson (1982), in most cases these textures are readily distinguished from true spinifex textures. The metamorphic olivine grains are normally bladed, non-skeletal and randomly oriented (Figure 3.8). The grains are elongate parallel to the *b* crystallographic direction, unlike magmatic spinifex olivines, which are elongate parallel to *a* or *c*. These crystals are usually set in a matrix of secondary hydrous minerals such as talc, amphibole, chlorite, anthophyllite or serpentine and they commonly cut across the metamorphic fabric. In some cases their compositions are similar to those of spinifex olivines; in others, they are decidedly richer, or poorer, in iron. Further discussion is found in Chapter 4.

Another problem lies in the use of the term spinifex for textures visible only with a hand lens or microscope. It is true that textures in the chilled margins of komatiite flows are simply finer-grained versions of the macroscopic random olivine spinifex textures developed deeper in the flows, and that the acicular pyroxenes in flow margins or pillows or fragments of komatiitic basalts differ only in grain size from macroscopic pyroxene spinifex textures. Yet to accept

---

[1] In addition to its use as the name of a texture in komatiites, the term 'spinifex' has been adopted, in various parts of Australia, in the names of companies specializing in balloon flights, safaris, feminist publications, computer system interfaces, pigeon sculptures, stud horses, dogs and cattle, an outback arts project, a restaurant and numerous pubs, a wine maker, and 'essence of spinifex', a potion 'made from the grass species that is useful for those who have a sense of being a victim to and having no control over illnesses, especially those with persistent and recurring symptoms'. The logos of some of these enterprises are illustrated in Figure 3.5.

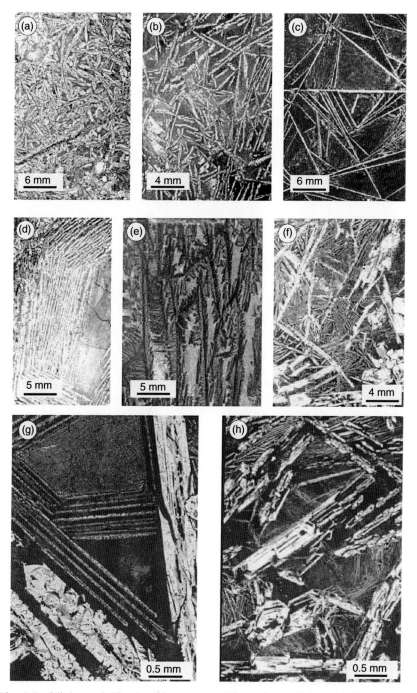

Fig. 3.7. Olivine spinifex in thin section: (a) random olivine spinifex texture from the Alexo komatiite flow; (b) random olivine spinifex from the northern part of Munro Township; (c) fine blades of olivine in platy spinifex of komatiite basalt in Munro Township; (d) platy olivine spinifex: a large skeletal olivine

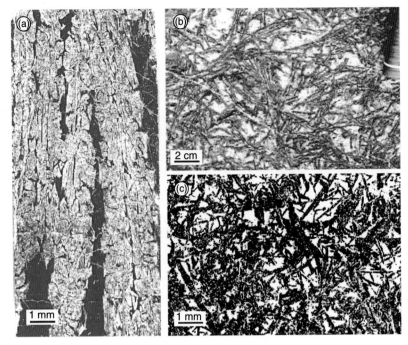

Fig. 3.8. Rocks with spinifex-like textures: (a) photomicrograph of harrisitic olivine from the Victor shoot at Kambalda; (b) platy olivines in an outcrop of the quenched margin of the Rum intrusion, Scotland; (c) photomicrograph of 'jack-straw' texture metamorphic olivine in a matrix of fine-grained talc and other hydrous minerals, Digger Rocks, Western Australia.

these microscopic textures as spinifex seems to provide licence for the term to be used for all manner of textures characterized by acicular minerals, be they in tholeiitic basalts, alkaline lavas, ocelli or even skarns. To avoid ambiguity and potential confusion, it seems safer to reserve the term spinifex for the macroscopic textures found in komatiitic lavas, or, more specifically, those textures that grow in the thermal gradients in the margins of flows and sills.

Fig. 3.7 (cont.)
blade is seen on the upper left; finer parallel plates fill most of the cell between the larger olivine grains and still-finer acicular needles of pyroxene fill the interior of the cell on the right; (e) vertical section through branching olivine grains of the 'chevron' olivine texture of the Alexo komatiite flow; a horizontal section is shown in Figure 2.14. (f) and (h); platy olivine spinifex from a komatiite of the Zvishavane area in Zimbabwe: note the presence of polyhedral olivine grains, an unusual feature of these rocks; (g) platy olivine spinifex in a Pyke Hill komatiite illustrating large solid 'master' blades and finer parallel secondary blades. The olivine is serpentinized in most samples, with the exceptions of (a), (f) and (h).

Yet another problem stems from the use of terms such as 'amphibole spinifex' to describe textures in which pyroxene needles are replaced by secondary tremolite or actinolite. Traditional usage has been that the name of the dominant *igneous* mineral is specified (as in 'olivine spinifex' or 'pyroxene spinifex') and that the types of secondary minerals that replace these textures are given separately. Although this usage is frowned upon by some petrologists who are accustomed to working with fresher rocks (Thompson (1983) coined the term 'komatispeak' for the practice), it is practical for most komatiite occurrences where excellent textural preservation allows easy recognition of the primary minerals. With this background, the use of the term 'amphibole spinifex' misleadingly suggests yet another type of spinifex in which igneous amphibole is the dominant component.

3.5 Textural variations in layered komatiite flows

Although textures and structures in layered komatiite flows have been documented and illustrated in great detail in many publications e.g., Barnes *et al.* (1973), Pyke *et al.* (1973), Arndt *et al.* (1977), Nisbet *et al.* (1977), Imreh (1978), Donaldson (1982), Arndt (1986a), Thomson (1988), Renner (1989), Wilson *et al.* (1989), some repetition seems necessary, to summarize some more recent findings, and to provide background for the discussion of these textures that follows in later chapters. For more detailed information the reader is referred to the original works.

A fully differentiated, layered, spinifex-textured flow contains the following units (from top to base): a chilled margin and aphanitic flow top (A_1), a spinifex layer (A_2), a thin layer of foliated skeletal olivines (B_1), an olivine cumulate layer ($B_{2,3,4}$), and a lower chilled margin.

Chilled margins (A_1)

A 1–10 cm thick chilled margin is found at the top of most komatiite flows. In the most common texture, small (~ 0.1 mm) polyhedral olivine phenocrysts are set in a matrix of elongate, platy, randomly oriented skeletal olivine grains, a type of microspinifex texture (Figure 3.9(a)). (Olivine habits are described using the terminology of Donaldson (1976, 1982) which is summarized in Chapter 4.) Examples of this texture are described by Pyke *et al.* (1973), Thomson (1988), Renner (1989) and Renner *et al.* (1994). Another type, observed in the Alexo komatiite flow (Arndt, 1986a), the flows from Zvishavane in Zimbabwe (Renner *et al.*, 1994) and in komatiite flows from Mt Clifford, Australia (Barnes *et al.*, 1973), consists of polyhedral phenocrysts, 0.1–0.5 mm across, in a glassy to microcrystalline matrix (Figure 3.8(b)). The olivine phenocrysts

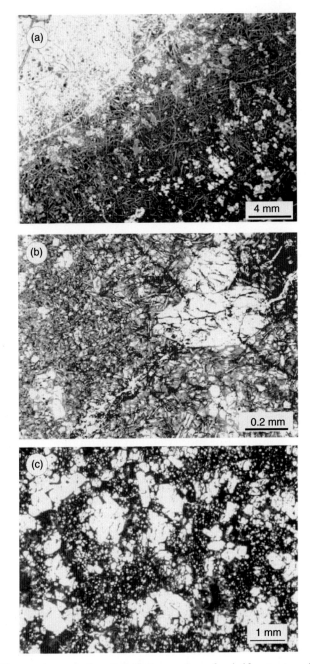

Fig. 3.9. Photomicrographs of chilled margins of spinifex-textured komatiite flows: (a) the upper chilled contact and chilled margin of a flow of the Zvishavane area, Zimbabwe (from Renner (1989)); the uppermost margin (top left) is composed mainly of glass altered to chlorite, then follows olivine microspinifex and olivine microphyric lava; (b) the lower chilled margin of a thin flow on Pyke Hill: large equant olivine phenocrysts are set in a groundmass of skeletal olvine, olivine microphenocrysts and altered glass (from Shore (1996)); (c) the bottom chill of another flow of the Zvishavane area, Zimbabwe, containing clusters of olivine phenocrysts in a matrix rich in olivine microphenocrysts (from Renner (1989)).

may be isolated or distributed in clusters. Their abundance varies from about 1–2% (in the chill of the Alexo flow, Figure 2.14(c)) to 15–20% (in less magnesian flows of Zvishavane and Mount Clifford).

Flow top (A_1)

Chilled margins form the uppermost part of what Pyke *et al.* (1973) called the 'chilled and fractured flow top', or A_1 unit. This unit ranges in thickness from 5 to 150 cm (average 15 cm) and is made up of aphanitic to very fine-grained lava cut by close-spaced (1–3 cm) polyhedral joints. Pyke *et al.* (1973) called these joints 'fractures', which is true in a sense, but they are comparable in many ways to the columnar jointing in basaltic flows, and probably result from contraction of the cooling solidifying lava. Between polyhedral joints, the lava is very fine-grained and contains randomly oriented, elongate skeletal olivines and a variable amount of polyhedral olivine phenocrysts, set in a very fine-grained or glassy groundmass.

Spinifex (A_2)

The size of olivine grains in the spinifex layer gradually increases from the top to the base of the layer. The top of the layer is taken as the level where skeletal olivine grains become visible in outcrop or hand samples. These grains are a few millimetres long and about 0.5 mm thick (Figure 3.6(a)). Deeper in the flow, the olivine grains become progressively larger and towards the base of some layers they form spectacular composite blades up to 1 m long (Figure 3.6(b)–(d)). Changes in the habit and orientation of the olivine grains accompany the increase in grain size. At the top, olivine occurs as simple, elongate, skeletal, randomly oriented tablets (random spinifex; Figure 3.6(a)); lower down the olivine blades lie parallel to one another within characteristic book-like crystals (plate spinifex; Figure 3.6(b)). With increasing depth in the flow, the books become larger and each book comprises numerous individual plates. In parallel with the increase in grain size, the orientation of the books changes from random to roughly perpendicular to the flow top. In some areas of good outcrop, the plates can be seen to form downward-splaying sheath or cone structures tens of centimetres across and up to a metre in length.

In many flows, the transition from random to platy spinifex is abrupt (Figure 2.12) and this led Gélinas *et al.* (1977) to subdivide the layer into an upper A_2 and a lower A_3 subunit. In the lowermost 5–20 cm of some spinifex layers, the size of the olivine grains decreases downwards and the texture reverts from platy to random spinifex. Olivine grains in certain samples from Kambalda, Australia (Thomson, 1988) or Dundonald Township in Canada

(Arndt *et al.*, 2004b), have nucleated on the underlying cumulate olivines and appear to have grown upwards where they interfere with the lowermost spinifex olivines. In other examples (e.g. in most of the flows exposed on Pyke Hill in Munro Township), perpendicular books of large plate olivines terminate abruptly at the contact with the B_1 layer.

Thomson (1988) recognized two generations of olivine in spinifex lavas of Kambalda. The first generation, which he called 'master blades', enclose volumes of melt within which crystallized a second generation of smaller, thinner, commonly more skeletal, 'secondary grains'. The two types are illustrated in Figure 3.7(d),(g). Some spinifex lavas, notably those in the Zvishavane area in Belingwe, contain equant polyhedral olivine grains distributed irregularly, often in clusters, amongst the skeletal spinifex blades (Figure 3.7(f), (h)).

The matrix between olivine grains in all but the coarsest spinifex lavas is composed of clinopyroxene, glass (always devitrified and commonly altered to secondary hydrous phases) and minor oxides (Figures 3.10, 3.11). Only in slowly cooled samples from deep within thick flows is plagioclase found, usually as irregular elongate grains intergrown with clinopyroxene. An exception is found in the komatiites of Gorgona Island where plagioclase is a common interstitial phase (Figure 3.10(c)). The habit of pyroxene grains interstitial to spinifex olivines is highly variable, ranging from large arrays of parallel needles (often complex, with cores of pigeonite or enstatite and mantles of augite) to very fine-grained sheaths, fronds and highly irregular intergrowths. A striking example is shown on the cover of the book. A remarkable feature of the habit of interstitial pyroxene, illustrated in Figures 3.10(a) and (f), is the manner in which the size and habit of the crystals changes dramatically from one interolivine cell to another. Compare, for example, the contrasting size and habit of pyroxene crystals in the three adjacent cells shown in Figure 3.10(a). Chromite occurs as minute (100 µm) equant grains, as inclusions in olivine or scattered through the matrix, and as skeletal grains that seem to have nucleated upon and grown outwards from the margins of olivine plates (Figure 3.11).

The modal proportion of olivine normally decreases from the top to the base of the spinifex layer (Figure 3.12 and Figure 4.1). In thicker flows, the variation may be extreme, from about 60% olivine in uppermost random spinifex lava to about 25% olivine in coarse plate spinifex. There are, of course, exceptions, such as in the Alexo komatiite flow, where we see an abrupt return to olivine-rich 'chevron spinifex' midway down the spinifex layer (Arndt, 1986a), and a less pronounced increase in the olivine content of the lowermost 5–20 cm where random spinifex reappears or where upward-growing

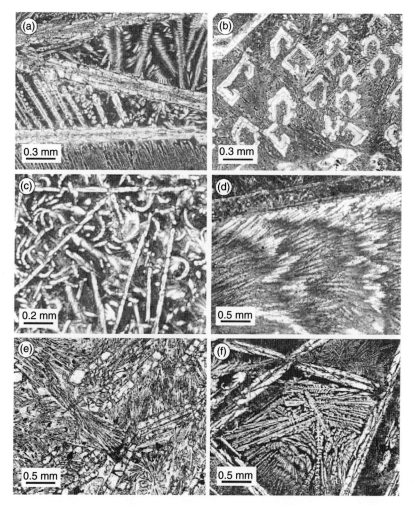

Fig. 3.10. Photomicrographs illustrating details of textures between the olivine blades of spinifex-textured komatiite: (a) clinopyroxene grains with contrasting sizes and habits in a komatiite from Zvishavane, Zimbabwe; (b) zoned hollow pyroxene needles with pigeonite cores (completely replaced by chlorite) and margins of augite, from olivine spinifex lava of the Alexo komatiite flow; (c) straight and curved clinopyroxene grains from a Gorgona komatiite; (d) complex fronds of augite crystals in a Munro Township komatiite; (e) augite–plagioclase intergrowths between olivine blades of a Gorgona komatiite; (f) augite grains with variable size and morphology interstitial to olivine blades of a Gorgona komatiite.

crescumulate olivine interferes with downward-growing olivine. Particularly in poorly differentiated flows, the size, habit and proportion of olivine grains change within the spinifex layer, to form alternating sublayers, or veins, patches and irregular blobs with contrasting texture and composition.

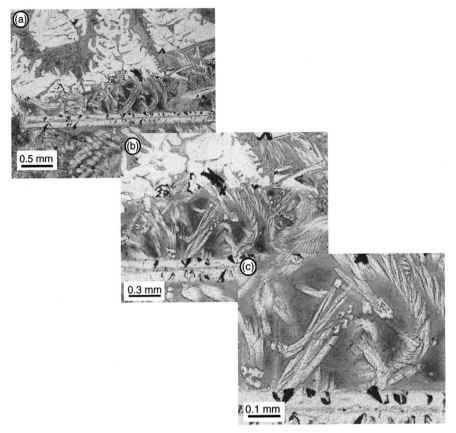

Fig. 3.11. Olivine–augite–chromite–glass textures in spinifex textured komatiite from Alexo. (a) The complex dendritic olivine grains, now replaced by colourless serpentine, in the upper part of the photo are parts of a much larger, several centimetre long, early formed olivine plate; the much thinner and less complex plate in the lower part of the photograph formed later. (b), (c) Details of augite and chromite morphologies. Chromite grains have nucleated on the olivine plate and grow outward into the interstitial liquid. Needle-like crystals and curved fronds of augite have grown within the interstitial liquid which has solidified to brownish glass. The glass is now devitrified but not strongly altered.

The B_1 layer

The B_1 layer is made up of elongate hopper olivine phenocrysts, typically 1–3 mm long, and these grains have a strong preferred orientation parallel to the flow margins (Figure 3.13). The phenocrysts lie in the usual pyroxene–glass matrix. B_1 layers in Archean komatiites are only 2–5 cm thick and the 20–150 cm thick layers in the Proterozoic komatiitic basalt flows on the Ottawa Islands (Arndt, 1982) are exceptional.

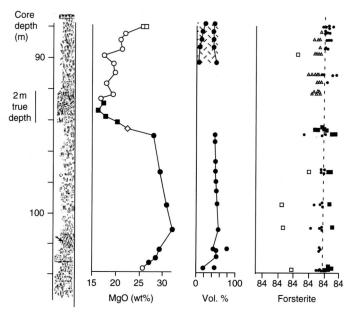

Fig. 3.12. Variation in modal olivine content from top to base of a layered, spinifex-textured komatiite flow from Zvishavane, Zimbabwe. Data from Renner (1989).

B_1 layers are absent from many layered komatiite flows and appear restricted to those exhibiting the most pronounced differentiation into spinifex and cumulate layers. The upper contact between the B_1 layer and spinifex lava is abrupt, except where the boundary is complicated by nucleation and upward growth of crescumulate grains. In examples where skeletal grains have grown upwards from the top of the B_1 layer (e.g. Figure 3.13(a)) it is apparent that the B_1 layer had consolidated before the growth of the spinifex layer was complete. In other examples, such as in most flows on Pyke Hill and in the Alexo flow, the degree of preferred orientation of olivine grains in the B_1 layer is greatest immediately below the contact with the spinifex layer (Figure 3.13(b)). This texture is thought to result from flowage of lava just below the spinifex layer, and it is taken as evidence that lava flowed through the B_1 layer after solidification of the spinifex layer. This type of structure has been related by Hill (2001) to breakout of lava through the crust of an inflated flow, which provokes lava flowage through the lava channel (see further discussion in Chapter 9).

The lower contact of the B_1 layer is gradational and corresponds to a gradual increase in the proportion of solid polyhedral grains and a decrease in the proportion of elongate hopper grains. A decline in the degree of preferred orientation accompanies this change. In the Alexo flow (Arndt, 1986a), the hopper grains become randomly oriented at a distance of only

Fig. 3.13. Photographs of B_1 layers in komatiites. (a) Polished slab of the base of the spinifex layer and the B_1 layer of the small komatiite sill in Dundonald Township (Arndt *et al.*, 2004b). Large skeletal grains of olivine, now replaced by dark-coloured serpentine, have grown upward from the top of the B_1 layer. (b) Scan of a large thin section of the B_1 layer of the Alexo komatiite flow. The base of the spinifex layer is seen at the top of the photograph. Immediately below, the short blades of olivine of the B_1 layer are aligned parallel to the contact. In the lower part of the B_1 layer the degree of preferred orientation is less pronounced. (c) Contact between the spinifex layer and the B_1 layer of a Pyke Hill komatiite. In this flow, as in the Alexo flow, the contact is abrupt and the olivine spinifex layer appears to have formed before the B_1 layer had fully crystallized.

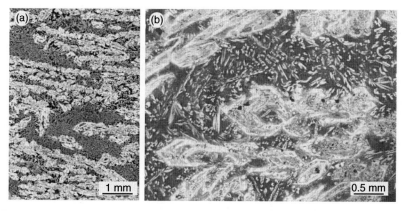

Fig. 3.14. The B1 layer of komatiitic basalts of Gilmour Island. (a) Stubby, slightly skeletal grains with clear preferred orientation are sparsely distributed in a clinopyroxene–glass matrix. (b) Detail of several skeletal B_1 olivine grains. The lower margin of each grain is lined by minute euhedral chromite grains and delicate skeletal augite grains are distributed through the glassy groundmass.

1–2 cm below the contact with spinifex lava (Figure 3.13(b)). The modal olivine content of the B_1 layer is intermediate between that of spinifex-textured and cumulate (B layer) rocks. An unusual characteristic of the B_1 layer, particularly in less-magnesian komatiitic basaltic flows, is an anomalously high chromite content (2–5% Cr_2O_3 (Hanski, 1979; Arndt, 1982)). In B_1 layers, it is common to see concentrations of equant, euhedral chromite grains, and in the unusually thick B_1 layers in the komatiitic flows of the Ottawa Islands, chromite grains have accumulated on the lower surfaces of olivine grains, as illustrated in Figure 3.14.

Cumulate layer (B_2–B_4)

Pyke *et al.* (1973), in their descriptions of komatiite flows from Munro Township, divided the lower olivine-enriched portions of spinifex flows into four subunits; B_1–B_4. The B_1 layer, although not always present, is a feature of well-differentiated flows world-wide. This is not the case for the B_3 or 'knobby peridotite' layer, which forms in the lower third of the cumulate layer and is distinguished from underlying and overlying rocks by the presence of 'knobs', or patches, 1 cm across, of pyroxene–glass matrix material. These patches resist weathering better than the olivine cumulate in which they are found, and stand out as knobs. The B_3 layer is a common feature of well-differentiated komatiites in the Pyke Hill outcrop, but is rarely recognized at other locations. The only report of a similar feature is in the komatiite flows of Zvishavane, where irregular blobs or patches of relatively coarse-grained pyroxene–plagioclase material are observed in thin sections taken from an intermediate level in the olivine cumulate layer (Renner, 1989; Renner *et al.*, 1994). These patches are far smaller than the Pyke Hill knobs – they are inconspicuous in outcrop – but may have a similar origin.

The remainder of the cumulate layer in Pyke Hill flows, and cumulate layers of most other thin komatiites, consists of olivine orthocumulate in which olivine grains are contained within a sparse pyroxene–glass matrix. The majority of the cumulus olivine grains (70–95%) are equant to tabular, usually subhedral with slightly rounded edges (Figures 3.15(a),(c)); the remainder are more elongate, tabular and moderately skeletal. Irregular veins and patches in which all the olivine grains are skeletal are present in many cumulate layers. And again, the komatiites on Gorgona Island are highly unusual in that the olivine grains in the cumulate layers are almost entirely highly skeletal, as illustrated in Figure 3.15(f).

Typical grain sizes of subhedral polyhedral grains range from 0.3 to 0.6 mm but grains up to 1 cm across are found in thicker flows. Grain size analyses by Arndt *et al.* (1977, 1989) and Renner *et al.* (1994) reveal reverse grading: in some Zvishavane flows, average grain length increases from ~ 0.2 mm in the cumulate layer immediately above the basal chill zone, to a maximum of ~ 0.35 mm about two thirds of the way up the cumulate layer (Figure 3.12). In other flows there appears no systematic change in grain size or modal olivine content.

The total modal olivine content of cumulate layers varies from flow to flow, from as low as 40–50% in poorly-differentiated flows with thin, irregular spinifex layers, to 80–98% in the interiors of thick komatiitic lava flows, e.g. Arndt (1977b), Barnes *et al.* (1988), Hill *et al.* (1995), Dowling and Hill (1993, 1998). In flows with normal dimensions, maximum modal olivine content (60–80%) is attained in the upper third of the cumulate layer. Higher in the flow there is a decrease to lower content in the B_1 layer; at deeper levels, modal olivine content first decreases gradually, then declines rapidly in a marginal olivine-depleted layer.

Basal chill zones

Basal chill zones form the lowest few centimetres of most komatiite flows. Their textures resemble those of the overlying cumulates in that they normally consist of polyhedral olivine grains in a finer-grained pyroxene–glass ground-mass. They differ in that the olivine grains are far smaller (0.1–0.3 mm) and far less abundant (typically 5–30%). Microspinifex texture, like that in many samples from upper chill zones, is rarely encountered in lower chills. The SASMAR drill core (Nisbet *et al.*, 1987), which penetrated the basal chill zones of several well-preserved Zvishavane komatiite flows, sampled the

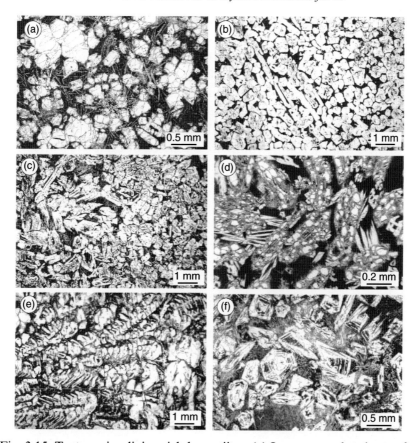

Fig. 3.15. Textures in olivine-rich komatiites. (a) Lower cumulate layer of a differentiated flow from Zvishavane, Zimbabwe: note the subhedral, equant, rounded habit, the variation in grainsize and the clustering of olivine grains. (b) Close-packed olivine grains from the interior of the olivine cumulate of the Alexo flow: most grains are equant and subhedral but some are elongate. (c) Olivine cumulate from just below the B_1 layer of the Alexo komatiite: most of the grains are skeletal. (d) Sparsely distributed, highly skeletal grains from just below the B_1 layer of a Gilmour Island komatiite. (e) Branching complex curved dendritic olivine grains in the upper part of the Murphy Well flow: the top of the flow is to the right. (f) Highly skeletal hopper olivine grains from the cumulate layer of a Gorgona Island komatiite.

unusual textures illustrated in Figure 3.9(c). In these rocks, clusters of poly-hedral olivine phenocrysts lie in a matrix containing numerous, exceedingly small (<0.1 mm), equant olivine microlites in a glassy groundmass.

Other unusual spinifex-textured komatiites

Komatiite flows in the 3.3 Ga Commondale greenstone belt in the southern Kaapvaal Craton have unusual textures and structures, as described by Wilson *et al.* (1989) and Wilson (2003). These flows lack the highly asymmetric layering of normal komatiite flows and instead show a more symmetrical distribution in which central olivine cumulate layers are flanked above and below by layers containing abundant microspinifex-textured veins and patches. Some of the spinifex is olivine dominated but more commonly acicular grains of orthopyroxene are the major component. Detailed petrographic descriptions of these rocks have not yet been published but their remarkable chemical compositions are discussed in Chapters 5 and 11.

Kareem and Byerly (2003) report similar textures in komatiite flows in the 3.3 Ga Weltevreden Formation, which is situated in the middle of the stratigraphic sequence of the Barberton greenstone belt. The spinifex textures of these flows are dominated by acicular grains of orthopyroxene, rimmed by augite, just as in the Commondale flows. Orthopyroxene is also a common interstitial phase in the olivine cumulates in the lower parts of the flows. The mineralogy of these spinifex-textured lavas resembles that of some pyroxene spinifex textures in komatiitic basalts, but the South African examples have very high MgO contents, greater than 30% in many cases, which classes them as komatiites. The high MgO is due in part to the unusually magnesian compositions of both the orthopyroxene and olivine, as discussed in Chapter 4.

Finally, the top 20 m of the komatiite flow in Murphy Well in the eastern Yilgarn in Australia is made up entirely of complex, branching skeletal dendritic grains, as shown in Figure 3.15(d).

Harrisitic textures

Harrisite is a type of texture that typically develops in the interiors of shallow-level intrusions such as the Rum intrusion, in Scotland, where the term was coined (Harker, 1908; Donaldson, 1974). The texture consists of very large branching olivine crystals, commonly tens of centimetres in length, with a complex, crudely skeletal habit (Figures 3.8 and 10.4). The texture is somewhat similar to platy olivine spinifex but is distinguished by the less skeletal, more solid morphology of the olivine grains and the coarser nature of the interstitial phases. The resemblance probably comes about because both textures form when olivine grows from a solid substrate into crystal-free liquid; downward from the crust of the flow in the case of spinifex, and upwards from the floor of the intrusion in the case of harrisite. O'Driscoll *et al.* (2007) have proposed that Rum harrisites formed when inputs of hot, picritic magmas were chilled against cooler layers of cumulate rock in the interior of the intrusion. Similar texture in the cumulate layers of thick komatiite flows in Western Australia probably formed in a similar manner, as discussed in Chapter 9.

3.6 Komatiite flows without spinifex textures

Massive flows without spinifex textures lie at the other end of the spectrum of komatiite flow types illustrated in Figure 3.2. Though less spectacular, they are far more common than spinifex-textured flows, making up 60–80% of most komatiitic sequences (Hill *et al.*, 1995; Dann, 2000). In their extreme form, these flows are almost undifferentiated and have a uniform olivine porphyritic texture throughout. Only the size and spacing of the polyhedral joints that pervade the flows change: at flow boundaries these joints are thin (1–3 mm), rectilinear, closely spaced and enclose polyhedral cells of lava 2–5 cm across (Figure 3.16(a)); deeper down towards the centre of the flow, they are thicker (5–10 mm), gently curved, and enclose polyhedral to irregular ovoid volumes of lava (Figure 3.16(b)). In areas of poor outcrop these structures have been mistaken for pillows. The polyhedral joints cut across the fabric of the lava, and in some thin sections are seen to bisect phenocrysts. Although they are commonly made more conspicuous during post-emplace-ment alteration (they appear to act as passageways to the fluids that passed through the komatiites during or well after their solidification), the joints themselves probably formed by thermal contraction during cooling and con-traction of the lava, in a manner akin to columnar joints. Exactly why the joints are polyhedral and not columnar, as in modern basaltic flows and sills and some Precambrian komatiitic basalts, is not understood.

Massive komatiite flows are generally thicker but not as laterally extensive as spinifex-textured layered flows. Detailed mapping in northern Munro Township (Figure 3.17) reveals a complex sequence of short stubby flows with steep, gently curved terminations; probably a two-dimensional section through a pile of lobate tongues of komatiite lava. This form of flow is developed to an extreme degree in the southwestern part of the Pyke Hill outcrop (Figure 3.18(a)), which is made up of lava lobes that are only slightly wider than they are thick. In the outcrops illustrated in Figure 3.17 there is a distinct polarity of morphological type, from 1–2 m thick spinifex-textured layered flows at the base, to 2–4 m thick massive flows at the top of this 50 m thick cycle. In other outcrops, the two komatiite types are interlayered (e.g. on Pyke Hill, Figure 2.11).

In the Barberton greenstone belt, massive sheet flows, up to 55 m thick, form the prominent resistant ridges that dominate the landscape (Figure 2.2).

Massive komatiite flows are texturally monotonous. Most contain polyhedral solid, equant, euhedral (rarely rounded) phenocrysts and sparse minute cubes and octahedra of chromite, within the ubiquitous augite–glass matrix (Figure 3.19(a)). Skeletal overgrowths are rare but a few flows contain elongate, platy skeletal

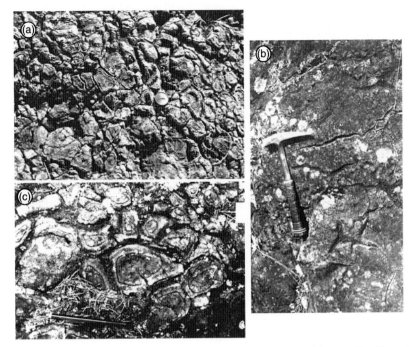

Fig. 3.16. Polyhedral jointing in massive, undifferentiated komatiite flows in northern Munro Township: (a) small polyhedral joints from near the top of a flow; (b) detail of polyhedral joins; (c) larger curved joints from the interior of a flow.

(hopper) grains (Figures 3.19(b),(c)). On a broad scale, the abundance of olivine phenocrysts varies little from the top to the base of the flow, or there may be a gradual increase in abundance to a maximum about two thirds of the way through the flow (Figure 3.4). No systematic studies of grain size variations appear to have been carried out on this type of flow. At the hand sample or thin section scale, the distribution of phenocrysts is seen to be irregular: the olivine grains commonly form clusters each containing 3–10 grains that are distributed through areas containing a much lower proportion of olivine (Figure 3.19(c)).

3.7 Flows with thin spinifex layers

This type of flow, which is distinguished by the presence of thin, irregular spinifex veins, has characteristics intermediate between those of differentiated flows with well-developed spinifex textures and massive flows. Spinifex textures in these flows are of the random type (platy spinifex is rarely present), B_1 layers are absent, the contact between spinifex and cumulate layers is gradational, and olivine in the cumulate layer is loosely packed. The modal abundance

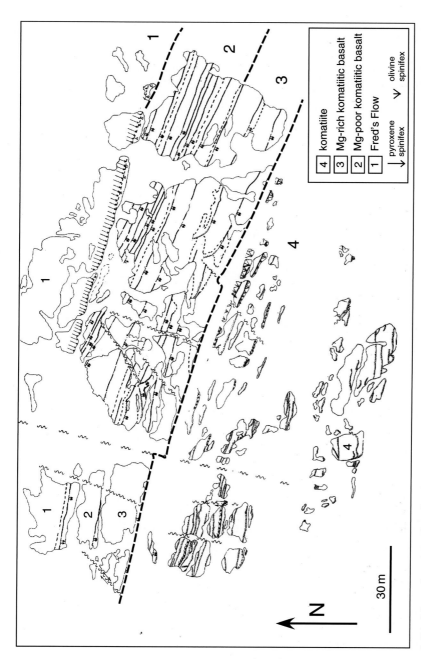

Fig. 3.17. Map showing the distribution of komatiitic basalts and komatiites in the northern part of Munro Township (Arndt, 1975).

Legend:
- 4 komatiite
- 3 Mg-rich komatiitic basalt
- 2 Mg-poor komatiitic basalt
- 1 Fred's Flow

→ pyroxene spinifex
⋎ olivine spinifex

30 m

N

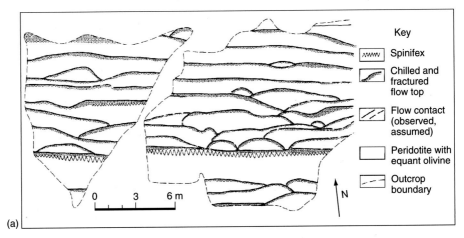

(a)

Key

ᶺᶺᶺᶺ	Spinifex
	Chilled and fractured flow top
⧅	Flow contact (observed, assumed)
	Peridotite with equant olivine
	Outcrop boundary

0   3   6 m

N

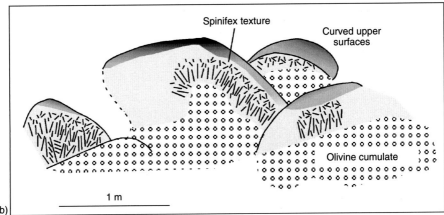

Spinifex texture

Curved upper surfaces

Olivine cumulate

1 m

(b)

Fig. 3.18. Sketches of pillow- or lobe-like structures in komatiites: (a) structures in massive komatiite in the lower part of the Pyke Hill sequence, from Pyke *et al.* (1973); (b) differentiated spinifex-textured structures in northern Munro Township, redrawn from Arndt (1975).

of olivine in cumulate layers is intermediate between that in cumulates of well-layered flows and the porphyritic lava of massive flows (Figure 3.4). Upper spinifex layers vary in thickness and are discontinuous along strike.

Where best developed, the veins of spinifex-textured lava form festoons of crescent-shaped, convex-upward veins that decorate the uppermost portion of the cumulate layers, as illustrated in Figure 3.20(c). In many such veins, the size and form of the spinifex crystals is symmetrical, varying from fine random spinifex at both margins to coarser platy spinifex in the centre. In other flows, irregular patches of spinifex texture are distributed haphazardly within the cumulate layer. In these patches the size and form of the spinifex grains varies considerably and contacts with surrounding cumulate lava are gradational.

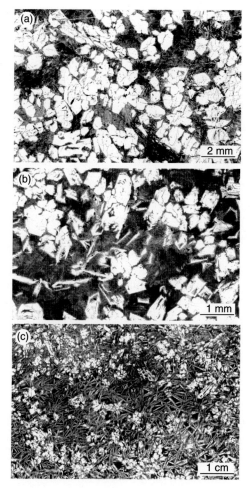

Fig. 3.19. Textures in massive komatiites, illustrated by three photomicrographs of samples from Munro Township. Note the variation in the size and habit of the olivine grains, from the solid polyhedral grains in (a) to the highly skeletal forms in (c).

The spinifex-textured lava in the veins and patches has a composition similar to that of the lower part of overlying spinifex layers. These features are believed to form by the congregation of intercumulus liquid in fractures or dispersed zones of the cumulate layer during the last stages of solidification of the flows.

The cross-cutting aspect of spinifex-textured veins in komatiite flows in the Barberton greenstone belt was cited by de Wit *et al.* (1987) as evidence that these units had formed as intrusive sills. In subsequent studies Dann (2000, 2001) reinterpreted them as internal features within extrusive units, part of the lava inflation process that is described more fully in Chapter 9.

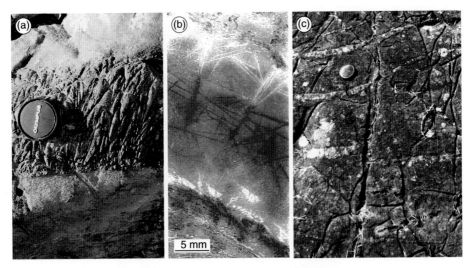

Fig. 3.20. Spinifex-textured veins in poorly differentiated komatiites: (a) small horizontal vein in the cumulate zone of a Pyke Hill komatiite; (b) scanned thin section of a vein in poorly differentiated komatiitic basalt of Gilmour Island; (c) upward-concave crescent veins in the cumulate zone of a komatiite in the northern part of Munro Township.

3.8 Thick dunitic units

Thick dunitic bodies within komatiitic sequences in the Yilgarn Block of Western Australia have attracted considerable attention, mainly because they are a common host of Ni sulfide ore deposits (Naldrett and Turner, 1977; Barnes *et al.*, 1988; Dowling and Hill, 1993; Lesher and Keays, 2002). Typical examples are the 220 m thick Honeymoon Well ultramafic body, which has a spinifex-textured top, and still thicker units without obvious volcanic features such as the >700 m thick Perseverance ultramafic complex and the still larger Six Mile and Mt Keith bodies (Figure 3.21). Thick dunitic units are not restricted to the Yilgarn Block but are also found in Canada (Jensen and Langford, 1985; Duke, 1986a, b) and South Africa (Anhaeusser, 1985).

The dunite bodies range from large but aerially restricted units with lenticular cross sections (about 0.5–2 km wide and 200–700 m thick) to extensive sheets such as in the Walter Williams Formation, which occupies an area of 120 km by 40 km between Ora Banda, Siberia and Ghost Rocks in the Yilgarn Craton of Western Australia (Figure 3.21). Many bodies pinch and swell along strike; the thinner intervening regions have been interpreted by Hill *et al.* (1995) and Dowling and Hill (1993) as in-filled lava tubes separated by contemporaneous flanking lava flows. Most of the dunitic bodies are steeply

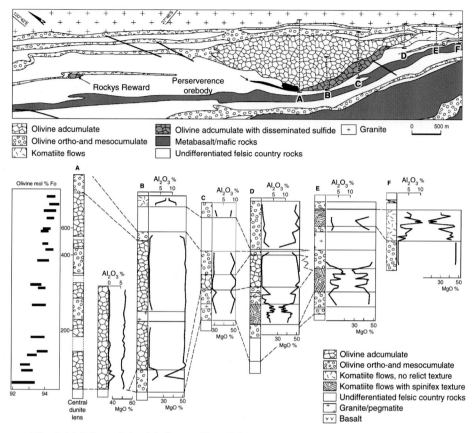

Fig. 3.21. Map of dunitic komatiite of the Perseverance ultramafic complex in the eastern Yilgarn, Australia (from Barnes *et al.* (1988)).

dipping and have conformable or only locally transgressive contacts with adjacent volcanic or sedimentary rocks.

The longitudinal extent of the lenses is unknown, but it is entirely possible that Mt Keith and the Yakabindie bodies (Six Mile, David, Goliath) are part of the same elongate channel, which would then be at least 25 km long. The Digger Rocks dunite unit in the Forrestania belt is probably an oblique section through a shallow plunging lens, at least 15 km long. It is possible to trace a dominantly orthocumulate, spinifex-free package containing lenticular dunite cores all the way from Yakabindie to Kingston, about 45 km.

The larger ultramafic bodies are zoned, with cores of massive, virtually monomineralic olivine adcumulate and margins of olivine ± pyroxene ± amphibole meso- and orthocumulates. Marginal zones are almost invariably serpentinized but interior dunitic pods and layers contain relict igneous olivine. Some of the dunite bodies also have internal layering: Six Mile has

alternations of adcumulate and mesocumulate, with layers of poikilitic orthocumulate towards the top, and Mt Keith contains discontinuous internal
layers of the so-called 'porphyritic olivine rock' which is probably a recrystallized harrisite lithology. Mount Keith contains layered pyroxenites and leuco-
gabbros in the upper part of a constricted zone flanking the main dunite lens
(Fiorentini *et al.*, 2007; Rosengren *et al.*, 2007). Layered orthocumulates,
coarse harrisites and in some places pyroxenites and gabbros are developed
above the central dunite layer in the sheet-like Walter Williams unit. Layering
on a centimetre to metre scale, defined by variations in grain size, texture and
olivine abundance, is common in marginal zones.

The adcumulates in the larger dunitic bodies are remarkable, virtually
monomineralic rocks made up of 95–99% coarse-grained (0.5–2 cm),
mosaic-textured, highly magnesian olivine with the composition Fo_{93-95}
(Figure 3.22(a)). Interstitial minerals are usually restricted to 1–5 mm lobate
interstitial or 0.1 mm euhedra of chromite, and in some cases interstitial Ni–Fe
sulfides. The content and morphology of chromite varies with the olivine
composition. Dunites with Fo > 93 mol% typically have no chromite at all,
while those with around Fo_{92-93} have lobate interstitial and sometimes poikilitic chromites up to 5 mm in size. More Fe-rich dunites typically contain more
abundant euhedral chromite (Barnes, 1998).

In meso- and orthocumulates olivine grain size is smaller and more variable,
and most grains are equant to moderately elongate, usually euhedral, sometimes
rounded. Intercumulus minerals are rarely preserved but in some samples poikilitic pyroxenes and minor aluminous amphibole and apatite have been recognized
(Hill *et al.*, 1987). Layers of harrisitic olivine (coarse, branching, moderately
skeletal grains) may be present, most commonly in upper marginal zones.

Centrally disposed disseminated sulfides are found in the Mt Keith and
Yakabindie dunite bodies. The Honeymoon Well dunite contains pods of disseminated sulfide within the lower 100 m or so, and in places has heavily
disseminated and minor massive ore at the contact. Perseverance has a substantial basal accumulation of disseminated and matrix ores within the dunite itself,
and basal massive ores, which were interpreted by Barnes *et al.* (1995) as being
hosted by precursor flows. Basal accumulations of disseminated and massive ore
are also found in the Digger Rocks dunite body (Perring *et al.*, 1996).

The exact nature of the large dunitic bodies is uncertain. Initially they were
interpreted as sills (e.g. Naldrett and Turner (1977)), then there was a period in
which they were thought to be extrusive and to have formed as lava lakes of
rivers. More recently the pendulum has swung back: Rosengren *et al.* (2005)
mapped apophyses of ultramafic rock that penetrated upward into overlying
volcanic units and used these, in combination with other features, to argue that

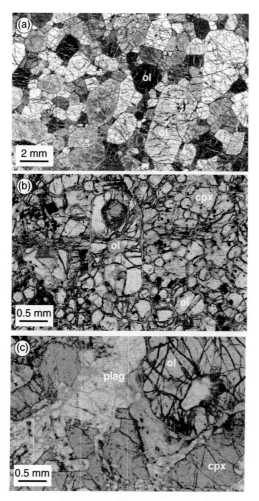

Fig. 3.22. Textures in intrusive komatiitic rocks. (a) Olivine adcumulate (dunite) from the Wildara area in the eastern Yilgarn, Australia. The rock consists almost entirely of unaltered olivine with an equilibrium texture. (b) Olivine cumulate from an ultramafic intrusion on Gorgona Island. The large olivine grain is cumulus; the smaller grains are poikilitically enclosed in augite oikocrysts. (c) Gabbro containing olivine, augite and plagioclase from a mafic intrusion on Gorgona Island. (ol – olivine; cpx – clinopyroxene; plag – plagioclase.)

the Mt Keith dunite was intrusive. The issue, still controversial, is discussed by Barnes (2007) and in Chapter 9.

3.9 Pillowed komatiites

Two types of pillow-like structure are known in komatiites. The first are large lobe- or tongue-like structures, such as have been mapped in the southwestern portion of the Pyke Hill outcrop (Pyke *et al.*, 1973; Shore, 1996; Figure 3.18) and in other parts of the Abitibi greenstone belt (Arndt *et al.*, 1977; Imreh, 1978; Jensen and Langford, 1985). The Pyke Hill structures have a bulbous, pillow or bun shape and range from 0.5 to 5 m wide and from 0.3 to 2 m high. They are composed of olivine porphyritic lava and are polyhedrally jointed. Structures in the northeast of the township, illustrated in the lower diagram of Figure 3.18, have similar dimensions (2–6 m wide) but have a convex upper skin of chilled lava, a central core of olivine cumulate and an intervening spinifex textured zone. The lower surfaces are concave and conform to the shape of underlying lobes. Both types of structure probably formed by the advance of lobes of lava at the front of a komatiite flow, in the manner of pillows in modern basalts. Similar lobe- or pillow-like structures are described in the Abitibi belt in Quebec by Imreh (1978) and Mueller (2005) and in the Barberton belt by Dann (2000).

The second type of pillow-like structure is smaller (20–100 cm) and more spherical. Examples have been described by Viljoen and Viljoen (1969b), Arndt *et al.* (1977), Nisbet *et al.* (1977) and Imreh (1978). In each case they are found in polyhedrally-jointed olivine porphyritic komatiites and many authors (e.g. Williams and Furnell (1979), Dann (2000)) are not convinced that these structures formed as distinct entities and are not simply cells between large, curvilinear joints; i.e. a well-developed form of the polyhedral jointing that characterizes many massive komatiite flows.

3.10 Volcaniclastic komatiites

In most greenstone belts, volcaniclastic komatiites are uncommon. In many Archean greenstone belts, outcrop of fragmental ultramafic volcanic rock is largely limited to thin (centimetre- to metre-scale) flow-top breccias and hyaloclastic rocks, as at the top of the Alexo komatiite flow (Barnes, 1983; Arndt, 1986a), Fred's Flow (Arndt, 1977b), or in several places at the top of the Walter Williams Formation (Hill *et al.*, 1987). Shore (1996) describes millimetre-scale breccias at the tops of Pyke Hill komatiite flows and distinguishes these units from the polyhedrally jointed upper chill zones of these flows. Similar units have also been recognized in drill cores of the Zvishavane spinifex flows in the Belingwe belt (Nisbet *et al.*, 1987; Renner, 1989), but not in the surface outcrop of the flows. The rubbly, soft breccias are easily eroded and may be far more common than the outcrop record would indicate. A thick persistent layer of fine fragmental komatiite in the lower Onverwacht Group of the Barberton greenstone belt had long gone unnoticed because it is strongly silicified (Thompson *et al.*, 2005).

In the Quebec portion of the Abitibi belt, ultramafic volcaniclastics are encountered more frequently and occur as discontinuous, thin tuffaceous units interlayered with komatiite flows and rare, thicker, coarser units with erosional contacts with underlying volcanic rocks. Gélinas *et al.* (1977) provided the first detailed description of fragmental komatiite. Other notable occurrences of volcanoclastic komatiites are at Steep Rock in the Wabigoon province of Canada (Barnes and Often, 1990; Schaefer and Morton, 1991), in the Lac Guyer region of Quebec (Stamatelopoulou-Seymour and Francis, 1980), at Scotia in the Yilgarn Craton (Page and Schmulian, 1981), at Ruth Well in the Pilbara of Western Australia (Nisbet and Chinner, 1981), in the Mendon and Kromberg Formations of the Barberton greenstone belt, South Africa (Byerly, 1999; Lowe and Byerly, 1999b; Ransom *et al.*, 1999), in northern Norway (Barnes and Often, 1990) and on Gorgona Island (Echeverría and Aitken, 1986). Only in Scandinavia and in French Guiana have abundant volcaniclastic komatiites been described. In the Sattasvaara area in Finland (Saverikko, 1985), ultramafic tuffs and coarse breccias are the most abundant type of ultramafic volcanic rock and predominate over komatiite flows; and at Dachine in French Guiana, poorly preserved ultramafic fragmental rocks are the hosts of a large deposit of microdiamonds (Capdevila *et al.*, 1999). Some of these units have been interpreted as pyroclastic, but they contain fragments of rock types that do not form pyroclastically (e.g., spinifex-textured or cumulate lava) and are therefore more likely epiclastic. Others contain only fine-grained glassy material and are limited in stratigraphic extent, so they are more likely hydroclastic. Komatiites *sensu strico* that contain unequivocal spatter and bombs have not been reported, but komatiitic basalts with such features have been described in the Möykkelmä area in Finland (Räsänen, 1996) and ferropicrites with such features have been described at Kotselvaara in the Pechenga belt (Green and Melezhik, 1999). Autoclastic lavas have been reported from Kambalda by Beresford *et al.* (2002), and are also observed at Sattasvaara in Lapland (Thordarson, pers. comm.).

In the following subsections I have chosen three examples that illustrate the main characteristics of fragmental komatiite.

Flow-top breccia of the Alexo komatiite flow

The top 1–2 m of the Alexo komatiite flow consists of uniform, unbedded breccia containing well-sorted, subangular to amoeboid fragments, 0.5–1 cm in maximum dimension. Textures in the fragments range from microporphyritic (1–3% olivine phenocrysts) through microspinifex to largely glassy (Figure 3.23(b)). Many fragments contain abundant olivine microlites and some contain small amygdules (1–2%). All primary minerals are replaced by secondary hydrous phases. A 2–5 cm thick chill zone delineates the contact between breccia and underlying spinifex lava. Textures of fragments within the breccia strongly

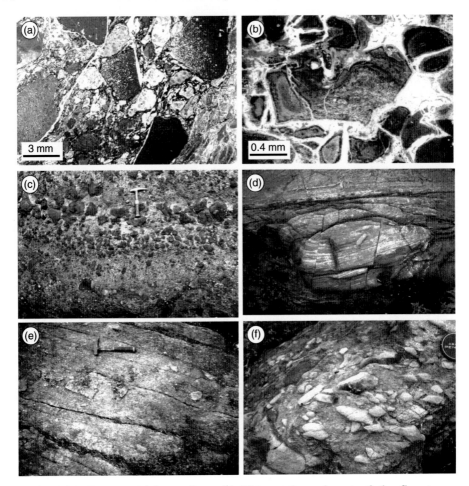

Fig. 3.23. Fragmental komatiites. (a) Thin section of part of the flow-top breccia at the top of Fred's Flow, Munro Township. Variably altered fragments of lava with variable grain size and texture lie in a matrix of finer, more altered fragments. Note the concave fractures at the margins of the larger fragments. (b) Hyaloclastite from the top of the Alexo komatiite flow. (c) Coarsely bedded hyaloclastite of picritic composition from Gorgona Island. (d) Fine, bedded komatiite tuff in a channel between two komatiite flows on Spinifex Ridge in the Quebec portion of the Abitibi belt. (e), (f) Unbedded fragmental komatiite, probably hyaloclastite, from Sattasvaara, Finland.

resemble textures in the uppermost part of the underlying flow, as do their chemical compositions (except for the effects of alteration – see Chapter 5). There seems little doubt that this breccia and others of the same type are hyaloclastites that formed during interaction of hot komatiite lava with seawater.

Such flow-top breccias are not to be confused with the polyhedrally-jointed A_1 layers of normal komatiite flows. The former are true fragmental rocks in which the clasts were mechanically generated and are displaced relative to one another; in A_1 layers the 'fragments' are cells separated by cooling joints in a once-continuous volume of lava. The latter structure formed *in situ* and the fragments have retained their original disposition.

It is not clear why hyaloclastitic flow-top breccias are not more common. The Alexo flow has no obvious morphological or chemical characteristics that distinguish it from komatiite flows without flow-top breccias. Part of the answer may lie in the low probability of exposure of these poorly-consolidated, soft rocks, as suggested by the presence of breccias in the drill core in the Belingwe belt and their absence in surface outcrop, but the lack of such breccias in regions of complete outcrop such as on Pyke Hill or in the underground workings at Kambalda demonstrates that this cannot be the complete solution.

Intraflow volcaniclastites, Spinifex Ridge, Quebec

Two types of ultramafic volcaniclastic rocks are found on Spinifex Ridge in the Quebec portion of the Abitibi Belt (Gélinas *et al.*, 1977b). The first type is a fine-grained tuff composed of small, chloritized shards, granules, globules and angular fragments of olivine grains. This rock is finely laminated, in places graded, and forms metre-thick beds between komatiite lava flows (Figure 3.23(d)). The second type is a coarser breccia made up of large angular fragments of lava from all parts of layered komatiite flows (chilled margin, spinifex, cumulate layer), more or less uniformly distributed in massive beds devoid of sedimentary structures. Beds of the first type of ultramafic volcaniclastite lie conformably on underlying komatiite flows; the second type fills an erosion channel. Both have compositions intermediate between those of spinifex and cumulate layers of komatiite flows (between 26 and 32% MgO).

Gélinas *et al.* (1977a) interpreted the second type of volcaniclastite as a channelized mass flow deposit containing fragments of detrital and/or volcanic origin. They proposed that the fragments in the first type of volcaniclastite formed either by hyaloclastic or pyroclastic processes, and were reworked and deposited by turbidity currents.

More recently Dostal and Mueller (2001) described komatiitic hyaloclastites in the Stoughton–Roquemaure group, in another part of the Abitibi greenstone belt in Quebec. Their descriptions make it clear that these rocks formed through fragmentation of hot komatiite magma as it came into contact with cold seawater.

Ultramafic pyroclastic komatiites at Sattasvaara, Finland

Saverikko (1985) described in great detail, and showed numerous photographs of ultramafic fragmental rocks from Sattasvaara in northern Finland. In this area he found thick, laterally extensive units that he described as tuffs, crystal tuffs, lapilli tuffs and agglomerates with compositions ranging from basalt (8% MgO) to komatiite (26% MgO). The ultramafic pyroclastic rocks form units from about 20 to more than 300 m thick within a sequence of massive, pillowed and blocky autobrecciated basaltic and komatiitic flows. The fragmental rocks are massive, poorly sorted and largely unbedded (Figure 3.23(e), (f)). The constituent fragments are angular except for some cored (accretionary?) lapilli. Vitric and lithic fragment clasts, the latter with olivine phenocrysts, some fragments of exotic rock types (amphibolite, calc-silicates, magnetite), and rare bombs are also recognized. The textures and structures in the fragmental rocks, the massive or blocky nature of amygdaloidal interlayered flows and the overall geological situation led Saverikko (1985) to conclude that the units are pyroclastic volcanics that had erupted in a largely subaerial environment. This interpretation has been questioned by R. E. T. Hill (personal communication) who cited features such as the relatively uniform composition and texture of the fragments, the presence of jigsaw-fit autoclastic pillow breccias, the lack of highly vesicular lavas, and aspects of the geometry and size distribution of the fragments as evidence that the rocks are hyaloclastites that formed in a subaqueous environment. The search for subaerial komatiite goes on.

Similarly, the fragmental picrites on the southern end of Gorgona Island most probably formed as hyaloclastites, rather than as pyroclastic rocks as envisaged originally by Echeverría and Aitken (1986). Mapping by Révillon *et al.* (2000) has shown that the margins of several picrite dykes that invade the fragmental units have been progressively transformed into blocks and fragments whose form, size and petrology closely resemble those in fragmental rocks themselves. In this example, the brecciation clearly came about as hot picritic lava invaded the water-charged tuffaceous sequence. The fragmental rocks were then reworked and redeposited to form the bedded units illustrated in Figure 3.23(b). Very probably similar processes produced the Sattasvaara fragmental komatiites.

3.11 Komatiitic basalts

To describe the range of structures and textures in komatiitic basalts requires first that the rock type be defined. As has been discussed in Chapter 2, the definition of komatiitic basalt is based partly on spatial relationships with komatiite, partly on chemical similarities and partly on the textures themselves. A distinctive type of magnesian basalt found in association with komatiite is characterized by an abundance of skeletal pyroxene. Commonly this pyroxene is very fine-grained and randomly oriented, but in some flows, the pyroxene grains are large (up to several centimetres long), and are oriented parallel to one another to form pyroxene spinifex texture. This is the principal diagnostic feature of komatiitic basalt: flows with this type of texture seem very commonly, although not invariably, to be spatially associated with komatiite, and in many examples a genetic relationship between the two rock types can be demonstrated using geochemical criteria. In some cases these basalts seem to be the simple products of fractional crystallization of parental komatiite, but frequently the komatiite seems to have assimilated country rocks and a more complicated petrogenesis is required.

Size, shape and morphology of komatiitic basalt flows

Komatiitic basaltic flows are thicker than komatiite flows, typically 2–20 m compared with 1–10 m for the komatiites. They also form far thicker units like the 120 m thick Fred's Flow (Arndt, 1977b) and the even thicker Ghost Range units (Jensen and Langford, 1985). Most of the thicker flows are internally differentiated and strongly layered, comprising olivine and pyroxene cumulate lower units, brecciated or spinifex-textured upper layers, and in some cases an intervening gabbroic unit (Figure 3.24). Spectacular layers composed of coarse, parallel needles of pyroxene, the so-called 'string-beef' spinifex (Figure 3.25), dominate the upper parts of some flows. Individual needles up to 50 cm in length form layers many metres thick, as in the upper part of Fred's Flow (Arndt, 1977b). Wilson *et al.* (1989) described giant cones, 1–2 m long and wide, built up of coarse pyroxene grains, in komatiitic basalts of the Nondweni greenstone belt in South Africa (Figure 3.25(a)). Other structures found in komatiitic basalts include pillows, polyhedral and columnar joints (the latter most commonly in olivine cumulate layers), hyaloclastites and other volcanic breccias. Pyroclastic komatiitic basalts can also be recognized from the acicular pyroxene textures in clasts and chemical similarities with associated basaltic and komatiitic flows. For more detailed accounts of the field characteristics of komatiitic basalts, the reader is referred to Viljoen and Viljoen (1969b), Nisbet *et al.* (1977), Imreh (1978), Arndt *et al.* (1979) and Wilson *et al.* (1989).

Textures

In more magnesian komatiitic basalts, those with 14–18% MgO, olivine is present and textures are similar to those in komatiites. Spinifex, porphyritic and cumulate textures are all developed. These textures differ from those in komatiites in that the olivine is less abundant: in spinifex textures, olivine occurs as wafers that are far thinner, though not necessarily shorter, than those

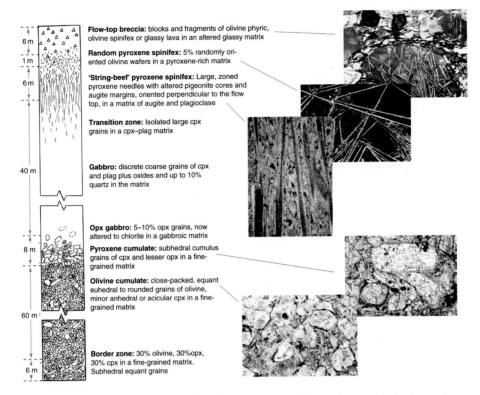

Fig. 3.24. Diagram illustrating the structure, dimensions and textures in Fred's Flow, a thick, layered komatiitic basaltic flow in Munro Township. (opx – orthopyroxene; cpx – clinopyroxene; plag – plagioclase.)

in the more magnesian rocks. In porphyries and cumulates, the olivine grains are less closely packed, less evenly distributed and less well sorted (Renner, 1989). Clusters of olivine grains are a common feature.

In less magnesian lavas, those with 10–14% MgO, subcalcic augite is the dominant mineral. Rapidly cooled samples from the margins of flows or pillows are composed of randomly oriented, thin pyroxene needles set in a once-glassy groundmass. Samples from the interiors of flows are coarser grained and the pyroxenes are more complex. Most consist of zoned hollow needles (0.2–0.5 mm wide, 1–50 cm long), some of which have cores of pigeonite or more rarely orthopyroxene, surrounded by mantles of augite (Figures 3.24, 3.25). The orientation of such needles varies from random near the tops of flows to parallel to one another and perpendicular to flow margins in interior spinifex layers. In thicker flows there is a definite progression of textures and mineralogies, from random olivine spinifex near the top to parallel pyroxene spinifex in the interior. The needles in spinifex textures are

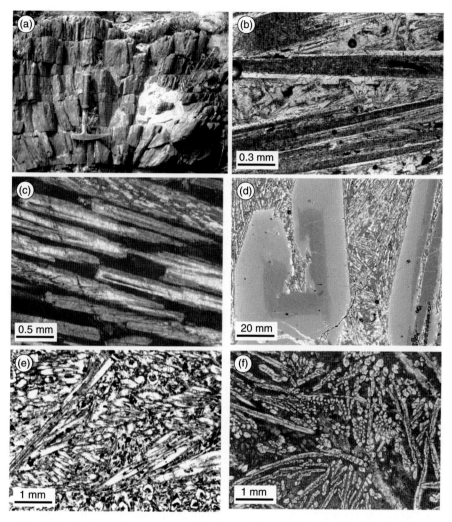

Fig. 3.25. Textures in pyroxene-rich komatiites. (a) Large sheaths of 'string-beef' pyroxene spinifex in an outcrop of komatiitic basalt in the Nondweni belt in South Africa. (b) Thin section of zoned pyroxene needles in 'string-beef' texture from the upper part of Fred's Flow, Munro Township. The needles are oriented perpendicular to the flow top (the photograph has been rotated by 90°). Each needle has a core of pigeonite, now altered to chlorite, and a margin of augite. They lie in a matrix of augite, plagioclase, iron oxides and quartz. (c) Thin section of zoned pyroxene needles from a cell interstitial to olivine blades in spinifex lava from the Alexo flow. Here the pigeonite cores of the zoned crystals have escaped alteration. (d) Backscatter electron image of a zoned pyroxene needle cut perpendicular to the long axis. The pigeonite core is darker than the augite margin. (e), (f) Komatiitic basalts from Munro Township consisting mainly of randomly oriented acicular grains of augite.

commonly in optical continuity and apparently are elements in far larger complex grains, each containing several hundred needles. The matrix interstitial to spinifex pyroxenes is made up mainly of smaller stubbier augite grains and plagioclase, with lesser amounts of brown amphibole, quartz, ilmenite and magnetite. Arndt and Fleet (1979), Wilson *et al.* (1989) and Parman *et al.* (1997) gave more detailed descriptions of the morphology and compositions of pyroxenes in spinifex-textured basalts.

The least magnesian komatiitic basalts are massive or pillowed and have textures that are not greatly different from those in tholeiitic basalts. In places where stratigraphic or chemical criteria can be used to identify the most evolved members of a komatiitic suite, as in Munro Township where basalts with ~7% form the interior portions of layered flows and share distinctive chemical features with more magnesian komatiitic basalts, these basalts are found to be somewhat coarser grained than typical tholeiites and to be characterized by early crystallization of pyroxene (Arndt and Nesbitt, 1982). Cameron and Nisbet (1982) discuss other textural, mineralogical and chemical criteria that can be used to distinguish komatiitic basalt from other types of basalt.

3.12 Intrusive komatiites

Komatiite is a dense magma that should commonly be trapped at various levels in the crust. Ultramafic intrusions of probable komatiite lineage were recognized from the very start of komatiite research. Viljoen and Viljoen (1969a) noted that a large discordant unit which cuts down through the type section of the lower Komati Formation of the Barberton belt shared many mineralogical and chemical features with the komatiite flows and they concluded that this unit, an irregular dyke, formed from komatiite magma. They also discussed the nature of the thick concordant olivine cumulate layers that form a conspicuous element of the Barberton landscape (Figure 2.2), and they concluded that these units were shallow-level sills. Mapping in southern Munro Township revealed the presence of a thick (>300 m) concordant unit dominated by olivine cumulates and of several smaller discordant dykes, all of which were clearly intrusive (Arndt, 1975; Arndt *et al.*, 1977). Other authors have described komatiitic intrusions of various sizes and dimensions in various parts of Australia, Canada and Zimbabwe e.g. Williams and Furnell (1979), Jensen and Langford (1985), Duke (1986a, b).

Investigations of komatiitic basalts in the Raglan belt in northern Quebec, important because they contain major Ni–Cu sulfide deposits, have yielded ambiguous results. These units invaded a series of poorly consolidated sedimentary rocks and opinion is divided as to whether the thicker olivine-cumulate-dominated units are mainly extrusive, as preferred by Lesher (1999, 2007) or mainly intrusive, as advocated by Trofimovs *et al.* (2003) and Rosengren *et al.* (2005). This issue is discussed in detail in Chapter 9.

A similar picture issues from studies of the 'Dundonald Beach' area in the Abitibi belt of Ontario. Houlé *et al.* (2002, 2008a) and Cas and Beresford (2001) described a series of thin (30–120 cm) concordant to moderately discordant units intercalated with graphitic shales. These authors used the presence of distinctive and diagnostic peperites to argue for a shallow-level intrusive origin of these komatiites. Arndt *et al.* (2004b) gave a detailed description of a 5 m thick, spinifex-textured layered komatiite in a superbly exposed glacially polished pavement in the same area. They used textural evidence and grain size variations at the upper contact to argue that this unit was intrusive as well; not into sediments but into other komatiitic units. An important aspect of this study was the subtle nature of the textures that had to be used to establish the intrusive origin: in normal, poorly exposed komatiites, these textures would not be preserved. To some extent the distinction is unimportant because even if these komatiites did reach the surface, they were emplaced at the very shallowest levels of the crust (Chapter 9).

At another scale, both spatially and in terms of overall importance (because they contain large Ni sulfide ore deposits, as described in Chapter 10), are the dunitic bodies in the Yilgarn of Western Australia. As discussed above, opinion is divided as to whether they were major extrusive lava pathways (Dowling and Hill, 1998; Hill *et al.*, 2004) or subvolcanic sills (Rosengren *et al.*, 2005).

4

Mineralogy

4.1 Introduction

The mineralogy of unaltered komatiites is remarkably simple. Only olivine crystallizes abundantly and its composition is tied closely to that of the komatiite liquid. During the early stages of crystallization, chromite, and in some cases Ni–Fe sulfides, accompany olivine, but they are normally very minor components. Other silicate minerals crystallize at a later stage, and they are normally confined to the matrix between olivine grains: this matrix consists of augite, chromite and glass in rapidly cooled lavas, and augite with or without orthopyroxene, pigeonite, plagioclase, chromite, Fe–Ti-oxides and in some cases quartz and amphibole in more slowly cooled lavas.

All komatiites are altered. In the least affected samples from Zvishavane in Zimbabwe and Gorgona Island in Colombia, pyroxene is fresh, olivine is altered to chlorite or serpentine, but only along grain margins and fractures, and fine-grained hydrous minerals patchily replace glass. In all other documented komatiites, secondary minerals replace the magmatic phases. Most if not all the olivine has been converted to chlorite or serpentine, pyroxene is partially replaced by tremolite and chlorite, and glass has completely altered to secondary hydrous minerals. In many cases the secondary minerals pseudomorphically replace the primary minerals, and little recrystallization is evident: in such samples the finest details of the habits and morphology of the original mineral grains can be recognized and the igneous precursor minerals are readily identified. In more altered samples recrystallization is more advanced and little remains of the original texture. Veins of serpentine, chlorite, carbonate and other secondary minerals, with varying sizes and abundances, cut through the samples. Under such circumstances it becomes impossible to determine the original mineralogy and textures, and difficult to establish whether or not the rock was originally a komatiite.

4.2 Primary minerals

Olivine

Olivine is the most abundant mineral in komatiites and typically makes up 40–80 vol% of the primary mineralogy of samples from normal lava flows, and very close to 100 vol% of some samples from the interiors of dunitic units. The habits and textures of olivine were described in broad terms in Chapter 3; here the morphological features are discussed in more detail and chemical compositions are presented.

Morphological characteristics

It has become common practice to describe the morphologies and arrangements of olivine crystals in komatiites using the terminology that Donaldson (1976, 1982) developed partly on the basis of his own observations of olivines in the Rum intrusion in Scotland and his own experimental work, and partly from the earlier descriptions and classification schemes of Drever and Johnston (1957), Naldrett and Mason (1968) and Nesbitt (1971). Donaldson (1976, 1979, 1982) recognized five main types of olivine morphologies in komatiitic rocks, illustrated in Figure 4.1 and described as follows:

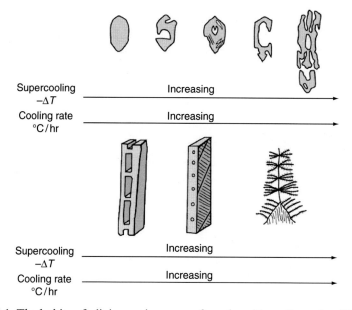

Fig. 4.1. The habits of olivine grains grown from basaltic melts under different cooling rates and differing degrees of supercooling (from Donaldson (1982)).

(1) 'Hopper olivine', which consists of equant to elongate crystals with hollow, or embayed, cores and euhedral to subhedral outlines.
(2) 'Plate olivine': two or more thinly tabular plates ($c{\sim}a{\gg}b$) or blades ($c > a{\gg}b$) stacked parallel, or nearly parallel, to one another, like a pack of cards, along the [010] direction. Individual plates may be complete, or have a lumpy interface, or be internally dendritic. In some instances the plates appear to diverge slightly from a common nucleation point to form a slight fan: depending on the extent of the divergence, these may be referred to as 'plate olivine' or 'radiate olivine' (Donaldson, 1976).
(3) 'Branching olivine', a variety of plate olivine in which more clearly divergent branches stem from multiple nucleation points.
(4) 'Polyhedral olivine': equant to tabular euhedral crystals with no embayments.
(5) 'Granular olivine': equant, subspherical crystals.' (Donaldson, 1982, p. 222).

Figure 4.1 illustrates the shapes of olivine grains that crystallized from mafic melts during experiments carried out by Donaldson (1976). Polyhedral olivines (top left) formed at slow cooling rates and/or low degrees of super-cooling, equant hopper olivines at larger values, and elongate hopper olivines at still larger values (top right). The olivines illustrated in the lower three drawings are 'chain', 'lattice' and 'feather' olivines, habits that, at the time of Donaldson's paper, had not been reported in komatiites. Shore (1996) and Faure *et al.* (2003) have subsequently reported grains with these morphologies in the rapidly cooled chill margins of komatiite flows of Pyke Hill.

The olivine habits shown in Figure 4.1 can be compared with those of olivine grains in the thin sections of komatiite illustrated in Figure 4.2 and in Figures 3.5 and 3.7. The habits of the olivine grains in the lavas depend on the composition of the silicate liquid and the conditions during their crystalliza-tion. To quote Donaldson (1976): 'The larger the olivine content of the liquid, the slower the cooling rate at which a particular olivine shape forms'. At a given, moderate, cooling rate, relatively stubby skeletal plate olivines grow from highly magnesian liquid, while finer, more elongate crystals grow from a more basaltic liquid. The influence of liquid composition is beautifully illu-strated by the textures in a vesicle in spinifex komatiite from Alexo (Figure 4.3). This vesicle was partially filled with interstitial silicate liquid as the lava cooled and the trapped gas contracted. Menisci formed during two separate influxes of liquid into the vesicle and each influx filled part of the vesicle with more evolved, less magnesian liquid: the inner meniscus has a lower MgO content than the outer one, which, in turn, has a lower MgO content than the host komatiite liquid. In the original liquid, which contained 25% MgO, olivine crystallized as stubby hopper grains; in the outer meniscus, the olivine crystals are fine and elongate; and in the inner meniscus, they are

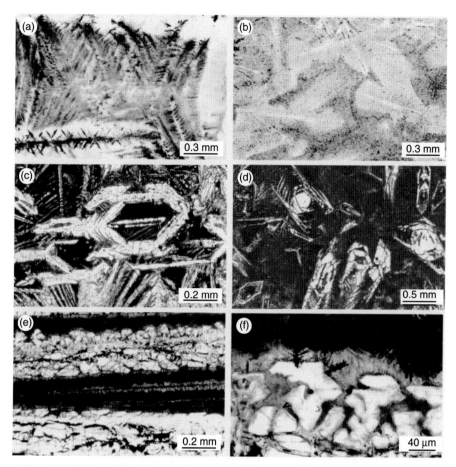

Fig. 4.2. Habits of olivine grains in rapidly-cooled komatiites. (a) 'Feather' olivine from the platy spinifex layer of a 1.5 m thick Pyke Hill flow. The olivine has been totally replaced by secondary minerals but the form of the original crystal is revealed by the distribution of minute chromite crystals that nucleated at the margins of the olivine (from Shore (1996)). (b) Acicular microlites of serpentinized olivine with multiple dendritic extensions set in an altered glassy groundmass, from the lower chilled margin of a flow from Pyke Hill. A photograph of the chilled margin is shown in Figure 3.9 (from Shore (1996)). (c) Hollow hopper olivine crystals from the spinifex layer of a komatiite from Zvishavane, Zimbabwe. (d) Hopper olivines in the lower cumulate layer of a komatiite from Gorgona Island. (e) Photograph looking down the a crystallographic axis of two parallel platy olivine crystals in the spinifex layer of a Pyke Hill komatiite (from Shore (1996)). (f) Detail of the margin of the upper olivine crystal in (e) (from Shore (1996)).

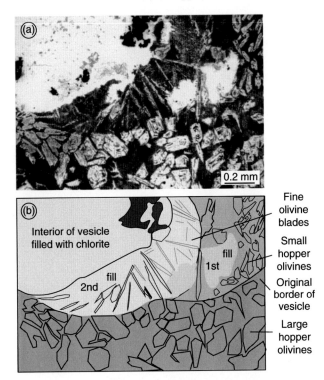

Fig. 4.3. (a) Vesicle in the olivine spinifex layer of the Alexo komatiite flow (sample M662) partially filled with interstitial liquid and showing contrasting olivine morphologies in successive menisci. (b) Sketch of the photomicrograph identifying the two menisci and highlighting the contrasting olivine morphologies. In the spinifex lava in the lower part of the image, skeletal hopper grains predominate, in the first infill of the vesicle the grains are smaller, more elongate and more skeletal, in the second infill, they form elongate blades. See text for further explanation.

still finer and still more elongate. The cooling rate must have decreased progressively as the vesicle filled with interstitial liquid but the effect of the diminishing cooling rate, which normally should have produced stubbier and less elongate crystals, was counterbalanced by the effect of the decreasing MgO content of the liquid.

The second factor that influences crystal morphology is the conditions during crystallization. These depend to some extent on the overall flow dimensions and morphology and on the eruptive setting but they are influenced more strongly by the position within the lava flow, because it is this parameter that controls the cooling rate and hence the manner of crystallization of the olivine. In the lower cumulate layers of differentiated flows, granular and polyhedral olivines predominate (tabular hopper olivines are always present, but in smaller quantities). These crystals grew during relatively slow cooling either during eruption and flowage, if they formed as phenocrysts, or in the lower part of the flow, if they formed by *in-situ* crystallization (see Chapter 9 for a more complete discussion of the process). In some porphyritic lavas and in B_1 layers, hopper olivines are the most common type (Figure 3.14); in spinifex lavas, a large range of skeletal forms are present (Figure 3.7). Their forms can be related both to cooling rate, which is highest near the surface of the flow, and to the manner in which the crystals grew. Faure *et al.* (2006) showed that the presence of a thermal and chemical gradient in the upper part of a komatiite flow has a strong influence on the morphology of the olivine and pyroxene crystals. The fine, highly skeletal crystals in the uppermost parts of the flow form through unconstrained growth that accompanied rapid cooling at the margin of the flow, and the larger, oriented plate olivines deeper in the flow formed through constrained growth during slower cooling, under conditions in which the thermal gradient had maximum effect.

For more detailed information about the morphology of olivine grains in komatiites, the reader is referred to Fleet and McRae (1975), Naldrett and Mason (1968), Nesbitt (1971), Donaldson (1976, 1982), Renner *et al.* (1994), Shore (1996) and Faure *et al.* (2003). Volcanic factors that influence the morphology and arrangement of olivine and other minerals in komatiites are discussed in Chapter 9.

Crystallographic features

Information about crystallographic orientations in skeletal olivine in komatiites is provided by Fleet and McRae (1975) and Shore (1996). In rapidly cooled lavas the olivine grains are strongly non-equidimensional, and crystal dimensions parallel to the *a* and *c* axes are greater than that parallel to the *b* axis. This difference produces tablet- or wafer-shaped crystals such as those shown in Figure 3.7(a),(b). In crystals with skeletal habits, the crystal faces {021} and {120} grow most rapidly. Differential growth produces U- or H-shaped crystals because growth on {021} and {120} planes is faster in the [100] direction than in the [001] direction. Sharp terminations of skeletal grains formed through growth along {041}.

Chemical compositions

Ubiquitous serpentinization often prevents complete analysis of olivine grains, particularly of their margins, but relict igneous cores are preserved in a large variety of komatiite types. Komatiites in regions of low metamorphic grade contain olivines that are clear and colourless and have normal optical characteristics, but olivine in komatiites from amphibolite-facies localities, notably

Perseverance in Australia, commonly has a red-brown colour. Binns *et al.* (1976) attribute this colour to exsolution of submicroscopic plates of opaque oxides during metamorphic heating. During this process, the igneous relicts of olivine interact and reequilibrate with surrounding metamorphic phases, with consequent changes in Mg–Fe ratios and trace-element abundances. This complication should be kept in mind when the compositions of such olivine grains are used to establish the compositions of the liquids from which they crystallized. Olivine grains in less metamorphosed komatiites seem not to have been affected (but see de Wit *et al.* (1983)).

Major elements and zonation

Compositions range from Fo_{94-95} in large unzoned olivine crystals in dunitic units, to less than Fo_{84} in the margins and tips of the finest olivine blades in spinifex-textured komatiitic basalts. Liquidus olivine in typical komatiite flows falls in the range Fo_{94}–Fo_{90}.

With the exception of some slowly cooled, very Fo-rich crystals in dunitic units, all olivine grains in komatiites are zoned. Some have cores with relatively uniform compositions and sharply zoned margins, as in the phenocrysts from the chill zone at the top of the Alexo komatiite flow, Ontario, whose composition profile is shown in Figure 4.4. The cores of these grains (Figure 4.4(a)) have the composition Fo_{94}, and probably grew in komatiite liquid containing $\sim$28% MgO during the ascent and eruption of the lava (see Chapter 12); the Fo-poor margins grew from more evolved interstitial liquid after entrapment of the olivine grains in a marginal chill zone. In most other komatiite lavas, the olivine grains show continuous zoning from core to rim (Figure 4.3(b)(c)), a consequence of the evolution of the composition of the liquid from which they grew, modified by diffusion (Bickle, 1982; Smith and Erlank, 1982; Arndt, 1986b; Silva *et al.*, 1997). The forsterite content of some grains in the well-preserved spinifex-textured Zvishavane lavas varies by up to 8 mol%, from Fo_{92} to $Fo_{<84}$. Such zoning is an inevitable consequence of the rapid cooling of komatiites, the pronounced changes in the composition of the liquid that accompany cooling, and the large temperature and compositional interval over which olivine crystallizes.

The highest forsterite contents of olivine in komatiite are reported in the unusual olivine–orthopyroxene-bearing units of the Weltevreden Formation in the Barberton greenstone belt (Kareem and Byerly, 2003) and in the Commondale greenstone belt (Wilson, 2003; Wilson *et al.*, 2003). In the latter area, the olivines are remarkably magnesian, containing up to 96.5 mol% forsterite, in keeping with the highly refractory and depleted major and trace element compositions of these peculiar komatiites.

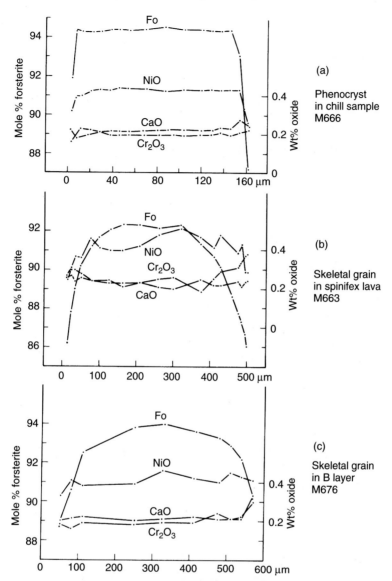

Fig. 4.4. Zoning profiles of olivine grains from three parts of the Alexo komatiite flow (from Arndt (1986b)): (a) the broad plateau and steep margins of the profile across a phenocryst in sample M666, the upper chilled margin; (b), (c) the hump-shaped profiles across grains in a spinifex lava, M663, and from the B₁ zone, M676.

Variations in olivine compositions within komatiite flows

The composition of olivine varies systematically from the top to the base of a komatiite flow. In many flows, the most magnesian olivine grains are found in the upper part of the cumulate portion, including in the B_1 layer, and in the uppermost spinifex-textured lavas. Figure 4.5 shows how the texture, mineralogy and olivine composition varies from the top to the base of the well-studied Alexo komatiite flow (Barnes, 1983; Arndt, 1986b; Lahaye and Arndt, 1996). The Alexo flow is a rare example in which olivine phenocrysts in the chill margins have escaped serpentinization (other examples include some komatiites in Munro Township, Zvishavane and Gorgona). The phenocrysts in the chilled margin of the Alexo flow have Fo contents as high as any in the flow (Fo_{94}) and this composition is that expected of olivine that crystallized from the erupted komatiite liquid, whose composition can be inferred from the composition of chilled flow margin itself (28% MgO). Another example is provided by the komatiites of Pyke Hill, Munro Township in which the most magnesian olivines have the composition $Fo_{93.5}$, which corresponds to that calculated for olivine in equilibrium with a liquid with $\sim 27\%$ MgO (Chapter 12).

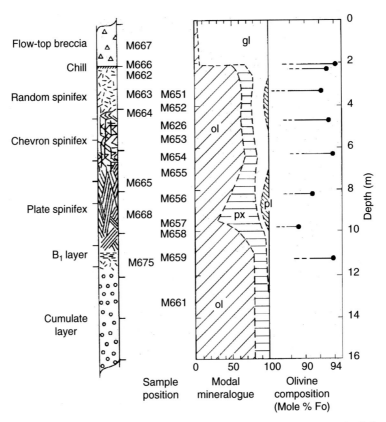

Fig. 4.5. Variation in modal mineralogy and forsterite contents of olivine in the Alexo komatiite flow (ol – olivine; px – pyroxene; pl – plagioclase; gl – altered glass). (From Arndt (1986b).)

From the base to the top of the olivine cumulate layer of thin komatiite flows in Munro Township (Arndt *et al.*, 1977), or the thicker flows in Zvishavane, Zimbabwe (Renner *et al.*, 1994; Silva *et al.*, 1997), the Fo content remains relatively constant. In thicker flows at Kambalda in Australia the Fo content decreases progressively upwards, as described in Chapter 9 (Lesher, 1989).

Certain komatiitic flows contain a second generation of olivine crystals, usually rounded polyhedral grains, which have distinctly higher Fo contents than that of the dominant olivine population. The Zvishavane komatiites, which erupted as liquids containing about 20% MgO and which crystallized microphenocrysts and skeletal olivine grains with the equilibrium composition $Fo_{91.5}$ (Renner *et al.*, 1994), contain a population of polyhedral to rounded phenocrysts with the composition $Fo_{92-93.5}$. Such grains are found throughout the flow, except within the lowermost portion of the spinifex layer. These grains must have grown from a more magnesian liquid that contained up to 26% MgO. Renner *et al.* (1994) explain these grains in a model in which an earlier, more magnesian komatiite crystallized in a crustal magma chamber, forming a cumulate layer that was remobilized during the passage of the magmas that erupted on the surface. The Fo-rich olivine grains were entrained in the less magnesian liquid and accumulated, together with primocrysts, in the lower parts of the komatiite flows.

In the komatiitic basalts of Gilmour Island, Hudson Bay (Baragar and Lamontagne, 1980; Arndt, 1982), the erupted liquid is estimated to have contained about 17% MgO on the basis of the Fo_{90} composition of olivines preserved throughout the cumulate part of the flow. Distributed irregularly through the lower part of the cumulate layer of these flows are isolated grains or clusters of polyhedral grains (Figure 4.6) with a more magnesian composition, Fo_{92}. As for the Zimbabwe flows, the more magnesian olivines are interpreted as phenocrysts or xenocrysts that had crystallized before emplacement from a more magnesian precursor magma.

Ni contents

Chemical analyses of olivine grains in several representative komatiite flows are listed in Table 4.1. The concentrations of many trace elements are far higher than in olivine from basaltic magmas. In the case of elements such as Ni and Cr, this is easily understood, being mainly a reflection of the high Ni and Cr contents of the parental ultramafic liquids. Ni contents vary as a function of the MgO content of the liquid, ranging from a maximum of $\sim 4000\,ppm$ in olivines with compositions in the range Fo_{91-93} (e.g. Nisbet and Chinner (1981), Smith and Erlank (1982), Arndt (1986b), Barnes *et al.* (1988)) to less than 1000 ppm in olivines that

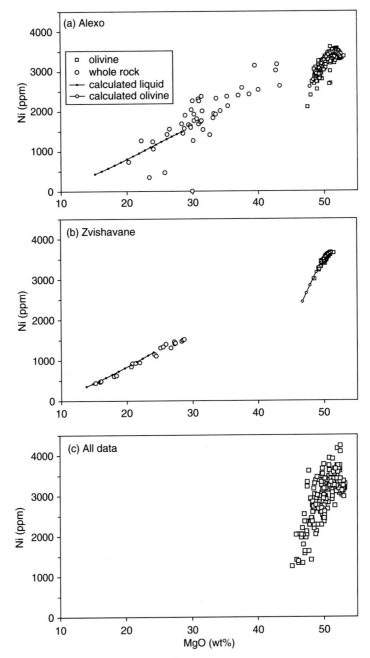

Fig. 4.6. NiO contents of olivine and whole rocks from komatiites from a Alexo, Canada, and (b) Zvishavane, Zimbabwe, and modelled fractionation trends for olivine and liquids. The latter were calculated using estimated parental liquid compositions from Arndt (1986b) and Renner *et al.* (1994). Olivine compositions reported in (a) and (b) are from A. Sobolev (unpublished data). Data in (c) are from numerous sources including Arndt (1986), Lahaye and Arndt (1996).

Table 4.1. *Compositions of olivine in selected komatiites*

	1	2	3	4	5	6	7
SiO_2	41.8[a]	41.6	41.9	41.0	41.27	39.85	41.61
Al_2O_3							0.05
Cr_2O_3	0.19	0.20	–	0.26	0.10		0.16
FeO	5.65	5.80	5.99	10.5	9.31	15.30	4.44
MnO						0.15	0.08
MgO	52.1	51.91	51.8	48.2	49.26	44.39	53.72
NiO	0.40	0.42	0.37	0.26	0.43	0.17	0.47
CaO	0.23	0.20	0.20	0.12	0.13	0.15	0.13
Fo	94.1[b]	94.1	93.9	89.1	90.4	83.8	95.6

[a] Oxides in weight per cent.
[b] Forsterite contents (Fo) in mole per cent.
1 Centre of phenocryst in chill sample M666, Alexo komatiite (Arndt, 1986b).
2 Centre of skeletal grain in matrix of chill sample M666, Alexo komatiite (Arndt, 1986b).
3 Centre of polyhedral grain in B_1 layer, sample M676, Alexo komatiite (Arndt, 1986b).
4 Centre of skeletal grain, spinifex sample M658, Alexo komatiite (Arndt, 1986b).
5 Centre of skeletal grain, spinifex sample Z2, Zvishavane komatiite (Bickle *et al.*, 1993).
6 Margin of skeletal grain, spinifex sample Z2, Zvishavane komatiite (Bickle *et al.*, 1993).
7 Grain of exceptionally Fo-rich olivine from a grain in komatiite from the Weltevreden Formation (Kareem and Byerly, 2003).

crystallized from basaltic residual liquids. When data from the literature are plotted in a MgO vs. Ni diagram, they plot in a diffuse cloud with generally positive slope (Figure 4.6(c)), but when precise analyses of olivines from single differentiated units are compiled (Sobolev *et al.*, 2007), the Ni contents are seen to pass through a maximum of about 4000 ppm in olivines with ca. 92 mol% forsterite (Figures 4.6(a),(b)). More magnesian and less magnesian olivines have lower Ni contents. This behaviour is explained as follows. Partition coefficients calculated using Ni concentrations in olivine and in estimated compositions of the liquid from which the olivine crystallized range from about 2 to 4 (Bickle, 1982; Arndt, 1986b; Smith and Erlank, 1982). These values agree well with partition coefficients derived from experiments with komatiitic starting materials (Chapter 9). The partition coefficients vary as a function of the MgO content of the liquid, as described by the following equation:

$$D_{Ni} = (124.3/MgO_{liq}) - 0.9$$

where D_{Ni} is the partition coefficient and MgO_{liq} is the MgO content of the silicate liquid (in weight per cent). The partition coefficient is lowest in the most magnesian liquids, which have the highest Ni contents, but increases as olivine fractionation drives down both the MgO and Ni contents of the liquid. It is the interplay between the increasing partition coefficient and decreasing Ni content of the liquid that causes the Ni content of olivine to pass through the maximum illustrated in the top diagram of Figure 4.6 (Smith and Erlank, 1982; Arndt, 1986b). Modelled trends for the variation of Ni and MgO contents of olivine crystallizing from two komatiite liquids (Alexo and Zvishavane) coincide with the measured olivine compositions. In the case of the Zvishavane flow the correspondence is remarkably good. In the Alexo case the measured data scatter about the modelled trend, probably as a result of variation in MgO or Ni content of samples from the flow. This variation could have arisen from minor element mobility during alteration and/or the presence of Ni–Fe sulfides in some samples. Samples from one section across the flow contained sulfide blebs and had slightly higher contents of Ni, Cu and PGE than samples from other parts of the flow (Brügman *et al.*, 1987).

More recent equations describing the partitioning of Ni between olivine and mafic–ultramafic liquids, such as those of Kinzler *et al.* (1990b), Beattie (1993) and Gaetani and Grove (1997) give results similar to those illustrated in Figure 4.6.

Cr contents

The Cr content of olivine in komatiites, shown in Figure 4.7(a), ranges from about 0.15 to 0.5% Cr_2O_3, values that are 10–20 times greater than that of olivine in typical terrestrial basalts. The reasons for these high concentrations are not immediately clear. Donaldson (1982) proposed five possible explanations: (a) delayed crystallization of chromite, (b) rapid crystal growth that led to disequilibrium partitioning of Cr, (c) a higher Cr partition coefficient due to the high MgO content or high temperature of the komatiite liquid, (d) high Cr content of the komatiite liquid, and (e) low oxygen fugacity (f_{O_2}) which increased Cr^{2+} in the melt and increased the partition coefficient. Shore (1996) gave a detailed appraisal of these explanations: he initially eliminated (b), because high Cr contents are found in olivines with various morphologies which grew at very different cooling rates, and (e), because experimental studies show that the partition coefficient does not vary with oxygen fugacity and because f_{O_2} in komatiites is not particularly low (Canil, 1997). Shore concluded that the other three explanations all had some validity. The high Cr content of olivine is directly linked to the high concentration in the komatiite liquid, and the Cr partition coefficient is higher than predicted, probably because of coupled substitution of Cr and Al for Mg and Si. The highest Cr contents are found in olivines that crystallized rapidly under conditions in which chromite did not nucleate, as illustrated in Figure 4.7. In this figure, the data

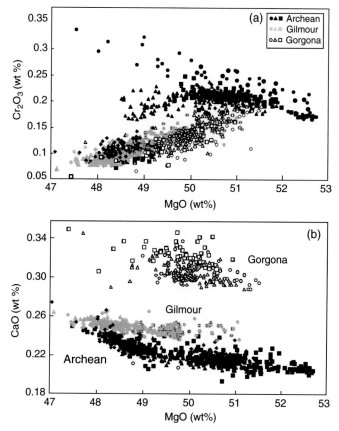

Fig. 4.7. (a) Cr_2O_3 and (b) CaO contents of olivines in komatiites (unpublished data from A. Sobolev). Olivine from Archean komatiites is shown as black symbols, olivine from Proterozoic Gilmour Island as grey symbols and olivine from Cretaceous Gorgona Island as open symbols. The different symbols represent analyses from different samples and/or different flows.

in the top left of the diagram are from olivine grains in sample M666, the chilled margin of the Alexo flow. The progressive increase in Cr with decreasing MgO coupled with petrographic observations indicates that chromite did not crystallize in this sample, probably because of the high cooling rates to which this sample was subjected. In olivine from all the other samples, which crystallized at lower cooling rates, the Cr content begins to decrease when the MgO content of the olivine drops below 50–51%, presumably because of the onset of chromite crystallization.

Ca contents

The CaO contents in reliable analyses of olivine in komatiites range from about 0.2 to 0.35 (Figure 4.7(b)). These values are broadly consistent with what would be predicted using partition coefficients of 0.02–0.04 as taken from experimental studies on basalts (e.g. Beattie *et al.*, 1994) or komatiites (Arndt, 1977a; Kinzler and Grove, 1985). Less predictable is the systematic variation in CaO content with age, as shown in Figure 4.7. Archean komatiites have the lowest values, followed by Proterozoic Gilmour komatiitic and Cretaceous Gorgona komatiites. The CaO contents of the Gorgona olivines are 50% higher than those in the Archean olivines. The reasons for this variation are not evident. If the partition coefficient were constant, CaO in olivine should reflect the CaO content of the liquid. The CaO contents of Gorgona komatiites are slightly higher, at a given MgO content, than those of Archean komatiites (Chapter 5), but the difference in the whole-rock CaO is about 10%, not 50%. Temperature no doubt plays a role, because the partition coefficient should increase as temperature decreases, but the eruption temperatures of the Gorgona komatiites are calculated to have been greater than those of Gilmour komatiitic basalts, yet the CaO contents of olivine from the latter are greater than those from the former.

Other elements

Lesher and Arndt (1995) noticed that the HREE patterns of olivine cumulates of the basal komatiite flow at Kambalda differed significantly from HREE patterns in spinifex-textured lavas of the same flow: chondrite-normalized HREE in olivine cumulates had a slight positive slope and the HREE in spinifex lavas a negative slope. If not a result of element mobility during alteration (see Chapter 5), the effect can be explained by olivine accumulation (a) if REE are highly fractionated by olivine, with $D_{Lu} \gg D_{La}$, and (b) if olivine–liquid partition coefficients in komatiites are significantly higher than in basalts. Analysis of an olivine cumulate from a Kambalda komatiite (Figure 4.8) shows this to be true. The chondrite-normalized pattern of the HREE is strongly sloping, concentrations are relatively high and calculated partition coefficients range from 0.01 for Gd to 0.1 for Yb – a factor of 10 higher than in basaltic liquids (Ray *et al.*, 1983). (The flatter slope and relatively high concentrations of the LREE are believed to be due to the presence LREE-enriched material in impurities or inclusions in the separated olivine.)

Chromite

Chromite occurs as delicate, dendritic or cruciform skeletal grains in spinifex lavas (Figures 4.9(a)–(c)), as small (10–100 mm) cubes or octahedra in spinifex, cumulate and porphyritic lavas (Figures 4.9(d), 3.12, 3.14), and as interstitial to poikilitic anhedral grains in dunitic bodies. Modal abundances rarely exceed 2–3% and are more commonly less than 1%. Chromite is more abundant in

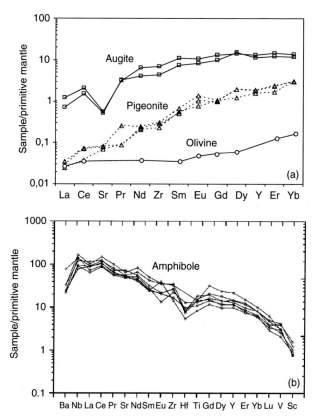

Fig. 4.8. (a) Mantle-normalized trace-element contents of olivine and pyroxenes in komatiites. The analyses for augite and pigeonite are new ion microprobe data measured by Bouquain (personal communication) in a sample of olivine spinifex (M105) from the Alexo komatiite flow; the olivine data are from a mineral separate from the Kambalda komatiite (Arndt and Lesher, 1992). (b) Mantle-normalized trace-element contents of amphibole in the Boston Creek Fe-rich komatiite (Stone *et al.*, 2003).

spinifex lavas than in the cumulate B layers of differentiated flows, and occurs relatively rarely as inclusions in olivine grains, a distribution consistent with experimental studies (Murck and Campbell, 1986; Kinzler and Grove, 1985) that show that chromite starts to crystallize significantly below the liquidus of the most magnesian komatiites (see Chapter 5). B_1 layers in komatiitic basalts with 15–20% MgO may have unusually high chromite concentrations. The data from a komatiitic basalt flow in the Siivikkovaara area in the Archean Kuhmo greenstone belt, Finland (Hanski, 1980) and some of the thick flows in the lower part of the Kambalda sequence (Lesher, 1989) show enormous concentrations of Cr in the upper part of the B layer. In flows on Gilmour Island in Hudson Bay, Canada, large numbers of minute chromite euhedra line the lower sides of elongate hopper olivines of the B_1 layer (Fig. 3.14, Arndt

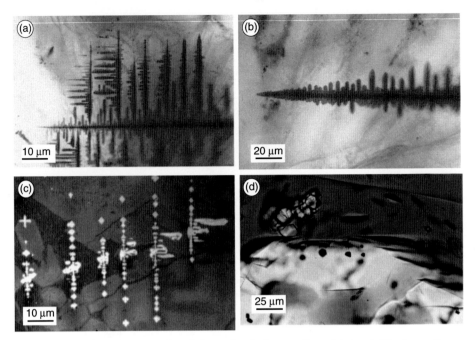

Fig. 4.9. Photomicrographs of chromites in komatiites: (a) dendritic chromite grain in spinifex-textured komatiite from Pyke Hill; (b) detail of the termination of the tip of one of the branches of the crystal illustrated in (a); (c) reflected light photomicrograph of a dendritic chromite crystal in a komatiite from northern Munro Township; (d) equant rounded grains of chromite included in an olivine crystal from a Pyke Hill komatiite. Photographs (a), (b) and (d) are from Shore (1996) .

(1982)). Ferrian chromite with a distinctive Al- and Mg-poor composition is also dispersed within or forms layers marginal to accumulations of massive sulfides at, for example, Kambalda (Lesher, 1989).

Table 4.2 lists some representative analyses and Figure 4.10 contains the compositions of the large collection of data compiled by Barnes (1998, 2000). As Barnes emphasizes in his two papers, which, together with Barnes and Roeder (2001) constitute the definitive work on chromites in komatiites, the interpretation of these compositions is complicated by the effects of metamorphism. The most obvious effect is the development of rims of chromian magnetite whose compositions are low in Ti and Al and plot along the Cr–Fe^{3+} join, as shown in Figures 4.10(a) and (f).

Chromites in komatiites generally have high $Cr/(Cr + Al)$ (Cr#), moderate $Mg/(Mg + Fe)$ (Mg#) and moderate to high concentrations of minor elements (Ti, Ni, V, Zn). They commonly are zoned, with margins richer in Fe and Ti but poorer in Cr than cores. The compositions of the cores fall in two

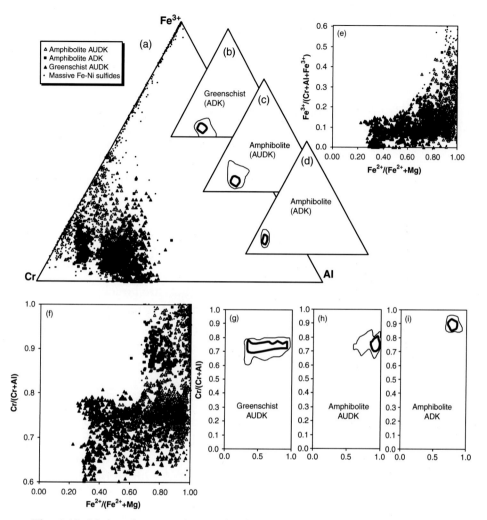

Fig. 4.10. Major-element microprobe data on compositions of chromites in Komatiites, all data plotted as atomic proportions within the spinel prism. Data shown as individual grain averages ((a), (e), (f)) and data density contours on 50th and 80th percentiles ((b), (c), (d), (g), (h), (i)). AUDK – Al-undepleted Komatiite (Munro-type); ADK – Al-depleted Komatiite (Barberton-type) (data from compilation of Barnes and Roeder (2001)).

broad fields in Figure 4.10(f), one with intermediate $Cr/(Cr + Al)$ and a wide range in $Fe^{2+}/(Mg + Fe^{2+})$ and the second with higher $Cr/(Cr + Al)$ and $Fe^{3+}/(Mg + Fe^{2+})$. Barnes (1998, 2000) attributed this variation to postcumulus and subsolidus Fe–Mg exchange, combined with the effects of metamorphism. Barberton-type Al-depleted komatiites from the amphibolite facies Forrestania Belt in Western Australia contain chromite with a significantly lower Al content, a feature that is probably inherited from the original magma composition. No analyses are available, however, from less metamorphosed Al-depleted komatiites (those from the Barberton belt are also at upper greenschist to amphibolite facies) and the relationship between chromite and komatiite composition remains unclear.

$Cr/(Cr + Al)$ in chromites from komatiites is significantly higher than in chromites from most types of basalt and peridotite and comparable values are encountered only in chromite from boninites (Cameron and Nisbet, 1982), from some island arc basalts and picrites, and from Alpine-type peridotites (Dick and Bullen, 1984). Mg# is lower than in chromites from abyssal basalts and peridotites, but comparable to values in boninites and arc basalts. The distinctive compositions of komatiite chromites are best illustrated in the diagram of $Fe^{3+}/(Cr + Al + Fe^{3+})$ vs TiO_2 (Figure 4.10(c)) (Barnes and Roeder, 2001), in which they plot in a separate field defined by low $Fe^{3+}/(Cr + Al + Fe^{3+})$ and TiO_2 contents higher than in chromites from boninites.

The compositions of chromites from komatiites are largely consistent with their crystallization from high-temperature ultramafic magmas: their high $Cr/(Cr + Al)$ derives from the high Cr and relatively low Al of the magma. The high Ti and V contents may indicate relatively high chromite-melt partition coefficients. For example, values of $D_{Ti}^{chromite-liq}$ between 0.6 and 0.8 can be calculated from the Ti contents measured in chromites from the chill zones of the Alexo komatiite flow and Fred's Flow. An explanation for the high Zn contents is less obvious. Groves *et al.* (1981) suggested that the high Zn may derive from sedimentary rocks assimilated by the komatiite, and that the feature may be genetically related to Ni-sulfide ores. However, Barnes (2000) showed the Zn enrichment is spatially associated with secondary magnetite rims and fracture fillings, and that the Zn enrichment is almost certainly secondary and hydrothermal in origin.

Pyroxenes

Three types of pyroxene – augite, pigeonite and orthopyroxene – occur in komatiites and komatiitic basalts. All three have distinctive habits and chemical compositions that depend on the composition of the host lava, and, more significantly, on the position within the flow at which they crystallized. Several different forms of pyroxene grains are observed: (a) large, elongate skeletal composite megacrysts, the dominant component of pyroxene spinifex lavas; (b) subhedral to euhedral, solid, prismatic cumulus grains, which occur in the lower portions of differentiated basaltic flows; and (c) small prismatic grains, needles, dendrites and delicate feathery intergrowths, which are found in the matrix of spinifex, cumulate and porphyritic lavas. Some of these habits were shown in Figures 3.11, 3.12 and 3.24.

Megacrysts in spinifex lavas

The most spectacular, characteristic and petrologically interesting form of pyr-oxene is the composite megacryst in pyroxene spinifex textures (Arndt and Fleet, 1979; Wilson *et al.*, 1989). When observed in outcrop, these grains appear as large, isolated or composite needles, as illustrated in Figure 3.25. In thin sections of well-preserved samples, the megacrysts are seen to have cores of pigeonite (or, very rarely, enstatite or bronzite) that are generally hollow and box-shaped in cross section. The pigeonite cores are mantled on the outside, and in some cases lined on the inside, by augite (Figure 4.11). Individual needles are 0.1–1.0 mm across and 2 cm or more in length. Close examination of thin sections demonstrates that such needles are the main element of much larger, highly complex grains.

In spinifex lavas, the needles are organized into colonies of individuals, oriented roughly parallel to one another with near-perfect optical continuity, or slightly branching in sheaths or columns that extend for distances up to several metres (Figure 3.25). The sheaths are oriented more or less perpendi-cular to the flow top, in some cases splaying out downwards. In acicular pyroxene textures, the composite pigeonite–augite needles are smaller

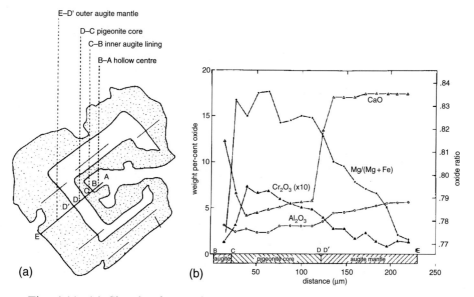

Fig. 4.11. (a) Sketch of a section perpendicular to the *c*-axis of a zoned pigeonite-augite crystal; (b) variations in the composition of the grain from the inner augite lining, through the pigeonite core to the outer augite margin (from Arndt and Fleet, 1979)

Table 4.2. *Compositions of pyroxenes in selected komatiites*

	Orthopyroxene			Pigeonite							Augite				
	1	2	3	4	5	6	7	8	9	10	11	12	13	14	15
SiO_2	55.69[a]	57.1	55.43	55.11	54.23	55.2	54.29	48.86	51.16	51.9	53.25	52.5	54.6	53.77	49.1
TiO_2	0.00	0.04	0.03		0.07	0.11	0.10	0.32	0.29		0.18	0.20	0.19	0.30	0.49
Al_2O_3	1.75	0.91	1.26	2.04	2.73	2.02	1.63	6.28	5.09	3.14	2.91	2.53	0.67	1.41	9.8
Cr_2O_3	0.84	0.50	1.04	0.61	1.17	0.63	0.85	0.53	0.53	0.39	0.09	0.81	0.23	0.75	0.35
FeO	8.10	7.9	5.96	9.70	9.37	9.4	7.84	8.56	7.70	7.6	5.71	5.91	5.59	5.35	7.8
MnO		0.21	0.18			0.26	0.24	0.17			0.19	0.15	0.18	0.10	
MgO	30.85	32.0	33.92	28.44	28.11	28.9	27.66	16.23	17.68	18.3	21.24	18.4	18.07	18.09	18.4
CaO	2.36	1.18	1.51	4.01	3.81	6.76	7.06	17.98	17.12	17.8	16.94	19.7	19.68	20.04	15.0
Na_2O		0.01	0.02		0.09		0.02		0.15		0.07	0.11	0.19	0.28	
Mg#	87.2[b]	88.0	91.0	83.9	84.0	85.0	86.3	77.2	80.0	81.0	86.9	85.0	85.2	85.5	80.8

[a] Oxides in wt%.
[b] Mg# in mol%.

1 Bronzite from the centre of a skeletal grain, spinifex sample Z2, Zvishavane komatiite (Bickle *et al.*, 1993).
2 Bronzite from the cumulate layer of Alexo komatiitic basalt (Arndt and Fleet, 1979).
3 Enstatite in komatiite from the Weltevreden formation (Kareem and Byerly, 2003).
4 Pigeonite from the interior of a skeletal grain, spinifex sample Z2, Zvishavane komatiite (Bickle *et al.*, 1993).
5 Pigeonite from the interior of a skeletal grain in pyroxene spinifex of Alexo komatiitic basalt (Arndt and Fleet, 1979).
6 Pigeonite from a composite needle interstitial to olivine blades, sample M105, Alexo komatiite (S. Bouquain, personal communication).
7 Pigeonite in komatiite from the Weltevreden formation (Kareem and Byerly, 2003).
8 Augite from the interior of a skeletal grain, spinifex sample Z2, Zvishavane komatiite (Bickle *et al.*, 1993).
9 Augite from the interior of a skeletal grain, pyroxene spinifex sample AX11d of Alexo komatiitic basalt (Arndt and Fleet, 1979).
10 Augite from a composite needle interstitial to olivine blades, sample M105, Alexo komatiite (S. Bouquain, personal communication).
11 Augite in komatiite from the Weltevreden formation (Kareem and Byerly, 2003).
12 Augite interstitial to needles in pyroxene spinifex of Alexo komatiitic basalt (Arndt and Fleet, 1979).
13 Augite from skeletal grain, olivine spinifex sample B95–15 of a komatiite from Barberton (Parman *et al.*, 2003).
14 Augite from the Boston Creek Fe-rich komatiite (Stone *et al.*, 2003).
15 Augite interstitial of olivine blades in spinifex komatiite from Gorgona Island (Aitken and Echeverría, 1984).

(typically 0.1–0.5 mm across and 1–5 mm long) and randomly oriented. The two types are analogous to platy and random olivine spinifex in more olivine-rich komatiites. In some flows, smaller and stubbier pyroxene needles are oriented parallel to the flow boundaries within a layer immediately below the main spinifex zone to form a unit comparable to the B_1 layer in komatiite flows.

The pyroxene needles are elongated parallel to the *c*-axis and bounded by {110} or less commonly by {010} prism faces (Figure 4.11). The pigeonite is optically homogeneous, the augite frequently characterized by fine laminations or striations subparallel to (001). Fleet and MacRae (1975), Arndt and Fleet (1979) and Shore (1996) described the optical and crystallographic features of pyroxenes in komatiites in more detail.

Very commonly the pigeonite in the centres of zoned pyroxene needles is altered to fine-grained secondary minerals, mainly chlorite, as shown in Figure 3.10(b). In this example the needles are clearly hollow and irregular blebs or traces of augite can be seen to line the hollow interior of many of the needles. In many samples the altered pigeonite in the interiors of zoned needles is easily mistaken for hollow cores, as in the example shown in Figure 3.25(b).

The chemical changes within a composite pyroxene needle that crystallized in the Alexo flow are illustrated in Figure 4.11(b). From the pigeonite core to the outer margin of the augite mantle, CaO and Al_2O_3 increase in a step-like manner while $Mg/(Mg+Fe)$ and Cr decrease continuously. This zoning is explained by crystal growth outwards and inward from a starting point close to the inner margin of the pigeonite core, a point identified by the maximum Cr content (Figure 4.11). Augite in the inner lining has a composition similar to that of the outer mantle, but with lower Ca, Al and Cr contents and far lower $Mg/(Mg+Fe)$.

Pigeonite in the cores of the needles is highly magnesian ($Mg/(Mg+Fe)$ = 0.81–0.85), and also has high Al_2O_3 contents (Figure 4.12 and Table 4.3). The clinopyroxene that surrounds the pigeonite cores is an aluminous subcalcic augite. In Figure 4.12, the compositions plot in a broad field extending from the high Mg#, high Wo compositions of the early crystallizing augite adjacent to pigeonite in zoned crystals, to augites with lower Mg# and only slightly lower Wo at the margins. The Al_2O_3 contents of both pigeonite and augite are high, around 2–3% in the former and from 3 to 10% or even higher in the latter. Fleet and McRae (1975) explained the high Al_2O_3 by coupled substitution into tetrahedral sites occupied by $Al + Ti + Cr$, and attribute this behaviour to the conditions of rapid cooling that led to the suppression of plagioclase crystallization.

Shimizu *et al.* (2005) reported another occurrence in komatiites in the Belingwe belt in Zimbabwe, where pigeonite seems to have appeared late in the crystallization sequence: it forms rims with relatively Fe-rich compositions on the margins of augite grains in olivine spinifex lavas.

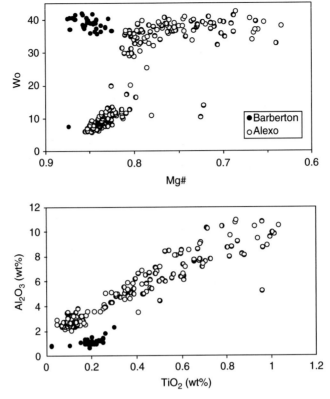

Fig. 4.12. Compositions of pyroxenes in the Alexo komatiite (S. Bouquain, personal communication) and in Barberton komatiites (Parman *et al.*, 1996, 2003).

The compositions of clinopyroxene in komatiites have come under scrutiny following the assertion of Steve Parman, Tim Grove and coworkers (Parman *et al.*, 1996,1997; Grove *et al.*, 1999; Grove and Parman, 2004) that the high Wo content of augite in Barberton komatiites is an indication that these pyroxenes crystallized at moderate temperatures from hydrous magma. Figure 4.12 shows that the compositions of augites in Barberton komatiites do indeed plot in fields separate from those of augites from other areas. However, the most evident distinction is not in their Wo contents, which are comparable to those of augites from the Alexo komatiite, but in their Mg/(Mg + Fe) ratios and in their Al_2O_3 contents. It is argued in Chapter 11 that the differences in augite compositions are related neither to the water contents of the komatiites, nor to the conditions of emplacement; but rather to differences in major element compositions of the Al-depleted Barberton komatiites and Al-undepleted komatiites from other regions.

Table 4.3. *Compositions of chromite and amphibole in selected komatiites*

	Chromite						Amphibole	
	1	2	3	4	5		6	7
SiO_2	0.06[a]					SiO_2	44.31	40.56
TiO_2	0.40	0.52	1.40	0.27	0.84	TiO_2	4.33	3.04
Al_2O_3	13.14	18.14	13.84	12.29	0.66	Al_2O_3	10.61	13.43
Cr_2O_3	54.00	53.3	47.23	51.37	51.93	Cr_2O_3	1.21	0.02
Fe_2O_3	6.29	1.69	8.47	4.55	12.74	Fe_2O_3	1.25	3.83
FeO	14.51	11.94	17.73	20.89	28.10	FeO	4.60	9.89
MnO	0.22	0.16	0.45	1.18	1.64	MnO	0.07	0.14
MgO	11.90	14.93	9.41	6.59	0.46	MgO	16.38	11.99
V_2O_3	0.40	0.22	0.31			CaO	11.47	11.72
NiO	0.08	0.08	0.16	0.06	0.04	Na_2O	2.81	3.07
ZnO	0.16	0.08	0.33	1.82	1.86	K_2O	0.59	0.26

[a] Oxides in wt%.
1 Chromite from a thin komatiite flow from Mt Clifford (Barnes, 1998).
2 Chromite from Mt Keith dunitic intrusion (Barnes, 1998).
3 Chromite from Mt Keith dunitic intrusion (Barnes, 1998).
4 Chromite from the olivine cumulate of a komatiite flow from Kambalda (Lesher, 1989).
5 Chromite in Ni–Cu-sulfide ore from Kambalda (Lesher, 1989).
6 Amphibole oikocryst from a komatiitic intrusion in the Abitibi belt (sample 27958) (Stone *et al.*, 2003).
7 Amphibole oikocryst from the Boston Creek Fe-rich komatiite (sample BCF1–60) (Stone *et al.*, 2003).

The orthopyroxene in spinifex textures of the unusual Si–Mg-rich komatiites from South Africa (Byerly, 1999; Kareem and Byerly, 2003; Wilson, 2003) and from crust-contaminated komatiitic basalts is unusually magnesian, but otherwise its composition is unremarkable.

Interstitial grains

Small acicular to dendritic grains of pyroxene are a ubiquitous component of the matrix of all rapidly cooled spinifex-textured komatiites and komatiitic basalts, and are also present as an interstitial phase in the olivine or pyroxene cumulate portions of layered flows. The habits of these pyroxenes are highly variable. Their form changes with decreasing grain size from straight to markedly curved, often composite, stubby skeletal prisms, laths and needles, to grains with fan, bow-tie and feathery spherulitic habits. Fleet and MacRae (1975) suggested that many of the acicular grains are transverse sections through spherules. Some spectacular examples of these textures in thin section have graced the covers of several journals (e.g. *Geology* **8**, 1980) and other examples are illustrated in Figures 3.11 and 3.25.

In most komatiites the pyroxene interstitial to spinifex olivine blades lies in a matrix of altered glass. However, in coarse platy olivine spinifex from lower parts of the spinifex layer in thick komatiite flows, and in most komatiites on Gorgona Island, plagioclase is intergrown with pyroxene (Figure 3.10(e)).

Many of the coarser interstitial pyroxenes have subcalcic augite compositions with unusually high Al_2O_3 contents, up to 8 wt% in some cases (Table 4.3). In some coarse olivine spinifex textures, the volumes of melt trapped between olivine plates are large, and in these spaces composite zoned pigeonite–augite needles like those in pyroxene spinifex rocks, and clino- or orthopyroxene oikocrysts, are observed. Orthopyroxene has been analysed in the centre of a composite grain in a spinifex komatiite from the Zvishavane locality in the Belingwe belt. This orthopyroxene, which contains $\sim$2% CaO, is more magnesian than the surrounding pigeonite (Mg/(Mg + Fe) = 0.87), and also is relatively aluminous (Al_2O_3 = 1.4–1.9%). Other, smaller augites crystallized together with plagioclase and these grains have lower, more normal Al_2O_3 contents.

The morphologies and compositions of pyroxenes in spinifex lavas resemble those in rapidly cooled lunar basalts. Their habits, compositions and zoning have been reproduced in dynamic cooling experiments by Walker *et al.* (1976), Lofgren (1980) and Faure *et al.* (2006), and there is general agreement that many of the unusual aspects of their composition and morphology can be attributed to the conditions of rapid cooling in which they grew.

Pyroxenes in gabbros and cumulate layers

Pyroxenes in the gabbroic and cumulate portions of layered komatiitic basalt flows have habits and compositions more like those in normal mafic flows and intrusions. Most grains are subhedral to euhedral and have prismatic habits. Grain sizes vary from 0.5 to about 1 cm and correlate with the dimensions and the conditions of emplacement of the host unit. In general the grain size is larger than in tholeiitic units with similar dimensions. The interiors of even relatively thin (20–100 m) units are composed of mafic rock with a medium-grained gabbroic texture (Arndt, 1977b).

Within the gabbroic units and cumulates in layered flows of the Abitibi and Circum-Superior belts in Canada (Arndt, 1977b), and in the Belingwe Belt in Zimbabwe (Nisbet *et al.*, 1977), augite is the predominant pyroxene and bronzite is subordinate; in many units in the Yilgarn and Pilbara Cratons in Australia, the proportions are reversed (Sun *et al.*, 1989). A thin orthopyroxene cumulate is isolated in the middle of the olivine cumulates of the Alexo komatiitic basalt (Arndt and Fleet, 1979; Barnes, 1983). The augites in these rocks have compositions like those in tholeiitic sills, without the low CaO and high Al_2O_3 contents that characterize spinifex pyroxenes. Bronzites also have compositions resembling those of plutonic pyroxenes. Arndt and Fleet (1979) and Kinzler and Grove (1985) provided more information about the crystal-lographic, morphologic and compositional characteristics of these minerals.

Plagioclase

Plagioclase appears infrequently in Archean komatiites, being largely restricted to the groundmass of samples from the interiors of thick flows, but it is an essential component of many komatiitic basalts. As mentioned above, plagio-clase is present in cells between unusually coarse platy olivine grains in the lower part of the spinifex zone of the Alexo komatiite flow (Barnes, 1983; Arndt, 1986b). The plagioclase has the composition An_{75} and is intergrown with augite. Calcic plagioclase is also common in the matrix of olivine spinifex komatiites from Gorgona Island (Aitken and Echeverría, 1984). In the Alexo flow, the presence of plagioclase can be related to unusually low cooling rates that prevailed in the deep interior of the flow; at Gorgona, the crystallization of plagioclase results from the relatively aluminous compositions of these flows.

Plagioclase with lower anorthite contents is found in the matrices of olivine and pyroxene cumulates and in the gabbroic portions of layered komatiitic basalts (e.g. Fred's Flow in the Munro Township, and layered flows in the Circum-Ungava belt (Francis and Hynes, 1979; Baragar and Lamontagne, 1980). In komatiitic basalts with less than about 10% MgO, plagioclase is an ubiquitous component. In almost all komatiitic lavas, although plagioclase starts to crystallize after pyroxene (Cameron and Nisbet, 1982; Kinzler *et al.*, 1990a), late cotectic growth of the two phases produces a complex plagioclase–augite intergrowth that characterizes many komatiitic basalts (Figure 3.25(d)).

Other magmatic minerals

Magmatic amphibole is a rare mineral in komatiites. It appears to be absent from thin komatiite flows, though in many cases its identification is compli-cated by the difficulty in distinguishing magmatic amphibole from meta-morphic phases. The presence of minor magmatic amphibole in the matrix of olivine orthocumulates in the lower part of a thick, layered komatiite flow in Munro Township was noted soon after the discovery of komatiites (Arndt, 1977b). Since that time W. E. Stone and coworkers have written a series of important papers (Stone *et al.*, 1987, 1997, 2003) in which they describe in detail the occurrence, the composition and the significance of amphibole in komatiites. In the most recent of those papers Stone *et al.* (2003) provided major- and trace-element and stable isotope compositions of amphibole in tholeiitic and komatiitic sills from various locations in the Abitibi, and some more detailed information about the amphiboles in the Boston Creek Fe-rich komatiite. Here the mineral occurs in the basal chill zones and olivine cumulate layers, mainly as oikocrysts in the matrix but also as inclusions in olivine grains. Electron and ion microprobe analyses indicate that their compositions are those of pargasite or hastingite and show them to be enriched in incompatible trace elements such as Nb, Zr and LREE. Particularly important is the water content, which ranges from 1 to 3%. Stone *et al.* (1997, 2003) proposed that the amphibole formed by subsolidus reaction of hydrous silicate liquid with olivine and pyroxene, either in the matrix or in melt inclusions trapped in the ferromagnesian minerals. Using a combination of bulk-rock and amphibole compositions, they calculated that the residual liquid from which the amphibole had crystallized contained at least 2–3% H_2O and that the parental magma contained 1–2% H_2O.

Other magmatic minerals identified in komatiites include ilmenite, Al-spinel and Ni–Fe sulfides. In the Fe-rich komatiites, chloro-apatite has been identified in the groundmass. And the poorly preserved fragmental komatiites at Dachine in French Guiana contain a large deposit of apparently xenocrystic diamond (Capdevila *et al.*, 1999).

Glass

Although we have no right to expect a phase as unstable as glass to have survived in 2–3 billion year old ultramafic lavas, the unusually fresh Zvishavane lavas provide examples of: (a) devitrified but still anhydrous glass in the groundmass between olivine and pyroxene grains; and (b) extant, unaltered and still vitreous glass in inclusions within olivine or chromite grains (Nisbet *et al.*, 1987; McDonough and Ireland, 1993; Shimizu *et al.*, 2001). Essentially unaltered glass is also preserved in the matrices and in olivine inclusions in the Cretaceous Gorgona komatiites (Echeverría, 1980). The glass in olivine inclusions is pale brown and translucent but that in the matrix is riddled with minute opaque grains. Some of the inclusions are partially crystallized to pyroxene and contain abundant minute grains of oxides or sulfides; others are totally vitreous. Retraction bubbles are found in a small number of inclusions.

The composition of glass in the inclusions is unusual. The MgO content is low, around 2–11%, and contents of SiO_2 (50–58%), Al_2O_3 (11–16%), CaO (14–20%) and Na_2O (0.7–1.6%) are high. This type of composition is that of a

Table 4.4. *Compositions of fresh and altered glass in selected komatiites*

	1	2	3	4	5	6	7
SiO_2	52.91[a]	48.38	41.6	52.4	57.85	57.47	43.79
TiO_2	0.46	0.27	0.36	0.33	0.73	0.68	1.24
Al_2O_3	15.36	10.15	11.0	19.8	16.39	16.81	14.40
FeO	12.66	10.44	10.9		3.24	2.45	7.40
MnO	0.13		0.27	9.1	0.07	0.07	0.10
MgO	5.74	18.36	19.1	5.1	2.73	2.86	14.23
CaO	9.09	4.59	8.25	7.0	14.53	14.73	12.51
Na_2O	2.76	1.22	0.51	4.1	1.59	1.12	2.52
K_2O	0.14	0.00	0.16	0.11	0.08	0.06	0.12
Total	99.26	93.41	92.3	97.9	97.78	96.85	99.79

[a] Oxides in weight per cent.
1 Isotropic interstitial glass in olivine spinifex lava, Zvishavane, Zimbabwe (Bickle *et al.*, 1993).
2 Altered interstitial glass in olivine spinifex lava, Zvishavane, Zimbabwe (Bickle *et al.*, 1993).
3 Altered interstitial glass in olivine spinifex lava, Alexo, Canada (Lahaye and Arndt, 1996).
4 Altered interstitial glass in olivine spinifex komatiite from Gorgona Is (Aitken and Echeverría, 1984).
5 Inclusion in olivine in spinifex lava, Zvishavane, Zimbabwe (Nisbet *et al.*, 1987).
6 Inclusion in olivine in spinifex lava, Zvishavane, Zimbabwe (McDonough and Ireland, 1993).
7 Inclusion in amphibole from a komatiitic intrusion (Stone *et al.*, 2003).

komatiite liquid that evolved by extreme olivine fractionation: as seen in Figure 5.2, the compositions lie on the extrapolation of the olivine control lines defined by whole-rock samples of Zvishavane lavas. This evolution probably took place mainly through crystallization of olivine on the walls of the host mineral.

Falling on the same trends for many of the major-element oxides are the compositions of devitrified glasses from the matrix between olivine or pyroxene grains. Some have compositions virtually identical to that of the glass in the inclusions (e.g. analysis 1 in Table 4.4), but others have higher MgO (up to 18%) and lower contents of most other oxides. As shown in Fig. 4.13, there is a good correlation between MgO and H_2O (the latter inferred from the difference between the sum of the oxides measured with the electron microprobe and 100%). The glasses with the highest MgO contents have the highest inferred H_2O contents. Similar relationships are seen in matrix glasses analysed in komatiites from Alexo and Munro (Arndt, 1986b and unpublished data), the Yilgarn Block (Binns *et al.*, 1976) and Gorgona (Aitken and Echeverría, 1984).

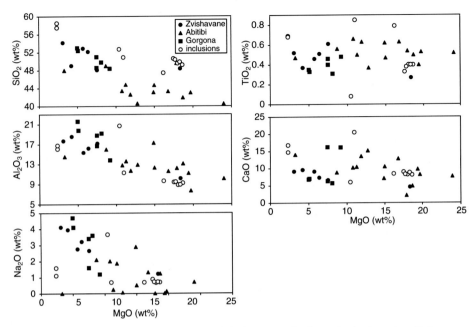

Fig. 4.13. Compositions of interstitial glasses and glassy inclusions in olivines from three regions.

Although many of the matrix glasses in komatiites from Gorgona Island are anhydrous and have MgO contents like those of Zvishavane inclusions, their compositions differ in other respects. As pointed out by Aitken and Echeverría (1984), their CaO contents are strikingly higher and SiO_2 contents somewhat lower. The variation of MgO contents in these glasses probably has two causes: (a) primary variation due to different extents of olivine fractionation; and (b) alteration during which the glass is converted to secondary Mg-rich phases. Discrimination between these two processes is not straightforward. Although it might be argued that the good correlation between MgO and H_2O is clear evidence for the latter process, the relationship can be explained equally well in terms of differing H_2O contents of the secondary phases that replace glasses of different compositions. An MgO-rich glass will be replaced by H_2O-rich phases such as serpentine and chlorite whereas an Mg-poor glass will be replaced by H_2O-poor phases such as tremolite, albite and epidote. In fact the effects of the two processes can be distinguished. For example, the wide variation in the compositions of glasses in a single sample of spinifex lava from Zvishavane is most probably due at least in part to alteration. In parts of the sample where the primary minerals (olivine and pyroxene) are freshest and the glass is clear and near-isotropic, both MgO and H_2O contents are low; near veins and in patches where primary minerals are altered and where the glass is clouded and has a distinct birefringence, MgO and H_2O contents are higher.

Another example of highly altered glass is that in the matrix of the Alexo spinifex lava M626. This material contains almost 18% H_2O and has been converted entirely to serpentine, a process that was apparently accompanied by an influx of Mg. On the other hand, the glasses in the Alexo chilled lava M666 appear to have preserved much of their original chemical compositions. Although their MgO contents are high, around 28%, their H_2O contents are only moderate. When their compositions are recalculated on an anhydrous basis, they lie on the olivine control lines defined by the whole rock compositions. There is no sign of the Ca loss that is a telltale indication of alteration in the MgO-rich relict glasses mentioned above. The glass in the Alexo chill sample appears to have preserved its high MgO content because it was quenched before significant amounts of olivine had crystallized. MgO contents in most other samples are lowered by olivine fractionation during the cooling of the interiors of the flows.

An interesting aspect of the composition of glasses in the Zvishavane lavas is their high Na_2O content. As discussed further in Chapter 5, Na_2O usually behaves in a mobile manner during the alteration of komatiites, but in these unusually fresh lavas, the primary Na_2O content appears to have been preserved. As is seen in Figure 4.13, the compositions of both glasses and whole rocks lie on a single line that intersects the composition of the olivine that crystallized within the lavas. This coherent behaviour suggests that the high Na_2O contents recorded in the Zvishavane lavas and glasses is a primary characteristic and that the lower Na_2O in komatiites from most other areas stems from Na loss during alteration.

4.3 Secondary minerals

The emphasis in this book is on the magmatic characteristics and history of komatiites, and space does not allow a full treatment of the effects of post-eruption alteration and metamorphism. These processes are considered from time to time, mainly because their effects must be taken into account if the igneous petrogenesis of the rocks is to be understood. For this reason the chemical consequences of the various alteration processes are discussed in Chapter 5, and the mineralogical changes are outlined below. For a more complete treatment of these processes the reader is referred to other sources. Jolly (1982), for example, gave a thorough account of the mineralogical effects of metamorphism, mainly at low to medium grades, on komatiites and related lavas in the Abitibi greenstone belt. Binns *et al.* (1976), Gole *et al.* (1987) and Barnes and Hill (2000) described the metamorphic mineralogy of komatiites in the Yilgarn belt, where rocks of higher grade than those in the well-studied parts of the Abitibi belt are found. They also discussed the influence of differing H_2O–CO_2 proportions in the metamorphic fluids. Descriptions of the secondary mineralogy of komatiites in the Barberton greenstone belt and other areas in South Africa are found in Viljoen and Viljoen (1969a), Smith *et al.* (1984), de Wit *et al.* (1987) and Wilson *et al.* (1989).

The compositions of the secondary minerals in komatiite depend not only on metamorphic conditions but also on the mineralogy and bulk composition of the lava. Thus, even in komatiites with the simple primary assemblage olivine–clinopyroxene–glass that have been affected only by hydration, the type of secondary mineral that replaces olivine is variable and depends on the bulk composition. The mineral is chlorite in lavas with less than about 25% MgO, serpentine in lavas with ~25–35% MgO, and serpentine + brucite in more magnesian rocks. During alteration, matter is exchanged between glass and silicate minerals. In Mg-poor komatiites, relatively abundant glass supplies Al to produce chlorite from olivine while the glass itself is converted to chlorite and tremolite; in komatiites with higher MgO contents there is insufficient Al to produce chlorite and the dominant hydrous phase is a serpentine mineral. In Mg-poor komatiites, the alteration is isochemical but for the introduction of water and the loss of alkalis and other highly mobile elements; in komatiites with higher MgO contents, Ca-bearing secondary phases are not produced and this element is lost from the rock (see Chapter 7).

Simple hydrated assemblages are common in most komatiite occurrences, and it is for this reason that the igneous characteristics – both textural and chemical – of these rocks can be determined with some confidence. However, more extreme types of alteration may be locally or widely distributed, and these processes have a more profound effect on the texture and composition of the altered rocks. Carbonatization, a relatively common process, produces the suite of secondary minerals. Quantitative treatments of the mineralogical and chemical effects of carbonatization are given by Smith *et al.* (1984) and Gole *et al.* (1987). Other types of alteration produce diverse mineralogical assemblages. In the Alexo area, for example, two flows, one komatiitic and the other basaltic, have undergone a type of alteration in which the early hydrous assemblage of serpentine–chlorite–tremolite plus relict igneous minerals was replaced by Ca-rich phases including grossular garnet and a Ca-rich pyroxene. Accompanying the Ca introduction was a loss of alkalis and Si. Still more extreme is the hydrothermal alteration suffered by some komatiites in the Barberton greenstone belt and Pilbara Craton, where spinifex and massive lavas have been partially and completely replaced by secondary quartz (Duchac and Hanor, 1987; Byerly, 1999).

4.4 Summary

(1) The primary mineralogy of komatiite is simple, comprising olivine, minor chromite, various types of pyroxene, (altered) glass and in some cases plagioclase or amphibole.

(2) The morphology of crystals of these minerals is complex. In the upper parts of flows mineral habits range from skeletal to dendritic, the form depending on the conditions of cooling and the composition of the liquid. Equant polyhedral or prismatic gains of olivine or pyroxene are found in cumulate zones.

(3) The contents of Ni, Cr and Ca in olivine are relatively high, due mainly to the high abundances of these elements in the komatiite liquid.

(4) Three types of pyroxene occur in komatiites: augite, pigeonite and bronzite. Characteristic of spinifex lavas in komatiitic basalts are zoned acicular grains with pigeonite cores and augite margins.

(5) Glass is present in the matrices of most komatiite samples. In the best-preserved samples from Zimbabwe and Gorgona, this glass is devitrified but remains essentially anhydrous. Fresh glass is also preserved in melt inclusions in grains of olivine and chromite.

5

Geochemistry

5.1 Introduction

Komatiites are extinct. The last eruption we know of took place on Gorgona Island some 90 Ma ago, and although sightings are regularly claimed in the literature and even more frequently in manuscripts submitted for review, none of these is convincing. When dealing with the chemical compositions of modern basalts, andesites, phonolites, even carbonatites, a natural step is to obtain samples of newly erupted lavas that are essentially unaltered and have compositions like those of the original magmas. We know of no newly erupted komatiites, and even the youngest examples – the Cretaceous Gorgona komatiites – have been affected by the circulation of hydrothermal fluids. They contain secondary hydrous minerals in veins and as pseudomorphic replacement of olivine and glass, and chemical analyses typically contain a small percentage of H_2O, and variable ferrous/ferric iron ratios. Such features would be grounds for rejection from many data banks of modern igneous rocks.

The freshest Archean komatiites, those from the Zvishavane area in Zimbabwe, are only slightly more altered than the Gorgona rocks, but most other known examples of Precambrian komatiites are less well preserved. It must be accepted that all komatiites are altered to a greater or lesser extent, and any investigation of their geochemistry has to deal with this alteration.

A special characteristic of komatiites allows us to do this, and indeed by using this characteristic, the chemistry of komatiites can often be interpreted more confidently than that of all other Precambrian or even early Phanerozoic volcanic rocks. In komatiites the only important liquidus mineral is olivine. Chromite also crystallizes, but in amounts so small that it influences only the Cr concentrations and has no significant effect on other elements. Because of the low viscosities of komatiites and the large interval between the liquidus and the temperatures at which other silicate phases appear (see Chapters 10 and

11), olivine crystallizes alone and segregates readily, and most komatiites display a wide range of compositions produced by olivine fractionation and accumulation. Thermal erosion and wall-rock contamination can upset such trends, but the effects of this process are generally secondary, as discussed in Sections 5.6 and 5.8. Olivine control lines produced by fractionation and accumulation provide a valuable tool for monitoring the effects of alteration. In Section 5.2 I discuss the effects of olivine fractionation and the use of the olivine-control-line tool; this is followed by a briefer mention of other techniques of identifying and quantifying element mobility. Then, having established which elements are immobile and reliable, I go on to a more general discussion of komatiite geochemistry.

5.2 Olivine fractionation in komatiites

MgO contents in samples from individual, differentiated, spinifex-textured komatiite flows vary by as much as 25%. In the Alexo komatiite flow, which has been studied by Barnes (1983) and Arndt (1986b) and described in Chapter 2, olivine-rich cumulates contain up to 45% MgO and the most evolved spinifex-textured lavas only 19%. Representative analyses of this and other komatiites are compiled in Table 5.1, and the variation in composition from top to base of the flow is shown in Figure 5.1(a). The lowermost flow in the Kambalda komatiite sequence (Figure 5.1(b)) has MgO contents ranging from 16 to nearly 50% MgO, and the thinner differentiated flows exposed in many greenstone belts contain 20–45% MgO. Because of these large variations, it is usually a simple matter to collect a suite of samples across a flow or suite of flows and use them as a basis for monitoring the extent of olivine fractionation and the effects of alteration.

5.3 Mobile and immobile elements: the olivine control line criterion

The most convenient way to represent chemical data from komatiites is to plot variation diagrams with the MgO content on the *x*-axis. MgO contents vary widely and systematically with olivine fractionation, and competing petrogenetic processes such as the fractionation or accumulation of other minerals, or contamination and alteration, produce trends oblique to olivine fractionation trends and are easily recognized. In Figure 5.2 modal olivine is plotted against the MgO contents in two suites of samples from well-preserved komatiites, one from Zvishavane, Zimbabwe and the other from Alexo, Canada. In both, the data correlate reasonably well and demonstrate the corelationship of MgO and olivine contents. In the case of the komatiites from Zvishavane,

Table 5.1. *Major-element compositions of selected komatiites*

	M667	M666	M663	M668	M661	B12	BH2	10-PPR-97	DLS OAC	94-19	94-35
SiO_2	45.9	46.1	46.1	46.9	44.4	47.6	49.9	45.8	41.5	46.1	47.0
TiO_2	0.34	0.36	0.35	0.45	0.19	0.44	0.11	0.70	0.01	0.63	0.37
Al_2O_3	6.75	7.11	6.92	8.80	3.25	4.49	7.56	7.24	0.3	10.8	10.7
FeO (tot)	11.0	10.9	10.8	11.6	8.24	12.06	5.5	11.1	6.2	11.1	10.4
MnO	0.18	0.17	0.16	0.16	0.12	0.20	0.12	0.16	0.1	0.18	0.17
MgO	30.9	28.4	28.9	24.0	42.8	27.1	30.8	26.9	51.6	21.0	22.2
CaO	4.88	6.61	6.31	7.98	1.01	8.00	5.39	7.92	0.1	9.17	8.50
Na_2O	0.03	0.29	0.43	0.09	0.01	0.05	0.34	0.15	0.0	0.90	0.56
K_2O	0.03	0.10	0.11	0.04	0.01	0.01	0.01	0.02	0.0	0.03	0.19
P_2O_5						0.03		0.03		0.05	0.03
(LOI)	10.6	6.8	7.7	6.5	12.9	5.7		7.03		2.39	3.26

Notes: Iron has been recalculated as FeO(tot) and then the analysis was recalculated to 100% on an anhydrous basis. LOI – loss on ignition.

Descriptions of samples and sources of analyses:

Alexo komatiite flow (Arndt, 1986b):
M667 – flow-top breccia
M666 – chilled upper margin
M663 – random olivine spinifex
M668 – coarse platy olivine spinifex
M661 – olivine cumulate

B12 – olivine spinifex lava from the Barberton greenstone belt (Lahaye *et al.*, 1995)
BH2–BH2 113.95–114.13 – aphyric chill from SiO_2-rich komatiite from the Commondale belt, South Africa (Wilson, 2003)
10-PPR-97 – Fe–Ti-rich komatiite from the Jeesiörova region, Finland (Hanski *et al.*, 2001)
DLS OAS – average composition of 40 samples from a dunitic lens in the Perseverence body, Australia (Barnes *et al.*, 1988)
94-19 – olivine spinifex-textured komatiite from Gorgona Island (Arndt *et al.*, 1997b)
94-35 – picrite from Gorgona Island (Arndt *et al.*, 1997b)

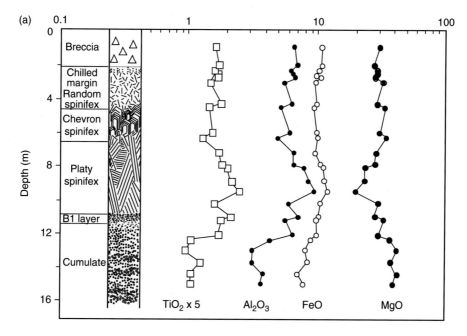

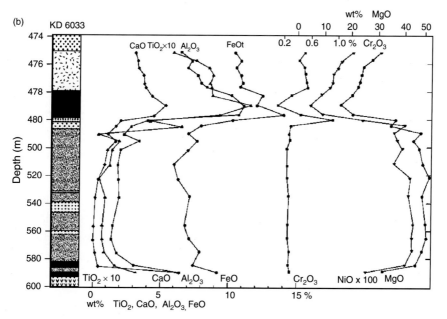

Fig. 5.1. Variations in the concentrations of major elements in sections across two spinifex-textured komatiite flows: (a) the komatiite flow from Alexo, Canada; (b) a flow in the lower part of the komatiite succession at Kambalda in Australia (from Lesher (1989)).

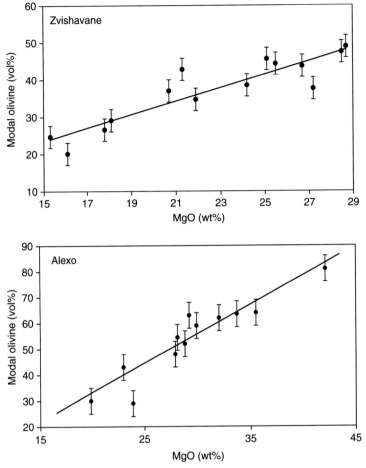

Fig. 5.2. Modal olivine vs MgO for Zvishavane, Zimbabwe and Alexo, Canada.

Zimbabwe, the initial liquid had 19–20% MgO: about 15% olivine fractionated from this liquid to produce the most evolved spinifex-textured lavas, and
about 25% olivine accumulated in the most magnesian olivine cumulates. The
scatter about the trend is partly due to errors in the modal analyses arising
from the difficulties in determining the modal mineralogy of coarse-grained
spinifex-textured samples, but is also influenced by variations in cooling rate
that affect the amount of occult olivine in the original glass and the alteration
of some samples. In the Alexo suite, the initial liquid contains about 28% MgO
(Chapter 12) and again, the more magnesian rocks are cumulates and the less
magnesian rocks crystallized from evolved liquids. Figures 5.3 and 5.4 are
variation diagrams for major and trace elements for the least altered samples
from the well-preserved suite of Zvishavane komatiites.

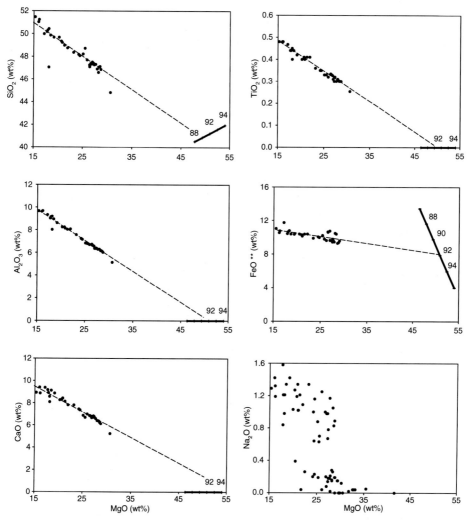

Fig. 5.3. Variation diagrams for major elements in fresh komatiites from Zvishavane, Zimbabwe. Al_2O_3, TiO_2 and SiO_2 behave as immobile elements and plot on tightly constrained trends that intersect the olivine composition. Na_2O is mobile and scatters broadly. FeO^{**} is Fe^{2+} calculated from the total FeO using the method of Barnes *et al.* (2007).

Immobile elements in Zvishavane komatiites

Consider the plot of MgO vs Al_2O_3 (Figure 5.3). The data fall on a well-defined linear trend ($r^2 = 0.998$) which intercepts the MgO axis at $49.4 \pm 0.5\%$. Because olivine contains insignificant Al_2O_3 (<0.3% – Chapter 4), the MgO intercept gives the MgO content of the olivine that fractionated and accumulated. The MgO value, which can be read off Figure 5.3, agrees within error

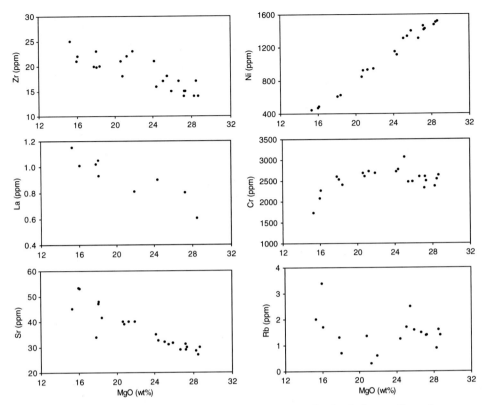

Fig. 5.4. Variation diagrams for trace elements in fresh komatiites from Zvishavane, Zimbabwe. Elements such as Zr, La and Ni define trends that intersect the olivine composition. Rb is mobile and scatters broadly.

with that in the average composition of olivine (Fo_{89}) that crystallized in the flow (Chapter 3). The amounts of fractionated and accumulated olivine, which can be calculated assuming incompatible behaviour of Al_2O_3, are consistent with the measured variations in modal olivine given in Figure 5.2. In all respects the variations of MgO and Al_2O_3 contents correspond to olivine control and it can therefore be concluded that in these komatiite flows *both* of these elements were essentially unaffected by hydrothermal alteration, metamorphism and other secondary processes.

Similar conclusions can be reached for other incompatible elements plotted in Figures 5.3 and 5.4 and listed in Table 5.2. These comprise the suites of elements normally thought of as impervious to hydrothermal processes – Al and Ti amongst the major elements, and trace elements such as most of the REE and high-field-strength elements (HFSE). Immobility for an element compatible with olivine such as Ni can be demonstrated by substituting it for MgO and comparing the *x*-axis intercept with the concentration of the

Table 5.2. *Trace-element compositions (ppm) of selected komatiites*

	M667	M666	M663	M668	M661	B-12	BH2	10-PPR-97	94-19	94-35
Sc	29.2	28.5	27	31.7	20.6		23.2	23.9		
V	163	154	150	188	76		140	206		
Cr	3011	2921	2919	2992	2199	2364	3776	3675	1830	2030
Co	116	119	124	103	123		64			
Ni	1656	1663	1874	1040	2977		1020	1413	1079	968
Cu	45	72	57	40			0.1			
Cs	3.25	1.83	1.55	1.2	0.33	0.40	nd	0.38	0.00	0.03
Rb	4.47	6.05	7.15	4.72	0.5	0.52	0.29	1.59	0.44	2.29
Ba	4.57	12.23	12.36	18.08	0.72	5.52	0.17	2.90	12.8	9.4
Sr	18.35	21.82	33.27	23.24	6.56	36.9	5.00	7.22	122.8	31.9
Nb	0.44	0.44	0.41	0.69	0.31	1.43	0.07	0.42	0.48	0.23
Zr	16.56	18.09	16.99	24.19	10.47	26	5.10	33.3	29.0	13.0
Y	7.98	8.65	8.11	10.54	4.56	8.9	5.50	10.1	14.2	15.0
Pb	0.35	0.11	0.22	0.08	0.04	0.15	0.087		0.19	0.01
Th	0.031	0.041	0.029	0.058	0.026	0.16	nd	0.06	0.15	0.03
U	0.012	0.0097	0.01	0.0175	0.01	0.03	nd	0.07	0.07	0.01
La	0.408	0.54	0.426	0.861	0.37	1.61	0.05	0.88	0.55	0.22
Ce	1.33	1.54	1.43	2.36	1.07	5.06	0.07	2.72	2.00	0.57
Pr	0.26	0.29	0.26	0.42	0.18	0.76	0.01	0.54	0.41	0.13
Nd	1.42	1.62	1.38	2.28	0.98	3.87	0.09	3.53	2.63	0.90
Sm	0.61	0.697	0.585	0.9	0.38	1.19	0.08	1.37	1.22	0.65
Eu	0.23	0.288	0.253	0.36	0.11	0.39	0.05	0.46	0.53	0.34
Gd	0.96	1.07	1.02	1.38	0.55	1.45	0.22	1.78	1.95	1.5
Tb	0.18	0.19	0.18	0.24	0.1	0.25	0.07	0.32	0.35	0.30
Dy	1.23	1.38	1.29	1.74	0.65	1.51	0.60	1.88	2.31	2.32
Ho	0.27	0.28	0.27	0.35	0.15	0.33	0.16	0.38	0.47	0.48
Er	0.81	0.899	0.84	1.13	0.43	0.93	0.51	1.06	1.34	1.35
Tm	0.12	0.13	0.12	0.15	0.07	0.13	0.09	0.13	0.21	0.23
Yb	0.79	0.888	0.816	1.1	0.43	0.86	0.60	0.83	1.22	1.42
Lu	0.12	0.13	0.123	0.165	0.07	0.13	0.10	0.12	0.18	0.21

Sample description as in Table 5.1.
nd – not determined.

element measured in olivine. Data from the Zvishavane flows plot on a tight trend in the Ni vs Al_2O_3 and Ni vs TiO_2 diagrams (Figure 5.5) with an intercept at 3400 ppm, a Ni concentration typical of those measured in olivine phenocrysts. This approach can only be used for elements abundant enough to be analysed with the microprobe. For others such as the platinum group elements (PGE), immobile behaviour can only be inferred on the basis of limited scatter

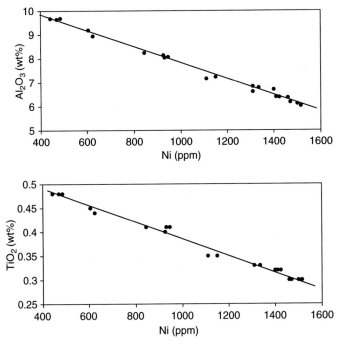

Fig. 5.5. Ni vs Al₂O₃ and Ni vs. TiO₂ diagrams for the Zvishavane suite. Normally the concentration of a trace element in a magma undergoing fractional crystallization should decline exponentially but, as explained in Chapter 4, this effect is counterbalanced by a variation in the Ni partition coefficient, which increases, and the Ni content of the liquid, which decreases.

in variation diagrams. In the case of Cr, simple olivine control can only be assumed in the most magnesian komatiites. In others chromite coprecipitates with olivine in sufficient quantities to control the behaviour of Cr, and possible effects of alteration are not easily monitored.

Mobile elements in Zvishavane komatiites

Element mobility takes two forms. Many elements scatter widely in variation diagrams and show no relationship whatsoever to olivine control lines. A good example is Rb whose concentrations range from 0.1 to 2.5 ppm and do not correlate at all with MgO (Figure 5.4). Isotopic studies by Hegner (reported in Nisbet *et al.* (1987)) demonstrated that Rb was lost from many samples from the Zvishavane komatiites. Potassium is similarly mobile. Some other mobile elements define tight linear trends but have MgO intercepts far removed from the appropriate olivine composition.

Table 5.3. *Summary of element mobility in komatiites from various localities*

Greenstone belt	Locality	Immobile	Mobile
Belingwe belt	Type area of Reliance Formation	Al, Ti, Mg; Ni, Cr(?), REE except Eu; HSFE such as Nb, Ta, Zr, Hf, Y	Si, Ca, Na, K; large-ion lithophile elements (LILE) such as Cs, Rb, Ba, Sr
	Zvishavane drill core	All major elements except Ca, Na, K; Ni, Cr(?), REE except Eu; HSFE	Ca, Na, K; LILE
Alexo	Non-rodingitized spinifex samples	All major elements except Ca, Na, K; Ni, Cr(?), REE except Eu; HSFE	Ca, Na, K; LILE
	Flow-top breccia and serpentized cumulates	Al, Ti, Fe(?); Ni, Cr(?), REE except Eu; HSFE	Si, Mg, Ca, Na, K; LILE
	Rodingitized samples	None	All major elements; all analysed trace elements
Barberton	Spinifex Stream	Al; Ni, Cr(?), REE except Eu; HSFE	Si, Ti, Fe(?), Mg(?), Ca, Na, K; LILE
	Schapenburg spinifex samples	Si, Ti, Al; Ni, Cr(?), REE except Eu; HSFE	Fe(?), Mg(?), Ca, Na, K; LILE
	Schapenburg cumulates	None(?)	Most major elements; REE; HFSE and LILE

CaO, for example, has an anomalously high intercept at about 56% MgO (Figure 5.3), and Ba (not shown) and Sr have intercepts that are too low (Figure 5.4). The results of the survey of element mobility in the Zvishavane komatiites are summarized in Table 5.3, which contains a list of elements whose abundances were measured, and identification of these elements as mobile or immobile.

The behaviour of Na is particularly interesting. The concentrations of this element in the suite of least-preserved samples from the SASMAR drill core in the Zvishavane region (Nisbet *et al.*, 1987) are much higher than those in more altered samples from elsewhere in the Belingwe belt (Figure 5.6). In samples from the Abitibi and Barberton suites, Na_2O contents vary widely encompassing about the same range as in the Zimbabwe samples. The reasonable assumption, based on differences in the degree of alteration of the sample suites, is that the Na_2O contents of the original magmas were closer to those of the least-altered

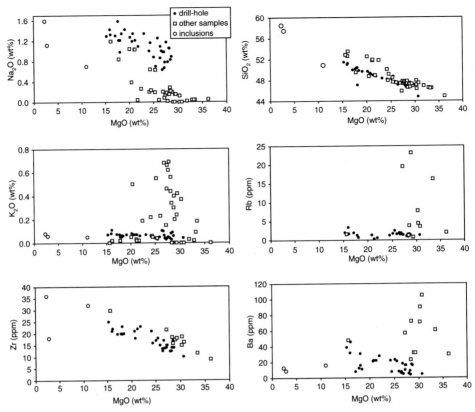

Fig. 5.6. Variation diagram of MgO vs Na$_2$O, SiO$_2$ and selected trace elements in fresh and more altered komatiites from the Belingwe greenstone belt. Most of the more altered komatiites appear to have lost 50–80% of their original Na$_2$O content. Data from Bickle *et al.* (1993) and Shimizu *et al.* (2004).

samples and that the other samples had lost a large proportion (50–80%) of their original Na contents.

The interpretation of Na mobility is confirmed by analyses of glass inclusions in olivine phenocrysts (McDonough and Ireland, 1993; Shimizu *et al.*, 2001; M. Gee, unpublished data). The glassy portions of such samples are essentially anhydrous and should have preserved their original magmatic compositions. In Figure 5.6 it is seen that glasses from the inclusions are collinear with the least-altered whole-rock samples from the SASMAR drill core. The glasses have low water contents and probably represent our best samples of near-magmatic compositions; a problem is to correct for the effects of post-entrapment crystallization and interaction with the surroundings, but this can be done using the reheating techniques developed by Sobolev and Kostyuk (1975).

More altered komatiites

The behaviour of major and trace elements in more altered komatiites is similar in many respects to that in the Zvishavane komatiites, as can be seen in diagrams such as Figure 2.6 (Barberton komatiites) and Figure 2.16 (Abitibi and Gorgona komatiites). The results are much the same as for the Zvishavane komatiites: elements such as Al, Ti, Zr, most REE, Ni are immobile and Rb, Ba, Si, Ca, Na, Si, Eu etc. are mobile.

5.4 Mobility of Mg

In general MgO seems to have been relatively immobile in Archean komatiites. This is shown by the tight trends in diagrams of MgO vs various immobile elements, and by the appropriate intercepts with MgO axes in variation diagrams. Several authors (Echeverría, 1982; de Wit *et al.*, 1983, 1987; Parman *et al.*, 1997, 2001) have suggested, however, that the high MgO contents of some komatiites may be due to Mg gain during interaction with seawater or hydrothermal fluids. De Wit *et al.* (1987) supported their interpretations with several arguments. They showed a positive correlation between MgO and H_2O, which they claimed to be a result of MgO uptake during hydration. In fact, this relationship probably arises from differences in the H_2O contents of the secondary minerals that form at the expense of magmatic minerals in komatiites of different compositions: serpentine and chlorite, which replace olivine in Mg-rich samples, have $H_2O > 10\%$ whereas amphibole and sodic plagioclase, which replace pyroxene and glass in Mg-poor samples, have $H_2O < 2\%$. The H_2O content correlates with the original olivine content, which in turn is a function of MgO content: a positive correlation between H_2O and MgO is entirely to be expected.

De Wit *et al.* (1987) also plotted Mg–Fe ratios of olivine grains and host komatiites. In their diagram, only samples from Gorgona Island and Alexo had the relationship between olivine and host magma compositions predicted from Mg–Fe partitioning: in samples from Barberton and Munro Township, the komatiite samples appear to have anomalously high Mg/Fe ratios, a factor that the authors attributed to Mg gain during alteration. This argument is called into question by the compositions of Barberton komatiites in variation diagrams such as those shown in Figure 2.6. If all these samples had gained a similar amount of MgO, the trend of the altered samples would have intercepted the *x*-axes at MgO values higher than that measured in olivines. Had some samples gained more MgO than others, either the data would scatter, or the trend would rotate. The only way of producing the correct MgO intercept

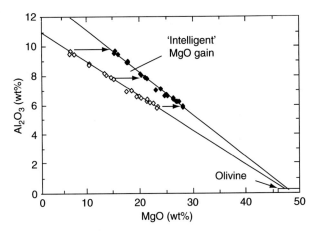

Fig. 5.7. Variation diagram illustrating the effects of MgO gain and showing that MgO gain will produce a tight trend that intercepts the *x*-axis at the olivine composition only if the amount of added MgO is inversely proportional to MgO content.

in a suite of Mg-metasomatized samples is to have some form of 'intelligent' MgO gain, a process in which the amount of added Mg must correlate systematically with the original Mg content, with less magnesian samples gaining more Mg through alteration than more magnesian samples. This type of behaviour, which is illustrated in Figure 5.7, seems most unlikely.

Further convincing evidence that most komatiites retain their original MgO contents can be obtained by reconstructing major-element compositions using modal analyses and estimated compositions of the igneous components. For example, the spinifex-textured komatiite M663 from Alexo contains about 50 modal per cent olivine (Arndt, 1986b). The olivine has the composition Fo_{88-92} and contains 46–50% MgO; it would have contributed about $0.50 \times 48 = 24 \pm 1\%$ MgO to the rock composition. The clinopyroxene (10–20% MgO) and interstitial glass (6–10%) in the matrix contribute another 5 wt%, to bring the total to 29% MgO. This value is very similar to the MgO content in the whole-rock analysis, which is 28.4% MgO. A similar procedure also works for more altered samples, such as the Crixas komatiites (Arndt *et al.*, 1989), and can be applied to any sample in which the modal composition can be measured, including the Barberton samples. Unfortunately, although numerous whole-rock analyses, normative compositions and mineral compositions of these komatiites have been published, nowhere could modal compositions be found.

The apparent discrepancy between olivine compositions and the Mg–Fe ratios of komatiites that was highlighted by de Wit *et al.* (1987) could have resulted from several different factors: (a) the presence of excess olivine, as phenocrysts in samples from chilled margins or skeletal crystals in plate spinifex lavas or cumulus crystals in samples from the lower parts could have increased the Mg/Fe of the whole rocks; (b) in more altered samples such as those from Barberton, the most magnesian olivines may not have been preserved; (c) the Mg/Fe of the olivine could have changed during metamorphism. Parman *et al.* (2004) reported whole-rock data for the chilled margins of Barberton komatiites, as well as new olivine analyses, and in these samples the rock compositions do correspond to those of the liquids that crystallized the most magnesian olivines (Chapter 12).

On the other hand, there is no doubt that individual samples in some komatiite suites have gained or lost a small amount of Mg. In the Alexo flow, for example, a sample from the flow-top breccia plots above the trends defined by all other samples in diagrams of MgO vs elements incompatible with olivine and an olivine cumulate plots below the trends. Both samples are very highly altered and contain absolutely no primary minerals; the breccia appears to have gained about 1–2% MgO and the olivine cumulate appears to have lost the same amount (Arndt, 1986b).

5.5 Other approaches used to demonstrate element mobility

Element mobility in komatiites can be identified by considering ratios of elements or oxides (Nesbitt and Sun, 1976). A good correlation between two elements can often be taken as evidence of immobile behaviour. For example, in Kambalda komatiites and basalts, Zr correlates well with Ti but not at all with Rb (Figure 5.8). On this basis it can be inferred that Zr and

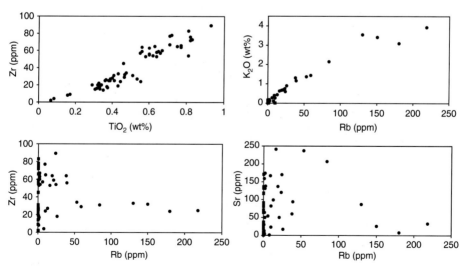

Fig. 5.8. Correlation between various major and trace elements in komatiites and komatiitic basalts from Kambalda; data from Arndt and Jenner (1986).

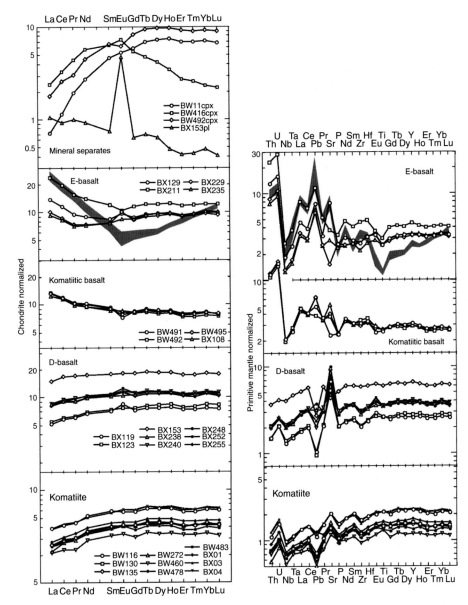

Fig. 5.9. Normalized trace-element patterns of whole rocks and mineral separates from komatiites and basalts from the Belingwe greenstone belt, Zimbabwe, from Shimizu *et al.* (2005). Trace-element data are normalized to the chondritic or primitive mantle values of McDonough and Sun (1995).

Ti are immobile and Rb mobile, as we concluded on the basis of the variation diagrams. Care must be taken, however, not to misinterpret correlation between two mobile elements that are affected in a similar manner during alteration, such as K_2O and Rb in the Kambalda komatiites and basalts. When

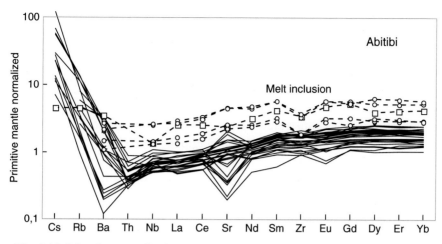

Fig. 5.10. Mantle-normalized trace elements in whole-rock samples of komatiite from Munro Township, Canada. Also shown are normalized trace-element concentrations in melt inclusions in olivine grains from Zvishavane, Zimbabwe (McDonough and Ireland, 1993) and from Alexo, Canada (Lahaye *et al.*, 2001).

plotted in the same diagram, these two elements correlate well but they scatter widely when plotted against MgO or against other trace elements. It appears that the two elements were similarly affected during alteration, but in this case their constant ratio does not indicate immobile behaviour.

Variations in element ratios are shown most clearly in diagrams in which element concentrations are normalized using chondritic or primitive mantle abundances – the familiar REE diagrams and spidergrams. The variable but often large Eu anomalies in the REE patterns of even the relatively fresh komatiites and komatiitic basalts from the Zvishavane region plotted in Figure 5.9 (Bickle *et al.*, 1993; Shimizu *et al.*, 2005) are clearly the result of alteration, because feldspar does not crystallize in rocks of such compositions. The spidergrams display more exaggerated anomalies of Sr, Pb and other large-ion lithophile elements (LILE) in the komatiites and basalts indicating strong mobility of these elements.

In Figure 5.10, mantle-normalized trace elements are plotted for komatiites from Munro Township in Canada. Again the mobility of elements such as Cs, Rb, Sr and Eu is clearly manifested by positive or negative anomalies. In Figure 5.10 trace-element analyses of melt inclusions in olivine grains (McDonough and Ireland, 1993; Lahaye *et al.*, 2001) are compared with the trace-element patterns in the whole rocks. The trace-element contents in the inclusions are higher than in the whole rocks because the analysed glasses represent liquids that fractionated olivine after entrapment. However, the overall patterns are parallel to those of the whole rocks, and the Cs, Rb, Sr and Eu anomalies are absent. The melt inclusions appear to have resisted the alteration that influenced the concentrations of these mobile elements.

Molecular proportion diagrams, introduced by Pearce (1968) as a tool for recognizing igneous fractionation trends in mafic rocks have been used by Beswick (1982) to monitor element mobility. The method depends on the recognition of departures from trends expected for the fractionation of igneous minerals, as illustrated in Figure 5.11. Crystallization of magmatic minerals such as olivine, pyroxene or plagioclase produces trends with characteristic slopes in these diagrams: the line with a slope of 1 in the lower diagram illustrates, for example, the effect of clinopyroxene fractionation. Element mobility either causes scatter or produces trends with slopes different from those that would result from crystallization of the minerals in the rocks. In general the approach yields conclusions similar to those obtained using the olivine-control-line criteria, but in certain cases the departures from igneous behaviour are not easily recognized. For example, Beswick (1982, 1983) and Jahn *et al.* (1982) claimed on the basis of trends in molecular proportion diagrams that Ca is immobile in komatiite suites, yet identical data plotted in simple MgO variation diagrams display some effects of mobility (e.g. Figures 2.6 and 5.3). The use of molecular proportion diagrams to monitor element mobility has been discussed in more detail by Butler (1986), Rollinson and Roberts (1986) and Pearce (1987).

Yet another approach is to compare moderately altered with somewhat less altered samples (e.g. Nesbitt and Sun (1976)). In Figure 5.6, the concentrations of selected major and trace elements in samples from drill core and from outcrop in the Zvishavane area of the Belingwe Belt in Zimbabwe are compared. The Zvishavane drill core samples are the freshest known Archean komatiites (Nisbet *et al.*, 1987), whereas the other samples are more altered and typical of Archean komatiites in general. The relatively large spread of SiO_2 values in the more altered rocks, compared with those from the least altered drill core samples, confirms this. As mentioned above, the Na_2O contents of the Zvishavane suite are higher than in the more altered samples, indicating that the latter have lost a large fraction of their original Na_2O contents. The mobile trace elements Rb and Ba are strongly enriched in the more altered samples, but for Zr, a relatively immobile element, the difference is smaller. This difference is due to either analytical bias, or perhaps a slight difference in the primary compositions of the different suites.

Although this approach yields useful information for the Zimbabwe komatiites, it is of limited use for more normal samples. There are two reasons for this: (a) the approach depends on the assumption that the samples being compared originally had the same composition, an assumption that may not

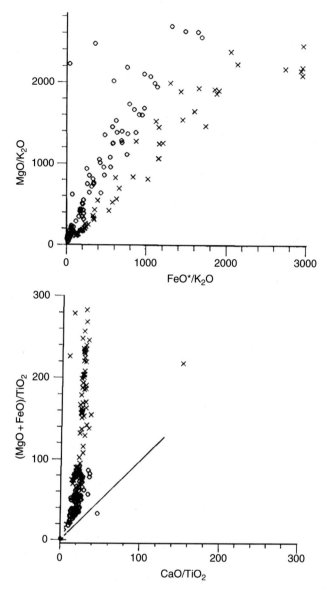

Fig. 5.11. Molecular proportion diagrams that ostensibly demonstrate immobile behaviour of TiO_2, $FeO + MgO$ and CaO (from Beswick (1982)). See text for explanation.

always be valid; (b) results from the Zvishavane komatiites provide evidence that in even the freshest komatiites many elements are mobile. Demonstration of similar element concentrations or element ratios in two samples, one moderately altered and the other highly altered, does not prove that the elements were unaffected by alteration.

Finally, in cases where melt inclusions in olivines or other minerals (chromite) are large enough to be analysed accurately with an electron or ion microprobe, they can provide samples of trapped komatiite liquid that probably retain much of their original composition. Valuable data have been obtained, for example, through the analysis of inclusions in olivine and chromite from the Zvishavane komatiites (McDonough and Ireland, 1993). The compositions of these inclusions are plotted in Figure 5.6 and in general the melt inclusions plot on the extensions of trends defined by the compositions of the least altered samples from the drill core. They confirm the interpretation that the altered samples probably have lost Na_2O and gained SiO_2, K_2O and Ba. Many new data have been obtained on these samples (e.g. Danyushevsky *et al.* (2002a), Woodhead *et al.* (2005)), but at the time of writing they have been published only in abstract form.

Analyses of melt inclusions in the Alexo komatiite flow by Lahaye *et al.* (2001) showed that in terms of their major-element compositions they plot around an olivine control line extending from the composition of the initial liquid in the flow, as estimated using the compositions of the chilled margin of the flow. In an MgO vs CaO diagram (their Fig. 4) the data scatter relatively widely suggesting that either MgO or CaO, or both, were slightly mobile. It would be interesting to plot MgO versus a less mobile element such as Al or Ti but individual analyses are not reported in the paper.

The effects on chemical compositions of hydrothermal alteration and metamorphism are clearly serious hurdles in any investigation of komatiite geochemistry. All komatiites are altered: all contain hydrous minerals, and all have seen an influx of water. The practice of normalizing analyses to 100% on an anhydrous basis is widespread (and has been followed systematically in this book), but it can be justified because in most komatiite suites the alteration is largely isochemical. Nonetheless, from the examples discussed above and considered in later parts of the book, it is evident that different suites are affected in different ways. Any study of komatiite chemistry must start with an investigation of the effects of alteration on the samples under consideration using a combination of the techniques discussed above. For each sample suite, the elements must be classified as mobile or immobile, and only the latter used in subsequent petrologic and petrogenetic interpretations.

5.6 Other types of mobile element behaviour

Certain studies of komatiite flows have demonstrated mobile element behaviour more sporadic and less systematic than that discussed above. Examples, drawn mainly from komatiites I have worked on, are listed below.

(a) In the central part of the Alexo komatiite flow and in parts of neighbouring basaltic flows (Barnes *et al.*, 1983; Arndt, 1986b), alkali elements, Sr and SiO_2 are strongly depleted and CaO is enriched (Lahaye and Arndt, 1996). This type of behaviour appears associated with a peculiar type of hydrothermal alteration that led to the formation of secondary calcic pyroxene and garnet. It resembles the rodingitization reported at the margins of ultramafic intrusions.

(b) In the random spinifex-textured portion of western outcrop of the Alexo flow, Ni, Cu and PGE have been added in the form of secondary sulfides that line the walls of amygdales (Brügman *et al.*, 1987).

(c) In volcanic rocks from the Kambalda region, Australia, mainly in komatiitic basalts but also in some komatiites, the formation of ocelli has resulted in drastic changes in the abundances of many elements. These structures, which are not to be confused with superficially similar spherical varioles, are small spherical to elliptical patches of relatively felsic material that lie in a more mafic matrix. They are strongly enriched in SiO_2, Na_2O, Sr and HREE and are depleted in MgO and FeO relative to the matrix. It is not clear whether the ocelli are primary structures (perhaps produced by liquid immiscibility) or whether they form during later alteration (Gélinas *et al.*, 1977; Arndt and Fowler, 2004).

(d) In other komatiites, alteration has resulted in whole-scale replacement of the original components. In many regions the influx of CO_2-rich fluids resulted in partial to complete carbonatization of the komatiite flows, a type of alteration that is accompanied by drastic changes in the chemical compositions of the samples without obliterating their volcanic textures (Pyke, 1975; Arndt *et al.*, 1989; Tourpin *et al.*, 1991).

(e) Particularly in greenstones belts subjected to intense hydrothermal circulation or in regions of strong weathering like Australia and southern Africa, some komatiites have been almost entirely replaced by quartz and sericite. The result is a rock that retains its spinifex texture but contains up to 90 wt% SiO_2 (Duchac and Hanor, 1987; Byerly, 1999). Thompson *et al.* (2005) described tuffs from the Barberton belt that have been almost completely replaced by silica but whose minor- and trace-element contents retain a record of their originally ultramafic compositions. In examples such as these the inference of essentially isochemical alteration obviously does not apply, and little use can be made of such material in petrogenetic discussions.

5.7 Igneous chemistry of komatiites

Major elements

Komatiites, by definition, have MgO contents greater than 18%. (In the following discussion, major elements will be quoted as weight percent oxide from analyses normalized to 100% on an anhydrous basis.) The abundances of

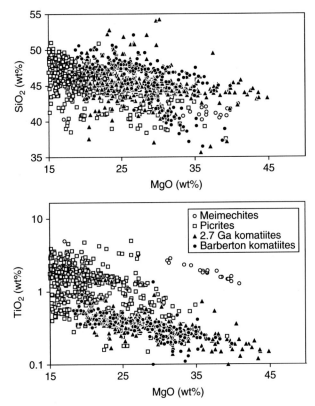

Fig. 5.12. Comparison of TiO$_2$ and SiO$_2$ contents of some picrites, meimechites and komatiites (data from Clarke (1970), Pedersen (1985), Arndt (unpublished) and sources listed in Figures 2.6 and 2.8).

Ni, Cr and other compatible elements are correspondingly high, and abundances of Al, Si, Ti, Zr and other incompatible elements correspondingly low (Table 5.2). Comparison with other igneous rock types is difficult, because few volcanic or non-cumulate plutonic rocks have such high MgO contents. Possible candidates are kimberlites and meimechites, but these have far higher concentrations of alkalis, P$_2$O$_5$ and TiO$_2$ and have far higher H$_2$O and CO$_2$ contents. Certain picrites, such as those from Baffin Bay (Clarke, 1970; Francis, 1985) and Disko Island (Pedersen, 1985), and the picrite–meimechite suite from Siberia (Fedorenko and Czamanske, 1997; Arndt *et al.*, 1998b), have comparable MgO contents, but these rocks have lower SiO$_2$ and significantly higher alkali and TiO$_2$ contents than most komatiites, as illustrated in Figures 5.12. These differences have been used by the International Union of Geological Sciences subcommission on the systematics of igneous rocks (Le Maitre *et al.*, 1989) to develop a geochemical definition of komatiite, as discussed in Chapter 1.

More generally, the major-element compositions of komatiites are simply those of a tholeiitic basalt that contains a high, but variable, proportion of added olivine. Only a few major elements, mainly Al_2O_3 and to a lesser extent TiO_2, Na_2O, K_2O, SiO_2 and FeO distinguish between komatiite, picrites, meimechite, boninite and other types of highly magnesian lava.

Variations in the chemical compositions of komatiites

Even within suites of rocks that fall clearly within the definition of komatiite, there are significant differences in the major- and trace-element compositions. Best known is the variation of Al_2O_3/TiO_2 and CaO/Al_2O_3, discussed below and illustrated in Figure 2.8, but also interesting are differences in the relative abundances of other elements. Gorgona komatiites, for example, have higher Al_2O_3 contents at a given Mg number than most Archean komatiites, and this characteristic is manifested petrographically in the relatively early crystallization of plagioclase in Gorgona komatiitic basalts (Echeverría, 1982; Aitken and Echeverría, 1984). Certain komatiitic rocks contain significantly higher Fe contents than typical samples from Munro Township or Barberton. Stone *et al.* (1987, 2003) described a layered mafic–ultramafic unit (probably a shallow-level sill) with a spinifex-textured upper portion from the Boston Creek area in Ontario that has FeO(tot) contents between 14 and 17%, values that are distinctly higher than in typical Archean komatiites. The high FeO contents of the Boston Creek unit resemble those of Theo's Flow, a mafic–ultramafic layered flow from Munro Township that Arndt *et al.* (1977) classified as tholeiitic, partly because of its high contents of FeO and other diagnostic elements, and partly because it lacked spinifex texture. This distinction cannot be made in the case of the Boston Creek unit, which is spinifex-textured. It seems more reasonable to recognize that komatiitic liquids may have a wide range of Mg–Fe ratios. The Boston Creek komatiite is an example of what Barley *et al.* (2000) have christened Karasjok-type Fe–Ti-rich komatiites as described in Chapter 2.

Another difference that perhaps has been given more attention than it merits is the SiO_2 contents of komatiites from South Africa, which are slightly elevated compared with those of komatiites from other regions. This characteristic is one of the few credible, though not indisputable, arguments for the notion that some komatiites formed in a subduction environment (Chapter 11).

Al_2O_3/TiO_2

The petrologically most important element in komatiites is aluminium, and variations of CaO/Al_2O_3 (a somewhat unreliable parameter because of the mobility of Ca) or Al_2O_3/TiO_2 (better, because these elements are relatively immobile) form an integral part of all komatiite classifications. Most komatiites from the type area in the Barberton greenstone belt are characterized by high CaO/Al_2O_3 (> 1) or low Al_2O_3/TiO_2 (10–15), and this characteristic formed part of the original definition of komatiite (Viljoen and Viljoen, 1969a, b; Jahn *et al.*, 1982; Nesbitt *et al.*, 1982). Correlated with relatively low Al_2O_3/TiO_2 is relative depletion of the HREE, a feature that is best expressed as high Gd/Yb (Figures 2.7 and 2.8).

Komatiites from most other regions have Al_2O_3/TiO_2 around 20, a value close to the chondritic value and distinctly higher than that of the rocks from the Barberton greenstone belt. Chondritic CaO/Al_2O_3 and flat HREE patterns accompany the chondritic Al_2O_3/TiO_2. The variations in these ratios form the basis of a chemical subdivision of komatiites into either two or three groups. Nesbitt and Sun (1976), Sun and Nesbitt (1978) and Nesbitt *et al.* (1979) divided komatiites into two groups: *Al-depleted komatiites* with low $Al_2O_3/$ TiO_2 and depleted HREE; and *Al-undepleted komatiites* with chondritic Al_2O_3/TiO_2 and flat HREE patterns. Jahn *et al.* (1982) recognized three types: *group 1*, which correspond to Nesbitt *et al.*'s Al-undepleted komatiites; *group 2*, the Al-depleted komatiites; and *group 3*, a complement of the Al-depleted komatiites, with high Al_2O_3/TiO_2 and relatively enriched HREE (low Gd/Yb).

I have always found these terms unsatisfactory. 'Al-undepleted' is an awkward term, and some Al-depleted komatiites owe their low Al_2O_3/TiO_2 to high TiO_2 rather than low Al_2O_3. In addition, I have never been able to remember which of Jahn *et al.*'s groups is which. In this book, I have therefore decided to follow a procedure being adopted by many authors and to call those komatiites with low Al_2O_3/TiO_2, and depleted HREE 'Barberton-type' komatiites, those with chondritic Al_2O_3/TiO_2 and HREE 'Munro-type' komatiites and those with high Al_2O_3/TiO_2 'Gorgona-type' komatiites. To this list could be added the Fe–Ti-rich 'Karasjok-type' komatiites. To help the reader who cannot remember whether the Barberton-type has normal or low Al_2O_3, I will often speak of Al-depleted Barberton-type komatiites, and so on.

It should be pointed out at this stage that several factors complicate the simple correlations between CaO/Al_2O_3, Al_2O_3/TiO_2 and HREE. One is the mobility of Ca during hydrothermal alteration and metamorphism, which leads to spuriously high or low $CaO–Al_2O_3$ values and renders the ratio an unreliable measure of magma composition. For example, the low CaO/Al_2O_3 ratios that worried Ohtani *et al.* (1988) and Walter (1998) probably can be attributed, at least in part, to Ca loss during metamorphism. A second factor is relative enrichment of moderately incompatible elements such as Ti and Gd by processes independent from those that cause the principal variation in

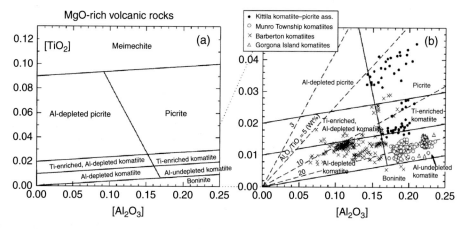

Fig. 5.13. Hanski diagram. [Al₂O₃] and [TiO₂] are the mole proportions of these elements, projected from the olivine composition using the procedure described by Hanski (1992): (a) the fields that define various types of highly magnesian lavas; (b) where various types of komatiite plot in the diagram.

Al_2O_3/TiO_2 or Gd/Yb. Certain komatiites from the Aldan Shield in Russia (Puchtel *et al.*, 1993) and throughout the Abitibi belt (Sproule *et al.*, 2002) have high Gd/Yb, low Al_2O_3/TiO_2 but chondritic CaO/Al_2O_3. There is little reason to believe that Ca has been lost systematically from all these rocks and it is more probable that the concentration of TiO_2 is higher, at a given MgO content, than that of other komatiites. Both Puchtel *et al.* and Sproule *et al.* interpret the enrichment of Ti and the more incompatible trace elements as source features. This and other aspects of the petrogenesis of komatiites are discussed in Chapter 13.

Perhaps the best method of classifying komatiites has been developed by Hanski *et al.* (2001). Figure 5.13 is Hanski's [Al₂O₃] vs [TiO₂] diagram in which [Al₂O₃] and [TiO₂] are the mole proportions of these components projected from the olivine composition. The values are calculated using the following equations: $[Al_2O_3] = Al_2O_3/(\frac{2}{3}-MgO-FeO)$, $[TiO_2] = TiO_2/(\frac{2}{3}-MgO-FeO)$, as described in detail by Hanski (1992). A major advantage of the diagram is that it eliminates the effects of crystal fractionation and accumulation: all rocks that lie on olivine control lines have constant [Al₂O₃] and [TiO₂] values and plot at a fixed position on the diagram. The diagram illustrates clearly both the compositional differences between komatiites and various types of picrites, and distinction between the various types of komatiite.

Komatiite type seems to correlate with age (Nesbitt *et al.*, 1982; Gruau *et al.*, 1987). As is also shown in Figure 2.7, Al-depleted Barberton-type komatiites predominate in ~3.5 Ga old terrains such as Barberton and Pilbara, and

Munro-type komatiites, although present, are relatively rare in both areas. Conversely, in 2.9 Ga and younger terrains, Munro-type komatiites are the norm and Barberton-type komatiites are known from only a few localities: Newton Township, Ontario (Cattell, 1987); Crixas, Brazil (Arndt *et al.*, 1989); Forrestania, Australia (Perring *et al.*, 1996). Al-enriched types (Gorgona-type or group 3 of Jahn *et al.* (1982)) are relatively rare. They occur sporadically in terrains of all ages but dominate the komatiites of Gorgona Island.

Incompatible elements

Trace-element concentrations vary widely in komatiites, and in normalized trace-element diagrams, the more incompatible elements range from strongly depleted to moderately enriched. On a broad scale there is some correlation between LREE and HREE ratios, and this correlation helps distinguish Al-depleted from Al-undepleted komatiites, but on the whole the variation in incompatible elements is not directly correlated with major-element data. Al-depleted komatiites from the Barberton greenstone belt normally are moderated enriched in LREE and other incompatible trace elements, and have REE patterns that show a gentle increase from Lu to La (Gruau *et al.*, 1990a,b). Komatiites from the Pilbara Block, which, like Barberton, is about 3.5 Ga old, show depletion of both LREE and HREE, and have hump-shaped patterns (Nisbet and Chinner, 1981; Gruau *et al.*, 1987). In the Fe–Ti-rich suite from Lapland (Hanski *et al.*, 2001), some komatiites have hump-shaped patterns while others are strongly depleted in LREE.

Barberton-type komatiites and the Fe-rich komatiites from Boston Creek often exhibit negative Zr and Hf anomalies in mantle-normalized trace-element diagrams. This feature has been attributed by Jochum *et al.* (1990) and Xie and Kerrich (1994) to garnet (or majorite) fractionation during partial melting. Positive Zr–Hf anomalies of variable magnitude in Munro-type komatiites were interpreted by the same authors as the result of the accumulation of a high-pressure phase, probably perovskite, in their deep mantle source. These anomalies are not reproduced in other high-quality thermal-ionization or inductively-coupled plasma mass spectrographic analyses (e.g. Blichert-Toft and Arndt (1999)) and in view of the difficulties in accurately measuring the ratios of HFSE to REE in variably altered komatiites (Lahaye *et al.*, 1995; Blichert-Toft *et al.*, 2004), these anomalies must be interpreted with caution.

A more robust characteristic of Munro-type komatiites is a moderate to strong depletion of the LREE and other immobile incompatible trace elements. The degree of depletion approaches, but rarely reaches or exceeds, that of modern N-type mid-ocean ridge basalt (MORB). Of the few examples of Al-depleted, post-3.0 Ga komatiites, those from Newton Township, Ontario, have flat to slightly enriched LREE, and those from Crixas and Ruth Well, western Pilbara Block, have depleted LREE and hump-shaped patterns. Jahn *et al.* (1980), in their description of komatiites from 2.7 Ga Finnish greenstone belts recognized all these types and Sproule *et al.* (2002) developed a more elaborate classification of komatiites from the Abitibi belt in Canada.

Other incompatible elements, if immobile, generally follow the LREE, or scatter widely, if mobile. Thus we normally see, in mantle-normalized trace-element diagrams of komatiites and komatiitic basalts, a relatively smooth progression from moderately incompatible elements such as Lu or Y to highly incompatible elements like La and Th, spoilt by erratic behaviour of the alkalis, Sr, U and Pb and by anomalies for those elements that are relatively enriched or depleted in the crustal rocks that contaminate some of these komatiites. The high-quality analyses of Shimizu *et al.* (2004), Figure 5.9, beautifully illustrate these features.

In most high-quality analyses of Munro-type komatiites, the HFSE – Zr and Nb – are concordant with the REE (Figure 5.9). Certain komatiites and komatiitic basalts have chondritic ratios or relative enrichment of the LREE, and negative anomalies of Nb and Ti. Notable examples include the so-called 'siliceous high-Mg basalts' from Western Australia (Arndt and Jenner, 1986; Cattell, 1987; Sun *et al.*, 1989; Barley *et al.*, 2000) and Canada (Cattell, 1987). These samples are also relatively enriched in SiO_2 and these characteristics are probably due to contamination of the komatiite by felsic crustal material (Huppert and Sparks, 1985a; Arndt and Jenner, 1986; Barley, 1986; Lesher and Arndt, 1995).

The concentrations of lithophile trace elements potentially provides important information about the tectonic setting in which they formed. As described earlier in the chapter, these elements are normally mobile during alteration, but their concentrations can be estimated using analyses of melt inclusions in olivine grains. McDonough and Ireland (1993), in their pioneering study of inclusions in komatiites from Zvishavane in Zimbabwe, showed that the concentrations of immobile elements such as Ba and Sr plotted on the same trends as the REE in mantle-normalized diagrams (Figure 5.10). On this basis they deduced that 'these magmas formed in sources similar to ... those of modern intraplate basalts and that komatiites are ancient analogues of modern plume-related magmas.' Lahaye *et al.* (2001) analysed melt inclusions in a komatiite from Alexo and they too found that potentially mobile elements such as Cs, Rb and Ba plotted on the same mantle-normalized spectrum as the REE. Strontium, on the other hand, was slightly depleted compared with REE such as Pr and Nd. The spectra display relative enrichment of the more incompatible elements and a small negative Nb anomaly, features that Lahaye *et al.* attribute to the assimilation of a small amount (~1%) of sulfidic sediment.

Compatible trace elements

The Ni contents of komatiites show an almost linear relationship with the MgO contents (Figure 5.14(a) and Figure 4.6). This is unexpected, because

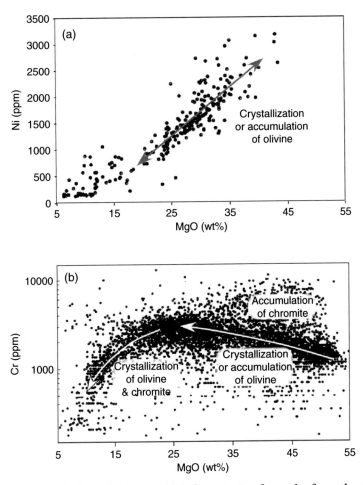

Fig. 5.14. (a) Variation of the Ni and MgO contents of samples from the Alexo komatiite flow (Lahaye and Arndt, 1996) and modelled evolution of the Ni contents in an evolving komatiite liquid and in the olivine that crystallizes from this liquid. The evolution was calculated using the following expression to calculate the partition coefficient $D_{Ni} = 124/MgO - 2.2$ (Arndt, 1986b). (b) Variation of the Cr and MgO contents of komatiites from all major greenstone belts. In most of the more magnesian samples (MgO > 25%), Cr increases with decreasing MgO because Cr is only moderately compatible with olivine, the sole cumulus mineral in these rocks. Another population of high MgO rocks have high Cr contents due to the presence of cumulus chromite. In most samples with less than 25% MgO, Cr decreases with decreasing MgO as a result of cocrystallization of olivine and chromite.

fractional crystallization of olivine, a mineral with which Ni is highly compatible, should cause rapid depletion of Ni in the evolving liquid and an
exponential relationship between Ni and MgO. In fact, as shown by Smith
et al. (1980) and Smith and Erlank (1982) and described in Chapter 4, the Ni
partition coefficient varies with the MgO content of the liquid, being low in
komatiite and higher in basalt (Hart and Davis, 1978; Kinzler *et al.*, 1990).
This variation causes the Ni content of olivine to follow a hump-shaped path
that reaches a maximum when the liquid contains about 20% MgO, as shown
in Figure 4.6. As a consequence, the Ni content of the evolving liquid follows
an open S-shaped curve as shown in the same figure. The compositions of
samples from a differentiated flow like the Alexo komatiite are mixtures
of evolved liquids and accumulated olivine, complicated by minor mobility
of Mg and Ni, and their compositions scatter about a near-linear trend, as
shown in Figure 5.14.

When an immiscible magmatic sulfide segregates, Ni, a chalcophile element,
is rapidly removed from the komatiitic liquid. The depletion of Ni, along with
other chalcophile elements like Cu and the PGE elements, is used as an
exploration tool during the exploration for Ni sulfide deposits, as discussed
in Chapter 10.

Figure 5.14(a) illustrates the rather complicated variation in the Cr content
in komatiites. Chromium is contained in two minerals. It is moderately compatible in olivine with a partition coefficient of around 0.3, and it is a major
component of chromite. Murck and Campbell (1986) showed experimentally
that during the cooling of komatiite at oxygen fugacity near the FMQ buffer,
olivine initially crystallizes alone and is joined by chromite only when the MgO
content of the liquid falls to about 25% MgO. The Cr–MgO variations in
Figure 5.14(a) are readily explained by the fractional crystallization and/or
accumulation of these phases. For example, in most of the rocks with high
MgO contents, the Cr content increases gradually with decreasing MgO
because these rocks consist mainly of olivine that had accumulated without
chromite from relatively magnesian liquids. The second population of high-
MgO komatiites contains high Cr contents because they contain both olivine
and chromite: presumably they are cumulates from less magnesian liquids that
had crystallized the two phases. Finally, in the samples with less than about
25% MgO, Cr decreases with decreasing MgO indicating that these rocks
formed as evolved liquids that had crystallized both olivine and chromite.
The mineralogy, composition and occurrence of chromite in komatiites is
discussed in detail in two comprehensive papers by Barnes (1998, 2000).

The compatibility of V with olivine depends on oxygen fugacity. Canil
(1997) showed experimentally that this element is moderately incompatible

in olivine, with a partition coefficient of around 0.03 at the FMQ buffer and that it becomes less compatible with increasing f_{O_2}. By analysing the covariation of V and other moderately incompatible elements such as Zr in several suites of Archean komatiites, Canil (1997) established the systematics of V partitioning between komatiitic liquid and olivine and used the results to show that the oxidation state of these ultramafic liquids was very similar to that of present-day oceanic basalts. He concluded on this basis that the oxidation state of the mantle has changed little through geological time.

Interest in the abundances of PGE in komatiites has been stimulated by the presence of Ni sulfide deposits in many komatiites (Chapter 10) as well as the application of the Re–Os isotope system (Chapter 6). The abundance of PGE in komatiites and other mantle-derived magmas depends on the nature and abundance in the residue of partial melting of phases such as sulfides, oxides and alloys that are capable of retaining these elements. Keays (1982, 1995) divided the PGE into two groups: the Pd group (Pd, Pt and Rh), which, in subsolidus peridotite, are hosted mainly by Cu-rich sulfides, and the Ir group (Ru, Ir, Os), which are hosted mainly by Fe–Ni sulfide and metal alloys. Platinum-group elements are very chalcophile ($D_{PGE}^{sulfide-silicate} \sim 10^4$–$10^5$ (Ballhaus *et al.*, 1994; Peach *et al.*, 1994; Fleet *et al.*, 1996)) and extremely siderophile ($D_{PGE}^{sulfide-silicate} \sim 10^5$–$10^6$ (O'Neill *et al.*, 1995)). If sulfides, oxides or alloys are retained in the residue during mantle melting, a significant proportion of the chalophile elements will remain in these phases and the concentrations of these elements in the partial melts will be low. Only when these phases are eliminated from the residue will the magma contain high abundances of elements such as Ni, Cu and PGE.

At low pressures, normal oxygen fugacities and relatively low degrees of melting (up to about 20%), all major silicate phases as well as oxides and sulfides contribute to the melt, which has a basaltic composition. As the degree of partial melting increases, the liquid becomes picritic or komatiitic and the sulfide phase progressively dissolves in the silicate liquid; when the degree of melting exceeds about 20%, sulfide totally dissolves into the melt (Wendlandt, 1982; Naldrett and Barnes, 1986; Rehkämper *et al.*, 1999). At low degrees of melting, when sulfide is retained in the residue, the PGE content of the silicate melt is low, and the more incompatible elements (Pd, Pt and Rh) are enriched relative to the more compatible Ru, Ir and Os. Only when the sulfide is eliminated from the source do PGE contents reach high levels and under these conditions the relative concentrations of the two groups of elements are more similar to those of their source. This behaviour is illustrated in Figure 5.15, which shows the chondrite-normalized PGE patterns of selected

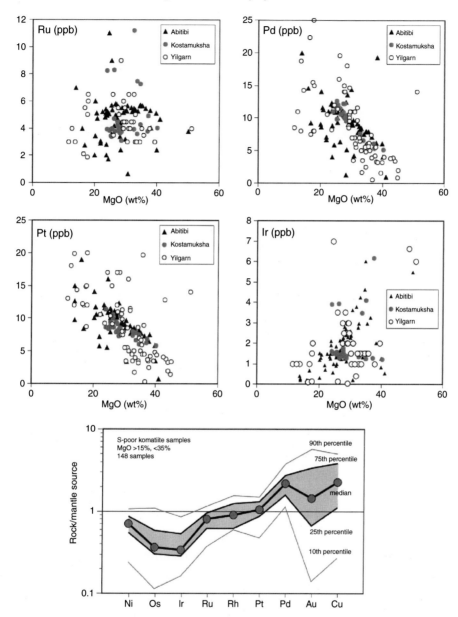

Fig. 5.15. PGE contents of S-poor komatiites (S < 0.2 %) compiled from the literature. Data sources: Crocket and MacRae 1986; Puchtel *et al.* 2004b; Barnes *et al.* 1995, 2004a; Brügmann *et al.* 1987; Puchtel and Humayun 2000. The bottom panel shows median and percentiles on combined data from all komatiite samples with MgO between 18 and 35%.

mafic and ultramafic rocks. The PGE content in komatiites is high and relatively unfractionated, a pattern that indicates that these magmas formed through high degrees of partial melting under conditions in which no sulfide was retained in the residue of melting (Naldrett and Barnes, 1986; Keays, 1995; Arndt *et al.*, 2004a).

Investigations of PGE in single komatiite flows or groups of flows have demonstrated that Pd, Pt and Rh are incompatible during the crystallization of komatiite magmas, while Ir, Os and Ru are compatible. Opinion was divided on whether the latter elements are accommodated in olivine (Brügman *et al.*, 1987), or whether they are withdrawn from the melt in alloys or oxides or sulfide phases (Naldrett and Barnes, 1986; Walker *et al.*, 1988). Puchtel *et al.* (2004b, 2007) used high-precision isotope-dilution data to show that an IPGE-rich phase or phases, in addition to olivine, controls the abundances of Ir, Os and Ru in komatiite flows.

Sulfide separates as an immiscible liquid during the cooling and crystallization of komatiite and this liquid is strongly enriched in chalcophile elements such as Ni, Cu and the PGE. If the sulfide accumulates in sufficient quantities (and close enough to the surface) it becomes an ore deposit. The Ni, Cu and the PGE content of the sulfide, and of the komatiite from which it separated, vary in a complex manner that depends on the initial composition of the komatiite magma, the way the magma erupted and the manner in which it interacted with wall and floor rocks. These aspects are discussed in Chapters 9 and 10.

Although not a PGE, Re behaves during magmatic and secondary processes in much the same way as Pd, Pt and Rh. The markedly different behaviour of the different groups of PGE results in large variations in element ratios such as Re/Os, the basis of Re/Os radiometric dating of komatiitic flows (Walker *et al.*, 1988).

5.8 Igneous chemistry of komatiitic basalts

Any discussion of the chemistry of komatiitic basalts is, to some extent, a matter of semantics. Komatiitic basalts are distinguished from other basalts and linked to the ultramafic members of the komatiitic magma series partly by field relationships and textural features, and partly by their chemical compositions. In some cases, major or trace elements provide a clear link between the ultramafic and mafic rocks. For example, certain basalts of the Abitibi belt share the distinctive chemical characteristics of the komatiites of that region (Figure 5.15). These basalts have a range of MgO contents, from 18% down to about 6%, high Al_2O_3 contents and contain relatively low concentrations of TiO_2 and incompatible trace elements (Table 5.4). They are distinguished from tholeiitic basalts by lower Fe contents and lower contents of HFSE and LREE. The more magnesian examples contain olivine or pyroxene spinifex

Table 5.4. *Major- and trace-element analyses of komatiitic basalts and related rocks*

	C11	C76	C9	C80	331/779	MC060 06	MC060 2B2	313-3203	9331	G26	94-7	92-12
SiO_2	45.85	49.29	51	50.77	54.54	51.30	55.97	49.41	49.57	49.0	48.74	48.69
TiO_2	0.49	0.71	0.68	0.75	0.51	0.61	0.72	1.36	1.76	0.82	0.48	1.15
Al_2O_3	9.36	12.2	13.16	13.94	4.49	8.67	10.49	11.86	9.96	11.3	12.99	14.44
FeO tot	11.76	9.53	11.52	12.6	10.16	15.59	9.70	13.20	13.00	11.4	10.37	11.13
MnO	0.24	0.17	0.21	0.23	0.18	0.01	0.16	0.33	0.21	nd	0.17	0.18
MgO	19.26	13.01	9.85	7.28	15.19	12.01	7.96	13.93	13.96	13.1	15.31	7.72
CaO	10.85	11.07	10.1	9.43	12.91	10.95	10.50	6.90	9.29	11.3	10.78	13.64
Na_2O	0.18	2.6	1.37	3.56	1.92	0.40	4.22	2.54	1.92	2.08	1.07	2.92
K_2O	0.03	0.03	0.69	0.07	0.05	0.01	0.09	0.37	0.19	0.10	0.05	0.06
P_2O_5	0.05	0.08	0.06	0.07	0.04	0.13	0.01	0.11	0.14		0.04	0.07
LOI	4.07	3.59	2.24	2.37	2.70	0.00	3.16	3.84	3.13	1.4	3.2	1.9
Sc	29	36	43	41			48	29.4	28.0	41		
V	175	225	259	270	104		196	295	310	283		
Cr	3000	1746	530	173	5099		3053	1201	792	1358	796	33
Ni	758	509	89	68	1473		1057	467	518	382	464	81
Cs						0.30		0.41	0.26		0.03	0.01
Rb	2.5	1	15.5	2.8	4.3	0.61	12.41	11.7	4.20	2	0.87	0.63
Ba	13	12	82	41	11	10	50	241	62.5	83	158	16.7
Sr	4.4	43	112	83		30	46	61.5	75.5		51	99.6
Nb	1.5	1.1	1.5	1.8	1.04	1.23	1.50	2.35	10.7	3	0.44	1.38
Zr	22.7	48	33	43	21.4	27.4	36.6	76	97.7	44	17.2	30.6
Y	12.5	19	16	20	6.40	7.50	15.26	14.2	17.6	17	15.4	25.2
Pb					3.86	0.30	0.63	0.647	0.651	0.74	0.11	0.43
Th					0.14	0.16	0.15	0.209	0.244		0.02	0.03
U					0.08	0.05					0.01	1.66

Table 5.4. (cont.)

	C11	C76	C9	C80	331/779	MC060 06	MC060 2B2	313-3203	9331	G26	94-7	92-12
La		2.49	1.47	2.07	3.43	1.67	3.86	4.52	8.10	3.54	0.36	1.67
Ce	2.79	6.36	4.54	5.81	3.35	4.25	7.92	11.4	21.0	8.52	1.20	5.04
Pr					0.94	0.68	0.95	1.48	2.75		0.25	0.93
Nd	2.77	5.14	3.74	4.94	3.63	3.28	3.90	8.19	14.8	5.69	1.65	5.28
Sm	1.05	1.76	1.37	1.78	0.96	1.05	0.99	2.41	3.79	1.69	0.90	2.31
Eu	0.41	0.57	0.61	0.67	0.36	0.32	0.33	0.62	1.44	0.64	0.42	1.18
Gd	1.55	2.51	2.13	2.6	1.23	1.08	1.35	2.86	4.05	2.23	1.74	3.53
Tb					0.17	0.17	0.28	0.46	0.615		0.34	0.64
Dy	1.89	2.97	2.61	3.2	1.09	1.17	2.13	2.85	3.63		2.36	4.08
Ho					0.20	0.26	0.53	0.55	0.683		0.51	0.84
Er	1.2	1.9	1.67	2.07	0.61	0.74	1.66	1.55	1.80	1.60	1.50	2.30
Tm								0.20	0.224		0.24	0.34
Yb	1.13	1.86	1.63	2.04	0.58	0.70	1.86	1.30	1.53	1.48	1.46	2.13
Lu		0.28	0.24	0.31	0.09	0.10	0.28	0.17	0.199	0.22	0.22	0.31

C11 – olivine–phyric upper zone border zone of Fred's Flow, Munro Township (Arndt and Nesbitt, 1984)
C76 – low-Fe olivine–phyric basalt from Munro Township (Arndt and Nesbitt, 1984)
C9 – pyroxene spinifex-textured komatiitic basalt, Munro Township (Arndt and Nesbitt, 1984)
C80 – evolved komatiitic basalt, Munro Township (Arndt and Nesbitt, 1984)
313-3203 – Fe-rich komatiitic basalt, Finnish Lapland (Hanski et al., 2001)
331/779 – komatiitic basalt, Barberton belt (Nesbitt et al., 1979)
MC060 06 – pillowed komatiitic basalt, Barberton belt (Cloete, 1999)
MC060 2B2 – pillowed komatiitic basalt, Barberton belt (Cloete, 1999)
9331 – Fe-rich picrite, Finnish Lapland (Hanski et al., 2001)
G26 – olivine-phyric top of komatiitic basalt flow, Gilmour Island (Arndt et al., 1987)
94-7 – olivine spinifex-textured komatiitic basalt from Gorgona Island (Arndt et al., 1997b)
92-12 – evolved komatiitic basalt (D-basalt) from Gorgona Island (Arndt et al., 1997b)

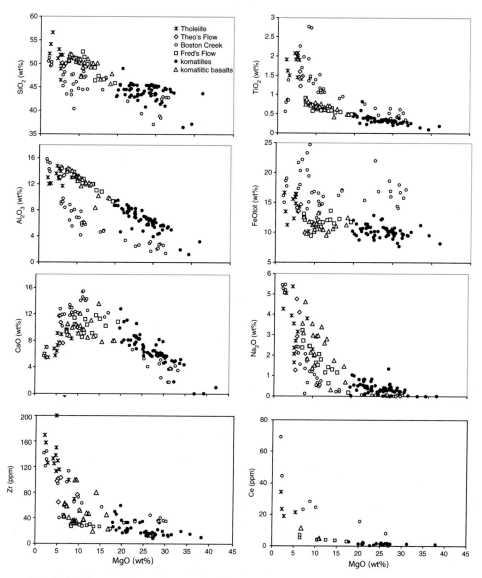

Fig. 5.16. Trace-element patterns of komatiites, komatiitic basalts, tholeiitic basalt and a siliceous high-magnesium basalt (a crustally-contaminated komatiite), from Kambalda; data from Arndt and Jenner (1986).

textures; they commonly occur in the upper parts of ultramafic–mafic volcanic cycles, and they plot on the extension of chemical trends defined by associated komatiites. For these basalts, a firm case can be made for their derivation by fractional crystallization of parental komatiite magmas.

The komatiitic basalts plotted in Figure 5.16 show a range of MgO contents from 18% (the limit with komatiite) to about 7%. Some of the more magnesian types contain olivine either as spinifex crystals or phenocrysts; others, however, owe their elevated MgO to the presence of abundant pyroxene. Some pyroxene spinifex lavas that are essentially free of olivine contain up to 16% MgO. In the komatiitic basalts plotted as Figure 5.16, Al_2O_3 contents increase progressively with decreasing MgO, indicating an absence of appreciable plagioclase fractionation. The CaO trend, on the other hand, levels out and then starts to decreases at an MgO content of around 12%. This inflection can be attributed to the onset of clinopyr-oxene fractionation. The inflection in the CaO trend is at a higher MgO than would be predicted from experimental data (Arndt and Fleet, 1979; Kinzler and Grove, 1985), probably because of a combination of clinopyroxene accumulation in spinifex lavas and Ca mobility. The more MgO-rich pyr-oxene spinifex lavas probably contain a large proportion of pigeonite and augite that accumulated during crystallization (Arndt and Fleet, 1979; Barnes, 1983).

The basalts that coexist with komatiites in many Archean and some Proterozoic greenstone belts have variable trace-element compositions. Some share the same geochemical features as associated komatiites, such as the LREE-depleted basalts which are associated with LREE-depleted komatiites of Munro Township (Arndt and Nesbitt, 1984) or the spinifex-textured basalts of the Belingwe greenstone belt (Bickle *et al.*, 1993; Shimizu *et al.*, 2005). Another example is the spinifex-textured basalts of the Barberton greenstone belt, which have the same Al-depleted characteristic as the komatiites of that belt (Jahn *et al.*, 1982). This is the case for the komatiitic basalts plotted in Figure 5.16, which are collinear with komatiites in most of the variation diagrams. On the other hand, other basalts in greenstone belts have geochem-ical characteristics that are quite unlike those of spatially associated koma-tiites. One group of basalts are relatively magnesian, with MgO from 18 to 5%, and have high contents of SiO_2 and incompatible trace elements such as Th and the LREE, and relatively low contents of elements such as Sc, V, Nb, Ta and Ti. Sun *et al.* (1989) have called these rocks *siliceous high-magnesian basalts*. Excellent examples are shown in Figure 5.9, which illustrates Shimizu *et al.'s* (2004) analyses of trace-element contents in komatiites and basalts from the Belingwe belt in Zimbabwe. The diagram shows that the concentrations of Zr, LREE and other immobile incompatible trace elements in the 'komatiitic basalts' and the so-called 'E-basalts' plot well above the trend defined by komatiites and depleted basalts. Other examples of the same type of basalt are known from various parts of Australia (Arndt and Jenner, 1986; Barley, 1986) and South Africa (Lahaye *et al.*, 1995), and have been reported in China (Polat *et al.*, 2005). In addition to their unusual chemical compositions, the

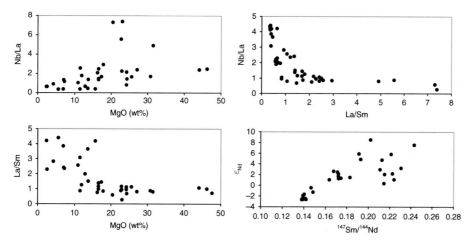

Fig. 5.17. Variation diagrams of crust-contaminated komatiitic basalts from Kambalda. Data from Arndt and Jenner (1986).

LREE-enriched komatiitic basalts commonly have distinctive isotopic compositions such as low ε_{Nd} values (Chauvel *et al.*, 1985; Cattell, 1987), and some contain xenocrysts of zircon (Claoué-Long *et al.*, 1988; Puchtel *et al.*, 1997; Shimizu *et al.*, 2004).

Figure 5.17 shows how trace-element ratios such as Nb/La or La/Sm correlate in a broad sense with Nd isotopic compositions in the komatiite-basalt suite from Kambalda in Australia. The ratios plotted in this diagram are sensitive indicators of crustal contamination, and most authors agree that these basalts owe their unusual compositions to the assimilation of granitoid crustal rocks of parental komatiite magmas. The processes involved are discussed more fully in Chapter 13.

Other basalts in greenstone belts may be relatively magnesian but have chemical characteristics that rule out any direct connection with komatiite. A series of olivine porphyritic basalts in Munro Township, for example, contain up to 17% MgO, but are markedly enriched in elements such as Ti, Zr, Na (Figure 5.15) (Arndt *et al.*, 1977; Arndt and Nesbitt, 1982). These basalts do not have spinifex textures, and although they form part of a stratigraphic sequence replete with komatiites, they clearly come from another source. Magnesian lavas from elsewhere in the Abitibi belt are enriched in the same group of elements, and have high rather than low FeO contents. The Fe–Ti-rich Boston Creek unit (Figure 5.15) is pyroxene-spinifex-textured, which points to a komatiitic affinity, but its chemical composition indicates a parental magma with a composition quite different from that which produced the Munro komatiites (Stone *et al.*, 1987). Clearly the range of basalt types preserved in greenstone belts is as diverse, if not more diverse, than in modern tectonic environments. To generalize about magmatic affinities and petrogenetic relationships is difficult, and it is usually wiser to treat each rock type individually.

6

Isotopic compositions of komatiites

6.1 Introduction

Komatiites were discovered just before the development of the Sm–Nd system
as a geochronological tool and geological tracer, and many of the earliest
Sm–Nd isochrons included komatiites. As with the earlier geochemical studies,
attention was focussed on these ultramafic volcanic rocks because they were
thought to provide reliable information about the composition of the Archean
mantle (e.g. Zindler (1982)). The argument was that these magmas represent
high-degree melts of the mantle and should therefore mirror the composition
of their source. Yet it soon became very clear that this was not true, as was
strikingly illustrated by the publication of a series of Sm–Nd isochrons with
anomalously old ages that were taken as evidence that these komatiites had
assimilated large quantities of material from old continental crust.

In this chapter I first review the progress that has been made in the application of five different radiogenic isotopic systems to the investigation of komatiites (Rb–Sr, Sm–Nd, Lu–Hf, Pb–Pb and Re–Os), and I describe how
successful (or unsuccessful) each of these methods has been as a means of
dating these rocks and as a source of information about their petrogenesis and
the nature and composition of their sources in the Archean mantle. I follow
this with a brief description of the analysis of stable isotopes in komatiites. For
the earlier work I draw heavily on the review papers by Zindler (1982) and
Smith and Ludden (1989).

6.2 The Rb–Sr system

The first attempts to date komatiites used the Rb–Sr method. Investigations of
mixed suites of ultramafic–felsic volcanic rocks from the Onverwacht Group
in South Africa (Allsopp et al., 1973; Jahn and Shih, 1974), from greenstone

belts in Zimbabwe (Hawkesworth *et al.*, 1975; Jahn and Condie, 1976), from Munro Township, Canada (Zindler, 1982) and from the Yilgarn Craton in Australia (Turek and Compston, 1971; Roddick, 1984) yielded little useful information on the time of eruption of the rocks. These investigations merely provided further evidence of the widespread mobility of Rb and/or Sr during hydrothermal alteration and metamorphism, and further support for the concept that Rb–Sr isochrons from whole-rock analyses of ancient volcanics either record the time of post-eruption metamorphism, or provide no useful geochronological information whatsoever. Virtually the only result that coincided with a likely eruption age was an isochron measured by Jahn and Shih (1974) on secondary minerals separated from a completely altered basalt from the Onverwacht Group. This isochron gave an age of 3430 ± 40 Ma, a result that agrees within error with ages obtained using more reliable methods. Subsequent U–Pb zircon, Pb–Pb and ^{40}Ar–^{39}Ar investigations (see below) provided an explanation for this largely fortuitous result: in the Barberton belt, unlike in many other greenstone terrains, the volcanic rocks were hydrothermally altered very soon after their eruption and were not strongly affected by subsequent major metamorphic events. The metamorphic minerals analysed by Jahn and Shih (1974) apparently record this alteration and were not influenced by the later events that resulted in the disturbances of the Rb–Sr system seen in the whole-rock investigations of Allsopp *et al.* (1973), Jahn and Shih (1974) and Jahn *et al.* (1982).

Strontium isotopic analyses of inclusion-free pyroxenes that had been carefully separated from differentiated komatiitic and tholeiitic flows and sills have provided more useful information. This approach, which was pioneered by Hart and Brooks (1977) and used by Zindler (1982) and Machado *et al.* (1986), provides a means of documenting the Sr isotopic evolution of the Archean and Proterozoic mantle. Virtually the only other source of such information comes from Sr isotopic analyses of alkaline intrusions (e.g. Bell and MacDonald (1982)).

Hart and Brooks (1977) and Machado *et al.* (1986) analysed pyroxenes from Fred's Flow, the layered komatiitic basalt from Munro Township, and from three tholeiitic layered sills from Munro and elsewhere in the Abitibi belt. The results are summarized in Table 6.1 and in Figure 6.1. Machado *et al.* (1986) found that the initial Sr isotopic composition of four pyroxenes from the layered sills was 0.70105 ± 0.00006 (2σ), and that the pyroxene from Fred's Flow had a higher value, between 0.70257 and 0.70273. This difference, which was consistent with earlier results of Brooks and Hart (1974), was unexpected, because on the basis of whole-rock analyses, the source of the komatiitic lava could be assumed to be chemically more depleted than that of the tholeiitic

Table 6.1. *Sr and Nd isotopic compositions of magmatic clinopyroxene in komatiites and related rocks*

Locality	Rock type	Sample	$(^{87}Sr/^{86}Sr)_o$	ε_{NdT}	Reference
Barberton	Komatiite	Px 355–1 (B)		2.0	Lahaye *et al.* (1995)
	Komatiite	Px 355–1 (A)		2.5	
	Komatiite	Px 10–6B (A)		0.7	
	Komatiite	Px 107–6B (B)		0.2	
Abitibi	Komatiite	N1	0.70220	2.3	Machado *et al.* (1986)
	Tholeiitic sill	N2	0.70092	2.8	
	Tholeiitic sill	N3	0.7010	2.6	
	Tholeiitic sill	DUN2	0.7010	2.2	
	Tholeiitic sill	LAO	0.70113	2.8	
	Tholeiitic sill	OP1	0.7011	2.4	
Alexo	Komatiite	M668 (A)		3.8	Lahaye *et al.* (1995)
	Komatiite	M668 (B)		3.1	
Cape Smith	Komatiite	VA13	0.70307	2.6	Zindler (1982)
	Tholeiitic sill	78–148	0.70170	3.5	
	Tholeiitic sill	78–151	0.70214	3.0	
Gorgona	Komatiite	GOR 502	0.702714	9.69[a]	Révillon *et al.* (2002)
	Komatiite	GOR 525	0.703348	10.0	
	Komatiite	GOR 539	0.702847	9.07	
	Picrite	GOR 515	0.703522	10.1	
	Gabbro	GOR 511	0.702924	9.27	
	Ol gabbro	GOR 509	0.703498	6.62	
	Dunite	GOR 503	0.702996	5.25	
	Wehrlite	GOR 507	0.702729	5.96	

[a] Whole-rock values.

sills, and a lower initial Sr isotopic composition might therefore have been anticipated. These results suggested that the sources of komatiitic and tholeiitic magmas differed in their composition and perhaps in their nature, a point discussed in more detail in Section 6.3. The initial $^{87}Sr/^{86}Sr$ for the pyroxenes from the tholeiitic sills plots slightly below an estimated bulk-Earth reference line, and that of the Fred's Flow pyroxene distinctly above the line (Figure 6.1).

Zindler (1982) reported initial $^{87}Sr/^{86}Sr$ for pyroxenes separated from a komatiitic flow and a layered sill in the Proterozoic Cape Smith belt in Quebec. The cleanest pyroxenes from the Summit Flow, a layered komatiitic basalt from the upper Chukotat Group, have initial $^{87}Sr/^{86}Sr$ from 0.70170 ± 10

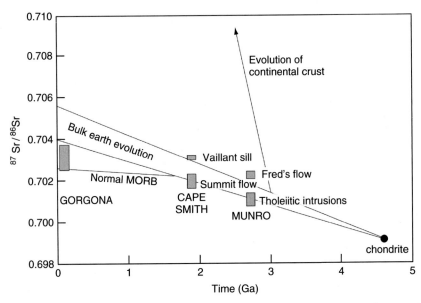

Fig. 6.1. Sr isotope compositions of clinopyroxenes from komatiites and related rocks. Also shown is the evolution of bulk earth, normal MORB and 3 Ga old continental crust. At Munro Township, the pyroxenes from tholeiitic intrusions plot on the curve representing the evolution of depleted mantle, but the pyroxene from komatiitic Fred's Flow plots above the curve. At Cape Smith, the komatiitic Summit Flow plots on the depleted mantle curve, but the Vaillant sill data plot above the curve. In each case, the position of the latter samples is attributed to contamination with older continental crust. The Gorgona pyroxenes plot between the depleted mantle and bulk earth curves. Diagram modified from Zindler (1982) with additional data from Machado *et al.* (1986) and Révillon *et al.* (2002).

to 0.702014 ± 7, values that lie at the lower boundary of the bulk earth evolution line (Figure 6.1). The cleanest pyroxene separate from the Vaillant sill, which intrudes the underlying Povungnituk Group, has a higher initial $^{87}Sr/^{86}Sr$ of 0.70307 ± 0.00007. This value lies slightly above the bulk earth evolution line and has been used by Francis *et al.* (1981) to support their petrological–chemical model of crustal contamination of Povungnituk Group magmas.

Much more recently, Révillon *et al.* (2002) analysed clinopyroxenes separated from komatiite flows and from comagmatic gabbros and peridotites from Gorgona Island. The aim here was to investigate the Sr isotopic composition of these rocks, which is unusually radiogenic for rocks from ocean basins. Figure 6.2 shows that the compositions of Gorgona lavas (Aitken and Echeverría, 1984), and those of other rocks from the Caribbean plateau (Kerr *et al.*, 1996) are displaced well to the right of the 'mantle array' in a

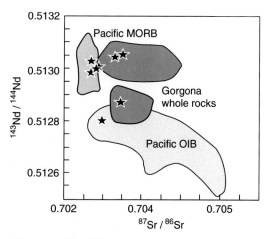

Fig. 6.2. $^{143}Nd/^{144}Nd$ vs $^{87}Sr/^{86}Sr$ diagram comparing the compositions of clinopyroxene from the Gorgona komatiites with the compositions of the whole rocks.

diagram of $^{87}Sr/^{86}Sr$ vs $^{143}Nd/^{144}Nd$. These isotopic compositions had been variously attributed to hydrothermal alteration of the lavas or to the assimilation of altered oceanic crust by Aitken and Echeverría (1984) and Kerr *et al.* (1996), or to an inherent component of the plume source by Hauff *et al.* (2000). The compositions of the magmatic pyroxenes plot to the left of most of the compositions of whole-rock samples, supporting the idea that the high $^{87}Sr/^{86}Sr$ is mainly due to interaction with seawater during or following eruption. However, the compositions of two pyroxenes plot away from the main group and are displaced to higher $^{87}Sr/^{86}Sr$, and another two pyroxenes have Nd isotopic compositions that are very different from those of the first group. These data provided further evidence for the heterogeneous nature of the source of Gorgona komatiites.

6.3 The Sm–Nd system

The first applications of Sm–Nd dating of Precambrian volcanic rocks seemed to have been successful at the time they were made. Hamilton *et al.* (1979a, b) obtained, from several greenstone belts in the Rhodesian Craton (as it was at the time), from Isua in Greenland, and from the Onverwacht Group in South Africa, linear trends in diagrams of $^{147}Sm/^{144}Nd$ vs. $^{143}Nd/^{144}Nd$ that met the criteria of isochrons. The ages calculated from these trends corresponded to those obtained using other methods and could be assumed to be correct. During the same period, DePaolo and Wasserburg (1979) measured a precise Sm–Nd

whole-rock and mineral isochron for the Stillwater Intrusion; Jahn *et al.*
(1982) duplicated the Onverwacht isochron; Zindler (1982) dated volcanic
and intrusive rocks from Munro Township and the Shibogamo Intrusive
Suite in the Abitibi belt, Canada; and McCulloch and Compston (1981)
published an isochron constructed using data from komatiites, basalts
and two felsic intrusive rocks from Kambalda, Western Australia. The Nd
and Sr isotopic data from komatiites and related rocks available at the end of
1981 were summarized by Zindler (1982), who concluded that, unlike the
Rb–Sr system, 'Sm–Nd systematics provide reliable age estimates for suites
of variably metamorphosed komatiites and associated rocks from
Precambrian greenstone belts'.

Subsequent work has shown that the early encouraging results were mis-
leading: many of these early results are now known to be either incorrect, or
when they gave the correct age to have done so largely by accident.

Early dating attempts

The first indications of something amiss came with the publication of new
Sm–Nd results by Claoué-Long *et al.* (1984) from Kambalda komatiites and
basalts, by Cattell *et al.* (1984) for a similar suite of volcanic rocks from
Newton Township in the Swaze Belt, Canada, and by Hegner *et al.* (1984)
for volcanic and intrusive rocks from the Pongola and Ishishwana Suites in
South Africa. All three sets of Sm–Nd data gave ages older than expected.
Claoué-Long *et al.*'s (1984) data plotted as a relatively well-defined linear
array in the Sm–Nd isochron diagram, which, if interpreted as an isochron,
gave an age of 3262 ± 44 Ma (Figure 6.3(a)). This age was some 500 Ma greater
than almost every other age obtained from the Kambalda area and surrounding Norseman–Wiluna belt ($\sim$2.5–2.7 Ga), but, because the other ages were
obtained largely from Rb–Sr or Pb–Pb dating, and because the Sm–Nd
isochron seemed internally consistent, Claoué-Long *et al.* interpreted the
3262 ± 44 Ma result as the age of eruption. Cattell *et al.*'s (1984) Sm–Nd
'isochron' gave an age of 2826 ± 64 Ma that was only slightly older than the
ages of rocks in neighbouring greenstone belts. These ages generally fell in the
range 2650–2750 Ma. However, because most of these ages had been
obtained using the reliable and precise U–Pb zircon method, the Sm–Nd
age was considered suspect. Cattell *et al.* checked the age of the komatiite
sequence by dating a felsic volcanic rock that underlay the komatiite-basalt
sequence using the U–Pb zircon method. This gave an age of 2697.0 ± 1.1
Ma, some 130 Ma younger than the Sm–Nd age. Hegner *et al.* (1984) also
suspected that their Sm–Nd age for rocks from the Pongola and Ishishwana

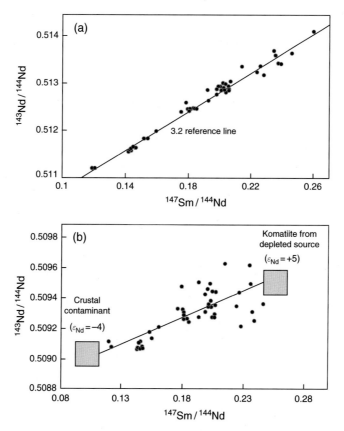

Fig. 6.3. Neodymium isotope compositions of komatiites and basalts from Kambalda, Australia: (a) measured values and (b) values calculated for an age of 2.7 Ga. These rocks have been dated by the U–Pb zircon method at 2.7 Ga but their present-day compositions (a) are aligned along a 3.2 Ga reference line (data from Chauvel *et al.* (1985), Claoué-Long *et al.* (1984) and Lesher and Arndt (1995)). The rotation of the array resulted from assimilation of older continental crust, which, at the time of eruption, had a lower ^{143}Nd/^{144}Nd, as shown in (b). The scatter about the reference line is far more apparent in (b); this probably results from some variation in the compositions of the komatiite magmas and the contaminants, compounded by disturbance of the Sm–Nd system during later metamorphic events.

Suites was incorrect because it too was higher than ages obtained using other methods.

Both Cattell *et al.* (1984) and Hegner *et al.* (1984) interpreted their linear Sm–Nd arrays as mixing lines between material from a depleted mantle source and an old enriched component, either enriched mantle or old continental crust, and both proposed that their Sm–Nd 'ages' were geochronologically meaningless (Figure 6.3(b)).

The Kambalda story

Convincing support for the interpretation of Cattell *et al.* (1984) and Hegner *et al.* (1984) came with the publication of new data from Kambalda. Chauvel *et al.* (1985) presented new Sm–Nd data, which also fell on an ~3.2 Ga 'isochron', but they were able to appraise this result using Pb–Pb data from volcanic rocks and sulfides (see below), and a Sm–Nd mineral isochron from a layered mafic–ultramafic sill, both of which pointed to an ~2.7 Ga emplacement age. Like Cattell *et al.* (1984) and Hegner *et al.* (1984), Chauvel *et al.* invoked contamination of komatiitic magmas with old enriched material to explain their results. Their interpretation was validated when Compston *et al.* (1986) published ion microprobe analyses of zircon xenocrysts separated from one of the Kambalda basalts. The ages of these zircons ranged from 2.69 to ~3.4 Ga: the younger age provided a maximum for the eruption age of the volcanics and left no doubt as to the inaccuracy of the Sm–Nd isochron age; and the older zircons provided clear evidence that the Kambalda volcanic rocks contained material derived from older felsic crust. Arndt and Jenner (1986) subsequently demonstrated that the chemical compositions of Kambalda basalts are consistent with contamination; Huppert *et al.* (1984) presented the model of thermal erosion, the likely contamination mechanism (see Chapter 9); and, finally, Claoué-Long *et al.* (1988) presented a new ion microprobe U–Pb zircon age (2692 ± 4 Ma) from a tuff interlayered with the komatiites, which pinned down the stratigraphic age of the volcanic rocks and provided the last nail in the coffin of the pre-3 Ga age.

Factors that disturb the Sm–Nd system

Crustal contamination

More recent attempts to measure Sm–Nd ages of Precambrian mafic and ultramafic volcanic rocks have also yielded equivocal results. Notable examples include Gruau *et al.*'s (1990b) and Lécuyer *et al.*'s (1994) investigations of komatiites in the Barberton greenstone belt, and Tourpin *et al.*'s (1991) studies of variably altered komatiite flows from Finland.

The problems associated with Sm–Nd dating of komatiites and other mafic rocks arise from the following set of circumstances. It has long been recognized (e.g. Zindler (1982), Cattell *et al.* (1984), Claoué-Long *et al.* (1984), Chauvel *et al.* (1985)) that the magmatic processes operative during the formation and crystallization of mafic and ultramafic magmas produce only minor variation in Sm/Nd and that this variation is generally too small to produce the spread of ratios necessary to give a precise isochron. During crystallization of komatiites, clinopyroxene is the only mineral capable of significantly changing the Sm–Nd ratio, and this mineral only appears close to the solidus (Chapter 4). The variation of Sm/Nd in comagmatic whole rocks is minimal even within strongly differentiated komatiite units such as Fred's Flow (Whitford and Arndt, 1977). Where larger variations are observed, they are most likely due to magma mixing or contamination (Arndt and Nesbitt, 1984; Kinzler and Grove, 1985). It is very likely that almost all published Sm–Nd whole-rock isochrons have been constructed from samples that either are genetically unrelated to one another, or have been affected by complex contamination or alteration processes that caused an abnormally large spread of Sm–Nd ratios. Contamination certainly has affected the compositions of komatiites and basalts from Kambalda (Chauvel *et al.*, 1985), Newton Township (Cattell *et al.*, 1984), Pongola (Hegner *et al.*, 1984) and Taishan in China (Polat *et al.*, 2005) and the process cannot be discounted for any suite of Archean mafic and ultramafic rocks.

The reason why contamination with old continental crust produces an anomalously high age is illustrated in Figure 6.3(b). The ^{143}Nd/^{144}Nd of old continental crust can be significantly lower than that of komatiite, particularly komatiite from a depleted mantle source, but the difference in ^{147}Sm/^{144}Nd of the two types of rock is not so large. When the compositions at the time of eruption are plotted in a normal isochron diagram, as in Figure 6.3(b), the komatiite samples plot in the top right of the diagram and old continental crust plots in the bottom left. Mixing komatiite and crust produces hybrid samples that define a crude linear array with a distinct positive slope. During the billions of years that have elapsed since the eruption of komatiite, radioactive decay has rotated the array, and increased the ^{143}Nd/^{144}Nd of every sample, but more so for samples with higher Sm/Nd. In so doing, the magnitude of the scatter about the linear array became small in comparison with the total spread of ^{143}Nd/^{144}Nd, and in many cases, the final array fulfils the criteria of an isochron.

Even samples from single flows do not necessarily give meaningful data: the isochrons from the Alexo komatiite flow (Dupré *et al.*, 1984) and from Fred's Flow (Zindler, 1982) give ages that agree with the apparent eruption age, as defined by U–Pb zircon data, only because the contaminants in both cases had a Nd isotopic composition similar to that of the komatiitic lava. Had these flows assimilated material with a different composition, as happened at Kambalda (Chauvel *et al.*, 1985; Lesher and Arndt, 1995), or had later alteration changed Sm–Nd ratios, as happened in the Tipasjärvi belt, Finland (Tourpin *et al.*, 1991), and in the Crixas belt, Brazil (Arndt *et al.*, 1989), the Sm–Nd ages would have been incorrect.

The effects of crustal contamination in suites of komatiites can be investigated by examining the correlation between ε_{Nd} values and trace-element ratios that are diagnostic of the continental crust. Many promising ratios such as Ba/La or Sr/Sm have to be excluded because the denominator elements are mobile during alteration, but ratios such as Th/Nb, Nb/La, Nb/Ta or La/Sm serve our purpose. The first three ratios are measures of the Nb–Ta anomaly, which is present in almost all granitoids or sedimentary rocks of the continental crust, and La/Sm is a measure of the degree of enrichment of incompatible elements, which is high in continental crustal rocks but low in most komatiites. In the examples shown in Figure 6.4, low ε_{Nd} values, which correspond to samples that have assimilated a large amount of old crust, correlate roughly with low Nb/La and high La/Sm. The correlation is

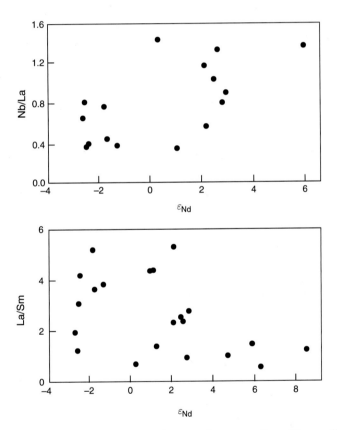

Fig. 6.4. Correlation between ε_{Nd} and trace-element ratios of komatiites and basalts from Kambalda, Australia. The correlations are poor, probably because of post-eruption processes that changed the trace-element ratios and/or perturbed the isotope system, but those samples with low ε_{Nd} values have low Nb/La and high La/Sm, characteristics that are consistent with contamination with old continental crust (from Arndt and Jenner (1985)).

evidently not good, probably because of variation in the nature of the contaminant or post-contamination fractional crystallization and alteration.

It is clear from the above discussion that in a series of variably contaminated
komatiites, any record of the isotopic composition of the mantle source is
obscured. When komatiites are used to track the composition of the Archean
mantle (as described in the following section), the contaminated samples must
be filtered out. Since certain types of komatiite, such as those from the
Barberton greenstone belt or Finnish Lapland (Hanski *et al.*, 2001), are
variably, often strongly, enriched in incompatible elements, La/Sm is not a
suitable discriminant, but in practice La/Nb or Th/Nb, which are high in
contaminated rocks, work well.

Post-eruption changes in Sm–Nd ratios

Of the numerous examples that exist in the literature of post-eruption disturbance of the REE contents of komatiites, the most striking is that of two
differentiated lava flows from the 2.7 Ga Tipasjärvi belt in Finland, studied by
Blais *et al.* (1987), Tourpin *et al.* (1991) and Gruau *et al.* (1992). The spinifex
lavas in the upper parts of the flows are strongly depleted in LREE, whereas
the olivine cumulates of the lower parts of the same flows have flat REE
patterns (Figure 6.5). Olivine fractionation would have had a minimal effect
on the REE patterns. The petrology and major-element geochemistry of the
spinifex lavas resemble those of other 2.7 Ga Munro-type komatiites and it is
probable that samples with this texture have preserved much of their original
chemical composition. The olivine cumulates, on the other hand, are strongly

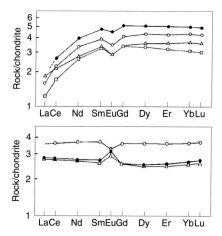

Fig. 6.5. REE profiles in a komatiite flow in the Tipasjärvi greenstone belt in
Finland (from Tourpin *et al.* (1991)). The upper diagram shows spinifex-
textured lavas; the lower diagram shows olivine cumulates.

altered, being pervaded by carbonate veins and carbonate replacement of the magmatic phases, and Gruau *et al.* (1992) proposed that LREE were added to these samples during the carbonate alteration. A whole-rock isochron generated by analysing samples from the flow has a slope that corresponds to an age of about 1.8 Ga, which is consistent with the time of regional metamorphism. Gruau *et al.* (1992) argued that the Sm–Nd ratios of the olivine cumulates had decreased during the metamorphism. When the data are plotted on a diagram of time vs ε_{Nd}, the more altered samples extrapolate back to anomalously high ε_{Nd} values at 2.7 Ga, the probable time of eruption.

Gruau *et al.* (1996) developed these ideas in their study of the oldest recorded supracrustal rocks from the Isua region in Greenland, which have been dated at about 3.8 Ga. Analyses of a series of metavolcanic, metasedimentary and granitoid rocks define a wide range of ε_{Nd} values, from just below zero to as high as + 5.3. Such high values are surprising for such old rocks, because they imply that the samples were derived from a strongly depleted mantle source. In order that the source had become so strongly depleted 3.8 Ga ago, enriched material of one type or another must have been extracted from the source very early in Earth's history. Gruau *et al.* (1996) noted that the range of calculated ε_{Nd} values was greatest in samples with low Nd contents and that in samples with Nd contents greater than 20 ppm, the ε_{Nd} converged to a value close to + 2. They argued that this value provided the best estimate of the composition of the mantle source and that other higher or lower values resulted from disturbance of the Sm–Nd isotopic system, either through change in the Sm–Nd ratio, as described above, or through the introduction of material with divergent isotopic composition. This example, together with other studies of Archean volcanic and granitoid rocks, highlights the susceptibility of the Sm–Nd system to post-eruption perturbations. Nonetheless, it is, together with Lu–Hf, the system least subject to such disturbance, and if we are to use komatiites as tracers of the composition and history of the Archean mantle, we have no choice but to exploit the Nd isotopic compositions of komatiites.

Nd isotopic constraints on the composition of the mantle source

Figure 6.6 is a compilation of ε_{Nd} values of komatiites plotted as a function of their age. In Figure 6.6(a), the data are plotted indiscriminately, and here it is seen that at many locations there is a very wide spread of ε_{Nd} values. In areas such as Kambalda, the spread to negative ε_{Nd} values results mainly from crustal contamination; in others such as in the Finnish greenstone belts, the variation results from post-emplacement alteration. Despite these complications,

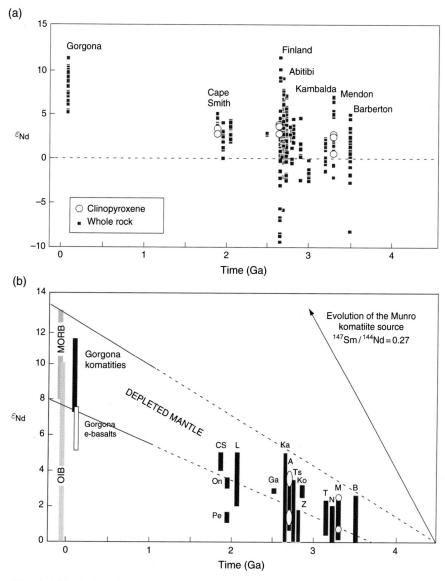

Fig. 6.6. Variation in the Nd isotopic compositions of komatiites and basalts through time. In (a) all data are plotted (CS – Cape Smith; On – Onega; Ve – Vetreny; L – Lapland; Ga – Gawler; Ka – Kambalda; A – Abitibi; Ts – Tipasjärvi; Z – Zimbabwe; Ko – Kostomuksha; T – Taishan; N – Nondweni; M – Mendon; B – Barberton); in (b) the vertical bars represent the compositions of samples with minimal contamination with old crust and that are little influenced by post-eruption alteration. See text for further explanation. Data are from numerous sources, including Echeverría (1982), Jahn *et al.* (1982), Zindler (1982), Cattell *et al.* (1984), Chauvel *et al.* (1985), Echeverría and Aitken (1986), Arndt *et al.* (1987), Gruau *et al.* (1987), Claoué-Long *et al.* (1988), Walker *et al.* (1988), Wilson and Carlson (1989), Chauvel *et al.* (1993), Lahaye *et al.* (1995), Lesher and Arndt (1995), Gruau *et al.* (1996), Ludden and Arndt (1996), Arndt *et al.* (1997b), Puchtel *et al.* (1997), Puchtel *et al.* (1999), Révillon *et al.* (2000), Hanski *et al.* (2001), Révillon *et al.* (2002).

it is seen that in general the ε_{Nd} value increases with age, from values that are slightly positive at 3.5 Ga to higher values in the Proterozoic komatiites and still higher values in the Cretaceous Gorgona komatiites.

In Figure 6.6(b), the data have been culled in an attempt to exclude samples contaminated with old continental crust or subject to alteration. This was done by excluding (a) crust-contaminated samples in which low ε_{Nd} values coincide with trace-element indices of contamination, and (b) samples whose petrographic or geochemical characteristics provide evidence of disturbance of the Sm–Nd system. The procedure is rather arbitrary but it probably serves to provide a clearer picture of the isotopic evolution of the komatiite source. Also shown in the diagram are the results from the few analyses of magmatic clinopyroxene from komatiites that have been performed: these analyses probably provide the best estimates of the composition of the komatiite magma and of its mantle source.

Although some authors have included the compositions of mafic rocks from the Isua belt in Greenland in diagrams of this type, they are not shown in Figure 6.6. Frei *et al.* (2002) have provided data on the Nd (and Os) isotopic compositions of ultramafic rocks from the ∼3.8 Ga Isua region in Greenland and because these rocks are directly associated with pillow basalts, it can reasonably be inferred that they formed as komatiites. However, following their formation, these rocks have been subjected to the protracted series of metamorphic and tectonic events that affected this region of old continental crust. The consequence is seen in the spectacular dispersion of REE data reported by Gruau *et al.* (1996) and also in the vast range of Nd isotopic compositions. Frei *et al.* (2002) and Frei and Jensen (2003) report ε_{Nd} values ranging from + 1.5 to + 12.4 in the ultramafic rocks and from –4 to + 6 in the basalts. Although the interpretation of positive ε_{Nd} values for the Isua and Akilia sequences has been subject of some debate (Vervoort *et al.*, 1996) and led McCulloch and Bennett (1994) to propose a model of progressive segregation of an enriched component from early Archean mantle, Frei *et al.* (2002) and Frei and Jensen (2003) concurred with Gruau *et al.* (1996) who suggested that the large dispersion of ε_{Nd} values resulted mainly from post-emplacement disturbance of the Sm–Nd system. This interpretation has subsequently been confirmed by Vervoort and Blichert-Toft (1999) by comparison of Sm–Nd and Lu–Hf isotopic data. There appears, therefore, to be little prospect of obtaining reliable initial isotopic compositions using the Sm–Nd (or any other) isotopic systems on ultramafic rocks from the oldest crustal sequences, and the best chance of obtaining constraints on the composition of the komatiite source in early Archean mantle probably comes from study of better preserved ∼3.5 Ga sequences in South Africa and Australia.

Despite these weaknesses, a number of important conclusions can be drawn from Figure 6.6: almost all ε_{Nd} values are positive, indicating that komatiites of all ages were derived from a source characterized by long-term depletion of incompatible elements. This applies not only to Munro-type komatiites, in which the trace-element patterns measured in the lavas are depleted, but also to komatiites with enriched trace-element patterns, such as the Barberton komatiites and Fe-rich komatiites.

The mantle source(s) of komatiites was certainly heterogeneous. In samples from many locations, ε_{Nd} values vary by at least 1–2 units and although some of this spread is no doubt due to open-system behaviour, covariation of ε_{Nd} values and trace-element ratios in areas such as the Abitibi belt (Lahaye *et al.*, 1995), Finnish Lapland (Hanski *et al.*, 2001) or most convincingly Gorgona Island (Révillon *et al.*, 2002) provides evidence that at least some of the variation is real.

In regions where data from both komatiites and tholeiitic basalts are available, the komatiites usually have the higher ε_{Nd} values. In Munro Township, for example, the ε_{Nd} values of komatiites range from $+2.5$ to $+4$, whereas in basalts the range is from $+1$ to $+3$; at Kambalda, the komatiites have ε_{Nd} values between $+4$ and perhaps $+8$ (Claoué-Long *et al.*, 1988), whereas apparently uncontaminated footwall basalts have ε_{Nd} values between $+1$ and $+2$; and at Gorgona Island, the komatiites and so-called 'd-basalts', which are depleted in incompatible trace elements, have significantly higher values than the trace-element-enriched 'e-basalts'. These differences provide evidence that the material that melted to produce the basalts had a different composition from the source of komatiites.

There is a crude increase in ε_{Nd} values, from close to zero at 3.5 Ga to a maximum of $+11$ in Gorgona komatiites, with perhaps a slight drop in Proterozoic komatiites. The rate of increase corresponds to that of depleted mantle, as estimated from the composition of mid-ocean ridge basalts, assuming that this portion of the mantle evolved through a linear increase in ε_{Nd}, starting around 4 Ga and reaching N-MORB values at the present time. The rate of increase is far less than that of mantle material with the Sm–Nd ratio of depleted Munro-type komatiite, as shown by the arrow in Figure 6.6(a). A similar discrepancy in the evolution of depleted upper mantle was recognized by DePaolo (1983) and Patchett and Chauvel (1984), who attributed the subdued growth of $^{143}Nd/^{144}Nd$ to the recycling of enriched material (continental crust or lower mantle) into the depleted upper mantle. In the case of komatiites, the explanation more likely relates to the melting mechanism that produces komatiite. As discussed in Chapter 13, Munro-type komatiites probably result from fractional melting and are derived from a source that had become depleted in incompatible trace elements through the preferential extraction of these elements during fractional melting. In such a situation, the Sm–Nd ratio of the material that melts to form the komatiite melts becomes much higher (more depleted) than the long-term ratio of the source (Cattell and Taylor, 1990; Arndt *et al.*, 1997b).

6.4 U–Pb and Pb–Pb methods

Age dating

Although minerals suitable for U–Pb dating (zircon, titanite, monazite, badellyite) do not crystallize in normal komatiitic magmas, the method has nonetheless been used to attempt to constrain the age of ultramafic volcanic sequences. For example, the age of the Kambalda volcanics was pinned down (a) by the U–Pb age of zircon xenocrysts in basalts (Compston *et al.*, 1986), which provided a maximum age of the lavas, and (b) by the U–Pb age of magmatic zircons in an interflow tuff (Claoué-Long *et al.*, 1988), which provided a precise stratigraphic age. More generally, U–Pb zircon dating of felsic volcanic rocks has bracketed the ages of komatiitic lavas in many Archean greenstone belts, particularly the Superior Province of Canada (e.g. Ayer *et al.*, (2002). Additional constraints come from zircons separated from pegmatoid gabbros in mafic–ultramafic intrusions, which in many cases have been shown, using petrological and geochemical criteria, to be contemporaneous with volcanism. Where suitable rocks are present, the U–Pb zircon method is certainly the most reliable way of dating Precambrian volcanic rocks.

The results of Pb–Pb dating of komatiites are not so clear-cut. A concerted effort was made in the 1980s to establish whether the method, which is relatively quick and straightforward, could give reliable ages for komatiites and other Precambrian rocks. Initial promise had been provided by work in the Abitibi belt in Canada (Dupré *et al.*, 1984) and the Belingwe belt in Zimbabwe (Chauvel *et al.*, 1993). For example, Chauvel *et al.* (1993) obtained a Pb–Pb isochron from samples from komatiites in the Belingwe belt in Zimbabwe (Figure 6.7) that coincided with U–Pb zircon and Sm–Nd ages from the same region. Dupré and Arndt (1986) then evaluated attempts to apply the Pb–Pb method to Precambrian ultramafic and mafic volcanic rocks, and they found that of 11 examples for which the age of the komatiite was known independently from other reliable techniques, the Pb–Pb method gave the correct age for only seven. In the other four cases the age was 100–300 Ma too low.

To obtain a precise Pb–Pb age, the samples being analysed must have had a range of U–Pb ratios at the time they erupted. The elements U and Pb are

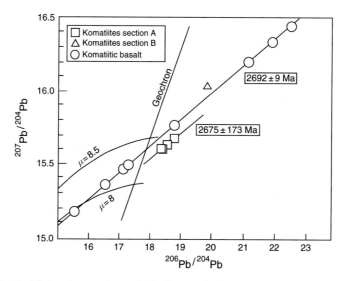

Fig. 6.7. Pb–Pb isochrons from the Belingwe belt in Zimbabwe, from Chauvel *et al.* (1993). The isochrons from the komatiites of sections A and B give ages that agree with Sm–Nd ages obtained on the same samples. The higher μ_1 value of the komatiitic basalts corresponds to lower ε_{Nd} values and this correlation is taken as evidence of crustal contamination. See Chauvel *et al.* (1993) for further details.

highly incompatible and are not fractionated during the emplacement and crystallization of mafic and ultramafic magmas: the spread of U–Pb ratios recorded in many komatiitic suites must therefore be due to another process. Dupré *et al.* (1984) suggested that the wide range of ratios in the Alexo komatiite flow was caused by hydrothermal alteration, which led to variable loss of Pb, perhaps accompanied by minor uptake of U. The alteration apparently took place on the sea floor very soon after eruption because the Pb–Pb age measured in this flow agreed within error with the U–Pb zircon age of underlying and overlying rhyolites. Dupré and Arndt (1986) proposed that all the Pb–Pb ages of komatiites obtained correspond to the times of alteration: in cases when this alteration was linked to hydrothermal processes on the sea floor, the age is close to the time of eruption and the Pb–Pb system gives an apparently correct result. In other cases, the alteration that reset the Pb–Pb system appears to have come later, often during subsequent metamorphism, and in these cases the ages are too low. Thin komatiitic flows seemed to give a correct age more commonly than thicker basaltic flows, perhaps because the ultramafic lavas were more susceptible to early hydrothermal alteration, but apart from this rather imprecise parameter, there seems to be no way of establishing whether a Pb–Pb age is correct or not. As a consequence, the Pb–Pb method has been largely abandoned as a dating method for old volcanic rocks.

Initial isotopic compositions

Initial Pb isotopic compositions cannot be calculated directly from measured isotopic ratios. In cases in which Pb isotopic compositions of magmatic sulfides have been measured, as at Kambalda (Roddick, 1984), Alexo (Dupré *et al.* 1984, Brévart *et al.* 1986), and the Cape Smith belt (Brevart *et al.* 1986), these can be used to estimate directly the initial compositions because such sulfides contain low concentrations of U and Th. When sulfides are not available, model initial isotopic compositions are calculated by making assumptions about the initial Pb isotopic composition of the bulk Earth and the age of the Earth, and by assuming single-stage evolution until the time of magma formation. This procedure yields μ_1 values, the time-integrated $^{238}U/^{204}Pb$ of the source. The Th–U ratios that the magmas had at the time of emplacement can be calculated from measured $^{208}Pb/^{204}Pb$ and $^{206}Pb/^{204}Pb$; and information about the time-integrated Th/U of the source is provided directly from the compositions of magmatic sulfides, or as model-dependent data using a procedure similar to that used to calculate μ_1.

For suites from the Abitibi belt and the Pilbara block, model μ_1 values fall within the range 7.8–8.0. In suites from other areas such as Kambalda and the Belingwe belt, μ_1 values are higher, between 8.2 and 8.4 (Figure 6.7). The latter results come from regions where the mafic–ultramafic magmas are thought to have been contaminated by significantly older continental crust, and the high μ_1 values are attributed to this contamination (Compston *et al.*, 1986; Dupré and Arndt, 1986; Chauvel *et al.*, 1993). In the Abitibi belt and Pilbara Craton, significantly older continental crust appears to have been absent, and the volcanic rocks probably have retained the Pb isotopic compositions of their mantle source. The Pb isotopic data, while yielding only ambiguous information about the composition of their mantle sources, provide important clues about the presence or absence and the age of continental rocks underlying greenstone belts.

When applied to younger, fresher rocks, such as the Gorgona komatiites and basalts, the results of Pb isotopic analyses are more convincing. Dupré and Echeverría's (1984) analyses of Gorgona samples showed that the different types of volcanic rock had distinct Pb isotopic compositions and provided the first indication of the heterogeneity of the Gorgona mantle plume.

Th–U ratios calculated from the same data range from about 2.9 to 4.1. Th–U ratios derived for the initial magmas correlate inversely with Sm–Nd ratios of the lavas. Dupré *et al.* (1984) and Allègre *et al.* (1986) used this correlation to obtain a value for the bulk Earth Th/U. Although the data used in these determinations included samples whose Pb isotopic compositions were dominated by crustal Pb, recalculation using suites from which the offending samples were excluded (e.g. Kambalda) yields an essentially similar result (4.0 compared to 4.1).

6.5 The Re–Os system

Age determination

^{187}Re decays to ^{187}Os with a half-life of 4.15×10^{10} years, a period that is comparable to those of ^{87}Rb and ^{147}Sm. As discussed in the previous chapter, Re behaves incompatibly during the fractionation of sulfur-undersaturated magmas, but Os is compatible either with olivine (Brügman *et al.*, 1987; Brenan *et al.*, 2003) or with minor phases – alloys or sulfides – that coprecipitate with olivine (Crockett and McRae, 1986; Walker *et al.*, 1988; Puchtel *et al.*, 2004b, 2007). As a result, the Re–Os ratio increases dramatically during the fractional crystallization of ultramafic magmas and measured ^{187}Re/^{188}Os values, even in single komatiitic flows, vary widely (e.g. from 1 to 10 in komatiites from the Vetreny belt in Russia (Puchtel *et al.*, 2001b)). There is always, however, some uncertainty as to whether the observed ranges of values are entirely magmatic or whether they have been increased during subsequent alteration. When using the Re–Os system as a dating method, or when the aim is to determine initial isotope compositions, magmatic olivine and chromite separated from the komatiites have been analysed because these minerals contain little Re and have much lower ^{187}Re–^{188}Os ratios.

The first Re–Os measurements on komatiites were made by J.-M. Luck (Luck and Allègre, 1984) using an ion microprobe, but this technique has now been supplanted by negative-ion mass spectrometry. Luck and Arndt (1986) reported, in abstract form, ion microprobe analyses of three whole rocks and a chromite separate from two komatiitic flows from the Alexo area. The data defined an isochron with an age that agreed with Sm–Nd, Pb–Pb and U–Pb ages obtained from similar or correlative samples. Largely through the work of Shirey and Walker (1995), who introduced the Carius-tube method of sample dissolution, and Creaser *et al.* (1991), who developed negative-ion mass spectrometry for the isotopic system, Re–Os isotopic analyses can be made routinely on komatiite samples, which have high concentrations of PGE.

Walker *et al.* (1988) analysed 12 komatiite samples from Pyke Hill in Munro Township for Re–Os as well as Sm–Nd and Rb–Sr isotopes. The Rb–Sr data scattered broadly around a 1 Ga reference line, the Sm–Nd data had a very narrow range of Sm–Nd ratios but mainly plotted on a 2720 Ma reference line, and the Re–Os data defined a good isochron with an age of 2726 ± 93 Ma.

It is now known that the Re–Os method has only limited potential as a dating method. Some useful results have been obtained, such as the isochron ages reported by Walker *et al.* (1988, 1999) for komatiites from Pyke Hill and Gorgona Island and the ages obtained by Puchtel *et al.* (2001b) for the Vetreny belt komatiites, but in most cases the Re–Os ages serve only to confirm results obtained by other methods; more commonly, the most useful application of the Re–Os isochron diagram is to establish the degree to which the samples have evolved as a closed system. The problem with the Re–Os method arises from the mobility of Re during alteration, a process that can result in the loss of a variable, but often very large, proportion of this element (e.g. Gangopadhyay *et al.* (2005)). The consequence is considerable scatter of data in $^{187}Re/^{188}Os$ vs $^{187}Os/^{188}Os$ isochron diagrams obtained for komatiites and other old mafic–ultramafic rocks.

Initial isotope compositions

The initial Os isotopic composition is expressed as γ_{Os}, which, like ε_{Nd}, gives the difference between the initial $^{187}Os/^{188}Os$ of a sample and that of a chondritic reservoir at the time of formation of the sample. Calculation of initial isotope compositions is not possible for samples in which the parent–daughter ratio has changed following their crystallization, but such samples can be identified by analysing the covariation of Re with other elements that are incompatible with olivine (Gangopadhyay *et al.*, 2005).

As shown in Figure 6.8, the majority of analysed komatiites have γ_{Os} values that range from near zero to strongly positive. The only suite with negative γ_{Os} is the Fe-rich komatiites from Boston Creek, in keeping with the other peculiar characteristics of these rocks. The positive γ_{Os} values pose an immediate problem because a positive γ_{Os} means that the sample came from a source that had long-term enrichment of the more incompatible element, Re in this system. The problem is that data from other isotope systems like Sm–Nd or Rb–Sr demonstrate that most komatiites come from a depleted source. The Re–Os system appears, therefore, to be decoupled from the isotope systems of lithophile elements such as Rb, Nd and Sm.

As pointed out by Walker *et al.* (1988, 1991), and Puchtel *et al.* (2001b, 2004a,b), the Os isotope compositions of komatiites fall into two groups. One group, which includes most Precambrian komatiites as well as a majority of samples from Gorgona Island, has near-chondritic compositions ($\gamma_{Os} \sim 0$).

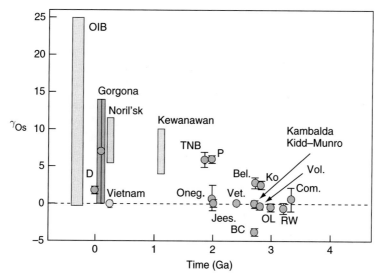

Fig. 6.8. Variation in Os isotopic compositions of komatiites through time. Principal sources of data are Walker *et al.* (1988, 1991, 1995, 1999), Brandon *et al.* (2003), Puchtel *et al.* (2001a,b, 2004a,b) and Gangopadhyay *et al.* (2005). (Locale names: OIB – ocean island basalt; D – Deccan (India); Gorgona Island (Columbia); Song La (Vietnam); Noril'sk, Siberian flood basalt province; Keweenowan basalts (Canada); TNB – Thompson Nickel Belt (Canada); P – Pechenga (Russia); Oneg – Onega (Russia); Jees – Jeesiörova (Finland); Vet – Vetreny (Russia); Bel – Belingwe (Zimbabwe); Kambalda (Australia); Kidd–Munro Abitibi belt (Canada); BC – Bostan Creek (Canada); Kostomuksha (Russia); RW – Ruth Well (Australia); Com – Commondate (South Africa).)

The second, more exclusive, group has higher γ_{Os} values. This group comprises Archean komatiites from the Belingwe belt in Zimbabwe (Walker and Nisbet, 2002), the Kostomuksha belt in the Baltic Shield (Puchtel *et al.*, 2001a), whose γ_{Os} values are between $+2$ and $+4$, Proterozoic komatiitic basalts from the Thompson belt, with γ_{Os} around $+6$, and some samples from the west coast of Gorgona Island, whose γ_{Os} values range from $+11.1$ to $+12.4$ (Walker *et al.*, 1999). Positive γ_{Os} are also reported in picrites of Proterozoic and younger ages, as well as in modern oceanic island basalts.

To explain these results, Walker *et al.* (1999) proposed that the komatiite source contained a high proportion of recycled oceanic lithosphere – the komatiites with near-zero γ_{Os} values could, for example, have come from the mantle portion of the lithosphere which had γ_{Os} values close to chondritic, whereas those with more strongly positive γ_{Os} values could have come from the basaltic portion, where Re–Os ratios are high and evolved to elevated $^{187}Os/^{188}Os$. Both portions of the oceanic lithosphere are depleted in incompatible lithophile elements and thus retained the depleted isotopic signatures for these elements. Following quantitative evaluation of this model, Walker *et al.* (1999) used mass-balance arguments to show that in order to explain the compositions of the most radiogenic Gorgona samples, the basaltic crust had to be improbably enriched in PGE or improbably old, or it had to constitute an improbably large proportion of the source. As an alternative, they suggested that the high $^{187}Os/^{188}Os$ was derived through interaction with the outer core, an argument supported by coupled $^{186}Os-^{187}Os$ data (Walker *et al.*, 1995; Brandon *et al.*, 2003). The idea that plume-derived magmas, including some modern oceanic island basalts, contain Os from the outer core, is currently the subject of active debate and the issue is far from resolved.

6.6 The Lu–Hf system

This system is based on the decay of ^{176}Lu to ^{176}Hf. Lu is the heaviest REE and Hf is a HFSE with a compatibility similar to that of Sm. Variations of Lu–Hf ratios coincide with changes in the relative abundances of middle and heavy REE. They therefore measure fractionation of ratios like Gd/Yb, one of the ratios used to distinguish Al-depleted Barberton-type komatiites from Al-undepleted Munro-type komatiites (Chapter 5).

Although Lu–Hf ratios might vary in suites of magmas derived through differing degrees of partial melting of the mantle, as long as the melting takes place under conditions in which garnet is retained in the source, they should not vary during fractional crystallization of komatiite, because both elements are strongly incompatible in olivine. There is little reason to believe, therefore, that the Lu–Hf system can be used to date komatiites and related rocks. Nonetheless, in some studies of the Hf isotope compositions of komatiites, linear arrays in the $^{176}Lu/^{177}Hf$ vs $^{176}Hf/^{177}Hf$ diagram have been obtained and in some cases these correspond to the presumed eruption age. In such cases, the cause is probably mobility of one of the elements, probably Lu, during alteration that took place during or very soon after eruption (Blichert-Toft *et al.*, 2004). In other words, as a dating method, Lu–Hf operates somewhat like the Pb–Pb method, and it suffers the same problems.

Like the Pb–Pb and Re–Os methods, the main application of Lu–Hf isotopic analysis of komatiites has become the use of initial isotope ratios to trace the history of the mantle source, and here the method is particularly valuable. Some important results obtained using the method are summarized in Table 6.2. Gruau *et al.* (1990a) built on preliminary results of Patchett and Tatsumoto (1980) to investigate the origin of the non-chondritic Al–Ti and Gd–Yb ratios that characterize Barberton-type komatiites. As recounted in

Table 6.2. *Summary of Hf isotopic compositions of komatiites and related rocks*

Locality	εHfT	Reference
Barberton, South Africa 3.5 Ga	+0.4 to +1.7	Gruau *et al.* (1990a)
Barberton, South Africa 3.5 Ga	−0.3 to +1.6	Blichert-Toft and Arndt (1999)
Various 2.7 Ga belts	+1.8 to +4.6	
Gilmour Is, Canada 1.9 Ga	+2.9 to +3.5	
Schapenburg belt, South Africa 3.5 Ga	+4.3 to +6.3	Blichert-Toft *et al.* (2004)
Gorgona, Colombia	+16.6 to +17.5	Thompson *et al.* (2004)

εHfT – the calculated Hf isotope composition, in epsilon notation, at the time of eruption.

Chapters 2 and 5, and discussed in more detail in Chapter 13, there are two competing explanations for these ratios: (1) they were derived from an ancient garnet-depleted reservoir within the mantle; and (2) garnet was extracted during the melting event that produced the komatiite magma. During melting in the upper mantle, Hf behaves like Gd and is moderately compatible with garnet whereas Lu behaves like Yb and is strongly compatible. As a consequence, mantle material with a high garnet content has a high Lu–Hf ratio and evolves to high ^{176}Hf–^{177}Hf ratios, whereas a source depleted in garnet has low Lu/Hf and evolves to low ^{176}Hf/^{177}Hf. Analysis of Lu–Hf isotopes should therefore be able to constrain the timing and the nature of the process that produced the high Gd/Yb, or low Lu/Hf, of Barberton komatiites.

The measurements of Gruau *et al.* (1990a) showed that Al-depleted komatiites from the Barberton greenstone belt had near-chondritic to slightly elevated ^{176}Hf/^{177}Hf (ε_{Hf} from zero to +1.5). In other words, these komatiites came from a source with the Hf isotopic characteristics of long-term garnet enrichment (positive ε_{Hf}) and not garnet depletion. The implication is that if the Al depletion and HREE fractionation resulted from the fractionation of garnet, this process took place only shortly before, or during, the melting event that produced the magmas. The results, illustrated in Figure 6.9, support the interpretation of Green (1974) and Ohtani *et al.* (1989), who argued that garnet separated at the site of melting during the komatiite-forming event.

Further light on the process was provided by the work of Blichert-Toft and coauthors (Blichert-Toft and Arndt, 1999; Blichert-Toft *et al.*, 2004) who used the newly developed magnetic-sector inductively-coupled plasma mass

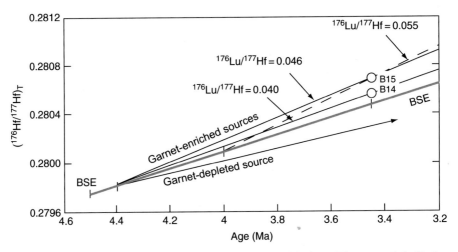

Fig. 6.9. Variations of Hf isotope compositions with time. The curve labelled BSE represents the evolution of the bulk silicate earth composition of Blichert-Toft and Albarède (1998). The line 'garnet-depleted source' shows the evolution of part of the mantle with a composition comparable to that of Al-depleted komatiite. B14 and B15 represent the extremes of compositions of Barberton komatiites, which all plot above the BSE reference. These compositions are reproduced by the evolution of garnet-enriched sources with $^{176}Hf/^{177}Hf$ only slightly above the BSE value (from Gruau *et al.* (1990a) and Blichert-Toft and Arndt (1999)).

spectrometer (ICPMS) to obtain far more precise analytical results than could be realized using the thermal ionization mass spectrometry employed by Gruau *et al.* (1990a). In addition to working on Al-depleted Barberton komatiites, Blichert-Toft and Arndt (1999) analysed Munro-type Al-undepleted komatiites from various Archean and Proterozoic localities. They found that the initial ε_{Hf} values were persistently positive, covering a relatively large range from $+2.6$ to $+7.8$. To explain the high positive ε_{Hf} values of the Barberton komatiites they proposed that these magmas, whose major- and trace-element characteristics pointed to garnet *depletion*, had come from a source with long-term *enrichment* of garnet. This is not as silly as it first seems because the retention of garnet in the residue of melting controls the chemical signature of the melts, and a source with a relatively high garnet content is more likely to retain garnet. Smith and Ludden (1989) had pointed out earlier that a garnet-enriched source seemed to be required to explain the high ε_{Nd} values of the Barberton komatiites. The picture that emerges is that the source of Barberton komatiites may not be normal mantle peridotite but could be eclogite or another type of rock that contains a significant garnet component. These ideas are explored further in Chapter 13.

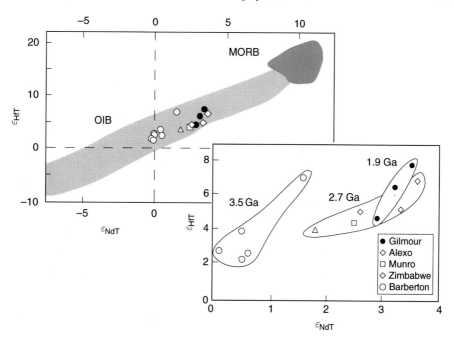

Fig. 6.10. Hf and Nd isotope compositions of komatiites and basalts (from Blichert-Toft and Arndt (1999)).

When the isotope data are plotted on an ε_{Hf} vs ε_{Nd} diagram (see Figure 6.10) they define a crude linear array that coincides with data from modern basalts and granitoids, and passes well above the composition of bulk silicate Earth. Two explanations have been proposed to account for the bias to high initial ε_{Hf}: (a) a reservoir with low Lu/Hf formed early in Earth's history and has remained hidden and separate from the rest of the mantle ever since; (b) the accepted values for the Hf isotopic composition of the bulk silicate Earth are inappropriate. The issue has not yet been resolved. New Hf isotopic analyses of chondritic meteorites have dispelled any doubt as to the reliability of these data (Blichert-Toft and Albarède, 1998; Blichert-Toft *et al.*, 2002): if the Earth formed from material with the Hf and Nd isotopic compositions of these meteorites, its bulk composition would have plotted below the trend defined by the compositions of all the rocks that can be sampled at the surface. Yet the possibility remains that the material from which the Earth was constructed did not have exactly the same Hf and Nd isotopic ratios of the chondrites. On the other hand, evidence is appearing through Hf isotopic studies of kimberlites (Bizzarro *et al.*, 2002; Nowell *et al.*, 2004), and from the search for [142]Sm anomalies in the oldest rocks (Boyet and Carlson, 2005) that indicates that at least some parts of the mantle have remained isolated from the rest for much of Earth's history, and the existence of a hidden reservoir with low ε_{Hf} remains entirely plausible.

The Hf isotopic compositions of komatiites and basalts from Gorgona Island are consistent with the Nd isotopic compositions discussed in Section 6.4. Thompson *et al.* (2004) have shown that ε_{Hf} values are moderate in the enriched basalts (12.8–13.3) and high in the depleted basalts, komatiites and picrites (15.8–17.7). These values provide further evidence that these magmas were derived from a very depleted part of the mantle which, according to Thompson *et al.* (2004), was part of the plume source of the Caribbean oceanic plateau.

6.7 Helium isotopes

Helium isotopic compositions are used to trace the origin of mantle-derived magmas and in particular to identify those that came from a deep source. Primordial ^{3}He was trapped during the Earth's formation and has not been significantly produced since that time, whereas ^{4}He is a mixture of trapped primordial component and radiogenic He produced by the decay of U and Th throughout geological history. The ^{3}He–^{4}He ratios of different parts of the mantle depend on reservoir residence times and ^{3}He–(U + Th) ratios. High ratios in oceanic island basalts (up to 32 times the atmospheric ratio (R_a) of 1.386×10^{-6}) indicate a deep mantle component rich in primitive ^{3}He. Lower, more uniform ^{3}He–^{4}He ratios of mid-ocean ridge basalts (8 ± 1 Ra) represent the more degassed and homogenized upper mantle. The isotopic data in theory allow us to constrain the depth of origin of the source of komatiites – whether in a mantle plume from deep in the mantle, or from a near-surface source such as convecting upper mantle or the mantle wedge above a subduction zone.

Three studies of He isotopic compositions of komatiites have been published (Table 6.3). Richard *et al.* (1996) analysed olivine separated from two Archean komatiites, from Alexo in the Abitibi belt in Canada and from the Belingwe belt in Zimbabwe, as well as from Proterozoic komatiite from the Ottawa Islands. Olivine from the Proterozoic localities had low ^{3}He/^{4}He values that were attributed by Richard *et al.* (1996) to alteration or the assimilation of old continental crust, as has been inferred for these komatiites on the basis of trace-element and Nd or Pb isotopic data (Arndt *et al.*, 1987; Chauvel *et al.*, 1993). The Alexo olivine, however, yielded ratios up to 40 R_a, the highest value recorded at that time in terrestrial lavas. After discussing, and dismissing, other possible sources of ^{3}He (radiogenic production in the olivine matrix, assimilation of crustal rock, incorporation of

Table 6.3. *Summary of He isotopic compositions of olivine in komatiites and related rocks*

Locality	Rock type	Sample	^{3}He/^{4}He (R/Ra)	Reference
Alexo, Abitibi belt	Komatiite	M712	26 to 39	Richard *et al.* (1996)
Zvishavane, Zimbabwe	Komatiite	Z11	0.05	
Gilmour Is, Canada	Komatiitic basalt	G21	0.14 to 0.17	
Munro Twp, Abitibi belt	Komatiite	MT-02	1.6	Matsumoto *et al.* (2002)
	Komatiite	MT-03	0.65	
	Komatiite	MT-04	29.2	
	Komatiite	MT-05	12.4 to 30.7	
Gorgona, Colombia	Dunite	GOR 503	7.8 to 9.3	Révillon *et al.* (2002)
	Wehrlite	GOR 507	12.4 to 13.1	
	Gabbro	GOR 509	13.6 to 18.2	

nucleogenic or cosmogenic ^{3}He), Richard *et al.* (1996) concluded that the data provide convincing evidence that komatiites form through melting in plumes derived from deep-seated mantle sources. They noted that the depleted trace-element and isotopic composition of the Alexo komatiite differed from the enriched compositions commonly associated with the plume sources of modern oceanic island volcanic rocks, but concluded nonetheless that the depleted material was part of the plume. A parallel can be drawn with the situation on Iceland, where alkali basalts are derived from an enriched portion of the plume and depleted picrites from a refractory depleted portion (Hémond *et al.*, 1993; Kerr *et al.*, 1995b). The He isotope data therefore provided evidence that the plume originated from somewhere deep in the mantle. The contrast between the undepleted character of primitive ungassed material and the depleted character of the komatiite source was taken as further evidence of decoupling of He isotopes from other isotopic systems.

Matsumoto *et al.* (2002), the authors of the second study, analysed olivine from komatiites from Munro Township, which is located 50 km to the east of Alexo and in the same stratigraphic unit; and they obtained very similar results. They found that the olivine preserved a high ^{3}He/^{4}He of about 30 Ra and they inferred that the original ^{3}He/^{4}He in the source might have been as high as 70. They pointed out, however, that even such a high ratio does not require that the source corresponded to primitive, ungassed Archean mantle. For example, some models for the ^{3}He–^{4}He evolution of the mantle produce Archean depleted upper mantle with ^{3}He–^{4}He ratios as high as those inferred for the Munro komatiites.

The third study, Révillon *et al.* (2002), reported He isotopic compositions of olivine separated from intrusive mafic–ultramafic rocks from the interior of Gorgona Island. According to Echeverría (1980), Kerr *et al.* (1995a) and Révillon *et al.* (2000), these rocks are comagmatic with the komatiites. These olivines yielded only modest ^{3}He–^{4}He ratios (~ 8–18) but they were significant for two reasons. First, the moderately elevated ^{3}He–^{4}He ratios were measured in samples with pronounced depletion of incompatible trace elements and correspondingly low ^{87}Sr/^{86}Sr and high ^{143}Nd/^{144}Nd, providing a further example of decoupling of He and other isotope systems. Second, the association of high ^{3}He–^{4}He ratios, MgO-rich rocks and forsterite-rich olivine constitutes a powerful argument for the hot plume origin of komatiite, as discussed further in Chapter 13.

6.8 Stable isotopes

Analyses of the stable isotopic compositions of komatiites, mainly oxygen, have been carried out with two specific aims. First, the isotopic compositions of magmatic minerals have been determined to provide information about the nature of the mantle source; and second, the compositions of partially to completely altered whole rocks and of separated secondary minerals have been used to establish the conditions of alteration and the composition and origin of the fluids that circulated through the volcanic pile.

Although some studies of altered komatiites and related rocks have included analysis of H isotope compositions, this approach is unlikely to yield useful data. O'Hanley and Kyser (1991) have shown that serpentines from Canadian localities with ages from Archean to Mesozoic have δD values (δD is the hydrogen isotope composition expressed using the delta notation) that correlate with their present latitudes. The implication is that exchange with modern waters has controlled the H isotopic composition of these minerals.

Analyses of primary minerals (olivine, orthopyroxene and clinopyroxene) have been undertaken by Hoefs and Binns (1978), Smith *et al.* (1984), Lahaye and Arndt (1996) and Révillon *et al.* (2002), as summarized in Table 6.4 and Figure 6.11. In large part the results are unsurprising and with few exceptions the compositions are similar to those obtained in analyses of fresh minerals from modern lavas. The conclusion reached following these studies is that the

Table 6.4. *Summary of oxygen isotope compositions of komatiites (whole rocks and minerals)*

Location	Yilgarn Craton	Munro Twp, Abitibi belt	Skead Twp, Abitibi belt	Barberton belt	Barberton belt	Tipasjärvi belt, Finland	Schapenburg, Barberton belt	Alexo, Abitibi belt	Gorgona, Colombia
	Hoefs and Binns (1978)	Beaty and Taylor (1982)	Beaty and Taylor (1982)	Hoffman et al. (1986)	Smith et al. (1984)	Tourpin et al. (1991)	Lécuyer et al. (1994)	Lahaye and Arndt (1996)	Révillon et al. (2002)
Whole rocks									
komatiite	+5 to +8.6[a]	+3.5 to +5.4			+3.6 to +7.6	+9.8 to +12.3	+3.2 to +4.7	+5.2 to +5.6	
komatiitic basalt	+6 to +9.4		+8.4		+5.9 to +7.3		+4.8 to +5.0		
other				+3 to +14					
Magmatic minerals									
olivine	+4.6 to +5.5							+5.2	
clinopyroxene	+4.6 to +5.5							+5.1	+5.18 to +5.35
orthopyroxene					+6.2				
Secondary minerals									
serpentine	+4 to +5.6				+1.7 to +7.1			+3.8	
chlorite								+3.1	
hornblende							+4.95 to +5.05		
magnetite							+0.08		
carbonate						+9.3			
'matrix'								+3.7 to +4.3	

[a] $\delta^{18}O$ values.

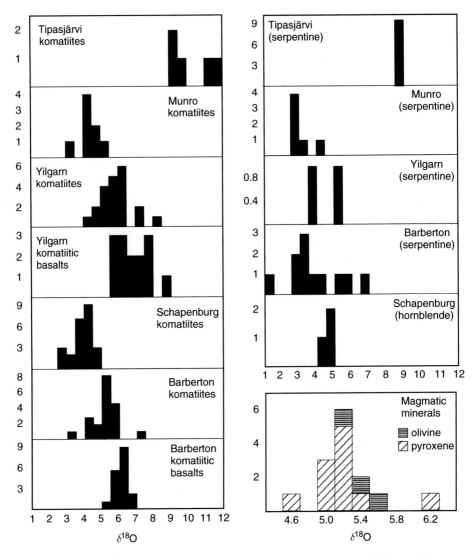

Fig. 6.11. Summary of oxygen isotopic compositions measured in komatiites. The histograms with black bars represent whole-rock data and the shadings in the last diagram represent the compositions of magmatic minerals from various localities. Data sources: Hoefs & Binns (1978), Beaty and Taylor (1982), Hoffmann *et al.* (1986), Smith *et al.* (1984), Tourpin *et al.* (1991), Lécuyer *et al.* (1994), Lahaye and Arndt (1996), Révillon *et al.* (2002).

oxygen isotopic composition of the source of Archean komatiites was very similar to that of present-day upper mantle. The only notable aspect of the results is the slightly elevated $\delta^{18}O$ value of Smith *et al.*'s (1984) analysis of orthopyroxene from the Stolzburg intrusion in the Barberton belt. They obtained a value of $+6.2$, which is distinctly higher than those of other magmatic minerals separated from komatiites. A possible explanation is that the magma that crystallized in this intrusion had assimilated rocks from altered (silicified?) oceanic crust, a process that led to the crystallization of orthopyroxene and changed the O isotopic composition. Using a crystallization temperature of 1450 °C, Smith et al. (1984) calculated that the parental magma of Stolzburg intrusion had $\delta^{18}O \sim 5.7$. This temperature is probably too high: the experimental studies summarized in Chapter 7 indicate that orthopyroxene starts to crystallize only at temperatures less than 1300 °C. Using 1250 °C as the equilibration temperature, Lahaye and Arndt (1996) obtained essentially identical $\delta^{18}O$ values for two magmatic minerals separated from the Alexo komatiite flow, $+5.1$ in the olivine and $+5.2$ in the clinopyroxene. This too was surprising because at normal magmatic temperatures, oxygen isotopes are fractionated between the two minerals. Lahaye and Arndt (1996) attributed the lack of fractionation to the high temperatures that prevail during the crystallization of komatiite. Finally Révillon et al. (2002) obtained $\delta^{18}O$ values in clinopyroxene separated from Gorgona komatiites and intrusion rocks that fell between $+5.18$ and $+5.35$, within the range reported in mantle-derived minerals.

The interpretation of the oxygen isotopic compositions of whole-rock samples of variably altered komatiite and of the secondary minerals separated from these rocks is more complicated (Figure 6.11). Although Hoefs and Binns (1978) reported preliminary data from Australian komatiites in their short, two-page report, the first detailed investigation of the oxygen isotopic compositions of komatiites and komatiitic basalts was that of Beaty and Taylor (1982). In their study, they analysed six whole-rock samples from a variably serpentinized komatiite flow from Munro Township and complemented this information with data from komatiites and serpentines from other parts of the Abitibi belt. The samples from the Munro flow had $\delta^{18}O$ values between $+3.5$ and $+5.4$ and all plotted on a mixing line between the inferred composition of fresh komatiite ($\delta^{18}O = +5.7 \pm 3$) and serpentine ($\delta^{18}O \sim +3$). The composition inferred for the komatiite was similar to that estimated for the present-day mantle on the basis of analysis of modern ultramafic rocks, and Beaty and Taylor concluded that the oxygen isotopic composition of the mantle had not changed through geological time. Using the compositions of the altered komatiites and the serpentine, and assuming a final metamorphic temperature of 300–330 °C, they calculated that the hydrothermal fluid that caused the alteration had $\delta^{18}O$ between -2 and $+2$ and that it was metamorphic pore fluid derived from Archean seawater. Their calculated water–rock ratio was high

(>1) and they proposed that the alteration resulted from the deep circulation of seawater through the volcanic pile.

Hoffman *et al.* (1986) reported whole-rock oxygen isotope data from a suite of diverse rock types – volcanic, cherty sediment and intrusive mafic–ultra-mafic – from the Onverwacht Group of the Barberton greenstone belt, South Africa. The $\delta^{18}O$ values of these samples lie between + 3 and + 14, a range that they compared with the compositions of Phanerozoic ophiolites and modern oceanic crust. They argued that when the samples are arranged in an 'ophio-litic pseudostratigraphy', the oxygen isotope profile of the Barberton samples is indistinguishable from that in Phanerozoic ophiolites. On this basis they defended the idea, advocated by de Wit *et al.* (1983) but now largely discre-dited (see Chapter 11), that the Onverwacht sequence represents a remnant of normal Archean oceanic crust. Like Beaty and Taylor (1982), they suggested that the Barberton rocks had been hydrothermally altered by fluids whose isotopic composition was indistinguishable from modern seawater.

In their study of komatiites from the Barberton greenstone belt, Smith *et al.* (1984) analysed a large suite (38 samples) of completely altered komatiites and komatiitic basalts, two types of serpentine, as well as magmatic orthopyroxene from the Stolzburg mafic–ultramafic intrusion. In the volcanic rocks they measured a large range of $\delta^{18}O$ values, from + 3.6 to + 7.6, with an average around + 5. They noted that the rocks were almost completely altered to secondary minerals and assumed that the alteration had taken place under high fluid–rock ratios. In their modelling of the alteration they considered two types of fluid, seawater with $\delta^{18}O = 0$ and magmatic water with $\delta^{18}O = +5$ to + 7. They found that the oxygen isotopic compositions of the komatiites and basalts could be explained by interaction with seawater at 100 °C, or with deep-seated fluids at the higher temperatures of 240–450 °C. Their general conclu-sion was that the alteration had taken place in an ocean floor environment.

This interpretation was called into question by Cloete (1994,1999), who pointed out that the volcanic rocks in any Archean greenstone belt would been subjected to at least two stages of alteration, the first on the sea floor and the second during accretion and orogenesis. In the Barberton region, the latter events followed soon after eruption, but in some other regions, deformation and regional metamorphism were much later, as documented in studies of komatiites from Finnish greenstone belts.

Tourpin *et al.* (1991) and Gruau *et al.* (1996) provided oxygen and carbon isotope data for several remarkably altered komatiite flows from the Tipasjärvi and Kuhmo greenstone belts in Finland; these flows were discussed in Section 6.3. The $\delta^{18}O$ values of whole-rock samples from the Tipasjärvi flow

are relatively high, from $+9.8$ to $+12.3$, and their carbon isotope compositions, expressed as $\delta^{13}C_{PDB}$, were low, from -2.3 to -3.4. These values coincided with those measured in a carbonate vein in the lower part of the flow, a
result that the authors took as evidence that the alteration that had changed
the mineralogy of the flow and disturbed the Sm–Nd and Rb–Sr systems
resulted from the circulation of carbonate-rich fluids.

Lécuyer *et al.* (1994) measured very different compositions ($\delta^{18}O = +3.2$ to
$+5$) in their study of a series of more highly metamorphosed (amphibolite-
facies) komatiites from the Schapenburg segment of the Barberton greenstone
belt. They established, using the ^{39}Ar–^{40}Ar method, that the metamorphism
took place at 2.9 Ga, long after the eruption of the komatiites at $\sim$3.5 Ga, and
they explained the isotopic compositions by assuming that the alteration took
place during metamorphism under conditions of high fluid–rock ratios and
high temperatures of around 450 °C. The fluid was assumed to be of metamorphic origin with $\delta^{18}O$ of $+5$ to $+7$ and δD of -65 to -50.

Lahaye and Arndt (1996) documented two alteration events in their study of
a 2.7 Ga komatiite flow from Alexo, Canada. In the komatiite samples the
measured $\delta^{18}O$ was in the range $+5.2$ to $+5.6$, values similar to those of Beaty
and Taylor (1982). Primary magmatic minerals are preserved in many parts of
the Alexo komatiite flow, which implies that fluid–rock ratios were low.
However, in other parts of the same flow there is evidence of major changes
in the concentrations of major and trace elements, the rodingitization
described in Chapter 5. To explain these changes requires that either the
fluid–rock ratio was high (contradicting petrographic observations) or that
the concentrations of these elements in the fluid were greater than those in
hydrothermal fluids of the type that circulate in modern oceanic crust.
According to Lahaye and Arndt (1996), the high concentrations required
that the trace elements were stabilized in F^-, OH^- or CO_3^{2-}complexes, in
which case the fluids were either metamorphic or magmatic and associated
with accretion and deformation, or that they were late magmatic and related to
the fluids that formed the gold deposits of the Abitibi belt.

The studies of the oxygen isotopic compositions of Archean komatiites and
related rocks provide evidence of at least two periods of alteration. An early
event accompanied the interaction between seawater and the volcanic rocks,
but it is by no means clear that the signature of this alteration is retained in the
rocks themselves. This being the case, it is not evident that the analysis of
komatiite can be used to estimate the composition and/or temperature of
Archean seawater. Nor does the analysis of cherts and other rocks in
Archean greenstone belts provide this information: Lécuyer *et al.* (1994), for

example, showed that the compositions of cherts in the Barberton belt are readily explained by fluid–rock interaction during the 2.9 Ga metamorphism that affected rocks that had been deposited several hundreds of million years earlier. More probably, the stable isotope compositions of komatiites are strongly influenced by fluid–rock interaction associated with the metamorphism, deformation and intrusion of granitoids that accompanied the accretion of the originally oceanic volcanic series to the continental crust, as Cloete (1994, 1999) realized during his very important study of the Barberton greenstones.

6.9 Conclusions emerging from the isotopic compositions of komatiites

(1) It is not easy to date a komatiite. A problem arises because parent–daughter ratios of most isotopic systems are not fractionated during the crystallization of ultramafic magma. In those cases in which the isotopic compositions of whole-rock komatiites define an isochron, the fractionation of parent–daughter ratios normally results from secondary processes that operated after crystallization. If this fractionation happened soon after eruption, the age is approximately correct; if it happened during later metamorphism, it can be very wrong. Another problem arises from irregular disturbance of the isotope system after crystallization. The loss of Re is common and this process limits the use of the Re–Os system, the only system in which parent–daughter ratios change during crystallization. U–Pb dating of zircon or baddelyite gives more reliable ages and the analysis of zircons extracted from komatiites has been used as a dating method. However, zircon does not crystallize in ultramafic magmas and either grains of this mineral in komatiites are xenocrysts or they grew during metamorphism. The best method of establishing the age of a komatiite is to date associated felsic volcanics using the U–Pb zircon method. The curious cohabitation of komatiite and rhyolite, discussed in Chapter 14, makes this possible in many Archean greenstone belts.

(2) The initial isotopic compositions of systems comprising lithophile elements (Rb–Sr, Sm–Nd, U–Th–Pb and Lu–Hf) indicate that the mantle source of komatiite was depleted in incompatible elements for a long period prior to komatiite formation. This is revealed by the low $^{87}Sr/^{86}Sr$ in magmatic pyroxenes, by low μ_1 values, and by the high ε_{Nd} and ε_{Hf} values measured in almost all komatiites. Given the uncertainty that surrounds the interpretation of the high ε_{Nd} and ε_{Hf} values measured in the oldest, polymetamorphosed parts of the Earth's crust, and the sporadic distribution of ^{142}Nd anomalies (Boyet *et al.*, 2003; Caro *et al.*, 2003; Boyet and Carlson, 2005), the coupled depletion recorded in the high ε_{Nd} and ε_{Hf} values of 3.5 Ga Barberton komatiites provides some of the best evidence of depletion in substantial portions of the early Archean mantle.

(3) The picture emerging from Re–Os isotopic studies is complicated. A small majority of measured komatiites do indeed come from a source depleted in Re, the more

incompatible element, but a substantial number of komatiites record near-chondritic Os isotope ratios. In addition, the Fe-rich komatiites from Boston Creek, whose Nd isotopic composition points to a source depleted in lithophile elements have negative γ_{Os} values, indicating derivation from a source with long-term enrichment in Re. Explanations that have been advanced to explain these variations in Os isotopic composition are: (a) the presence of a component of old recycled oceanic crust in the komatiite source or (b) the presence of Os from the outer core.

(4) All radiogenic isotopic systems record variations in initial isotope ratios that indicate that the komatiite source was heterogeneous. This variation is most apparent in the Cretaceous volcanic rocks from Gorgona Island, where there is evidence of two depleted components in the source of the komatiites and basalts and at least one additional enriched component in the source of the basalts. Similar variation is also present in the source of Precambrian komatiites, but here the message is garbled by variable contamination of the komatiite magmas with crustal rocks or by post-eruption disturbances of the isotope systems.

(5) The high He isotope ratios measured in komatiites from the Abitibi belt and from Gorgona Island are attributed to the presence in the komatiite source of material from the deep mantle. Together with the high temperatures required to form komatiite magma, this information strongly supports the hypothesis that komatiites formed in mantle plumes that rose from deep in the lower mantle. There is no evidence in any isotope system of isotopic compositions characteristic of a source in the mantle wedge above a subduction zone.

(6) The volcanic rocks of Gorgona Island have similar positive ε_{Nd} and ε_{Hf} values but a large range of Sm–Nd ratios, ranging from moderately low in depleted basalts to extremely low in picrites. This decoupling of trace-element ratios and isotope compositions provides strong evidence that the magmas formed by fractional melting (discussed further in Chapter 13). Similar decoupling in some Archean komatiites and the extreme depletion of incompatible trace elements in the Commondale komatiites (Wilson *et al.*, 2003) probably result from the same process.

(7) Oxygen isotopic compositions of magmatic olivine and pyroxene are similar to those in minerals from the modern mantle. The compositions of variably altered whole-rock komatiite samples and of secondary minerals record several alteration events in most komatiite suites. Initial alteration on the sea floor was followed by alteration during the metamorphism and deformation that accompanied accretion to the continent and subsequent tectonic events.

7

Experimental petrology

7.1 Introduction

Experimental work on komatiites has proceeded along three main lines. The first involved phase relation studies performed between 1975 and about 1984. These experiments, which were carried out at relatively low pressures, from 1 atm to about 4 GPa, were useful and provided valuable information about phase relationships and element partitioning, but they were not particularly exciting: they merely showed that komatiites behaved as might be expected – as a tholeiitic basalt containing 30–50% extra olivine.

More interesting is the second series of experiments, which started around 1984 and involved more extreme conditions. These experiments were carried out with ultra-high-pressure multi-anvil apparatuses and with shock-wave techniques, at pressures between ~4 and 20 GPa and temperatures from 1500 to 2000 °C, conditions corresponding to those in the deep interior of the Earth's mantle. It was apparent from an early stage of komatiite investigation that certain characteristics of komatiites – their high eruption temperatures and geochemical signature of garnet depletion – are best explained by melting deep in the mantle, and only such experiments are able to provide the information necessary to answer critical questions of komatiite petrogenesis.

The third line involved rapid-cooling experiments designed to understand the conditions in which spinifex and other textures form during the crystallization of komatiite flows. The more recent experiments of this type have succeeded in reproducing these textures and, when combined with fluid dynamic modelling, have provided valuable constraints on the mechanisms of komatiite eruption and solidification.

7.2 Experiments at 1 atm pressure

From 1975 to 1977, I carried out a series of experiments on natural and synthetic komatiites and komatiitic basalts (Arndt, 1976a) and obtained the

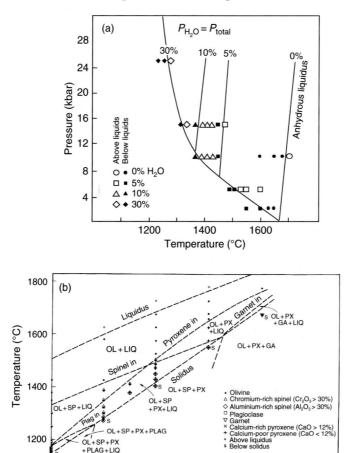

Fig. 7.1. (a) Anhydrous and hydrous liquidi and experimental data determined by Green *et al.* (1974) for the Barberton komatiite 49 J. (b) Phase relations of the Munro komatiite SA3091, from Arndt (1976a). (OL – olivine; SP – spinel; PLAG – plagioclase; PX – pyroxene; GA – garnet; LIQ – liquid.)

results summarized in Figure 7.1(b). In an olivine spinifex-textured komatiite from Munro Township containing 25% MgO (SA3091, Table 7.1), olivine is the 1 atm liquidus phase, appearing first at about 1500 °C. This mineral crystallizes over a very wide temperature interval, from 1500 °C to about 1190 °C. The only other phase present in this interval is chrome spinel, which crystallizes in minor quantities at temperatures below 1310 °C. Augite and plagioclase start to crystallize only below 1190 °C, relatively close to the solidus which is at about 1100 °C. Neither orthopyroxene nor pigeonite was observed.

Table 7.1. *Compositions of starting materials used in experiments on komatiites*

	SA3091	SKIB	comp A	comp B	M620	HSS-15	B-4
SiO_2	45.4	48.3	49.2	50.1	45.6	46.77	46.73
TiO_2	0.39	0.57	0.43	0.33	0.4	0.33	0.31
Al_2O_3	7.5	11.2	11.6	10.6	7.95	3.42	6.30
FeO	12.9	12.7	12.9	11.0	12.66	11.26	10.76
MnO	0.28	0.0	0.36	0.36	0.2	0.19	28.42
MgO	25.1	14.9	12.6	17.2	25.00	31.51	0.18
CaO	7.4	10.9	11.4	9.5	7.60	5.67	6.29
Na_2O	0.73	1.35	0.81	0.37	0.01	0.12	0.85
K_2O	0.15	0.0	0.13	0.07	0.02	0.08	0.13

Starting materials, the rocks they represent, and references to papers in which experiments are described:

SA3091 – Al-undepleted komatiite from Munro Township (Arndt, 1975)
SKIB – komatiitic basalt from Fred's Flow, Munro Township (Arndt, 1975)
comp A – komatiitic basalt from Fred's Flow, Munro Township (Kinzler and Grove, 1985)
comp B – komatiitic basalt from Fred's Flow, Munro Township (Kinzler and Grove, 1985)
M620 – Al-undepleted komatiite from Munro Township (Wei *et al.*, 1990)
HSS-15 – Al-depleted komatiite from the Barberton belt (Wei *et al.*, 1990)
B-4 – Al-undepleted komatiite from the Belingwe belt (Agee, 1998)

Essentially the same crystallization sequence was obtained with basaltic starting materials. In experiments on a natural pyroxene spinifex-textured basalt (FSX) and a mixture of oxides with the composition of this basalt (SKIB, Figures 7.2, 7.3(a)), both containing between 15% and 16% MgO (Table 7.1), olivine was again the liquidus phase at ~1350 °C and pyroxene first crystallized at 1190 °C from liquids containing ~10% MgO. This result was in apparent conflict with field observations and interpretations of chemical data. Pyroxene spinifex basalts, like their more magnesian counterparts, had at that time been thought of as non-cumulate rocks with compositions identical to those of the liquids from which they crystallized. The pyroxene-dominated mineralogy had been interpreted to indicate that pyroxene was the liquidus phase, in contradiction with the experimental result. Pyroxene spinifex lavas have MgO contents between ~10 and 18% MgO, and the implication from the mineralogy was that pyroxene crystallized in liquids with correspondingly high MgO contents. An inflection in the trend of chemical data plotted in the MgO–CaO–Al_2O_3 diagram (Figure 7.3(b)) supported this interpretation. As discussed in preceding chapters, it is now realized that many, if not all, spinifex lavas contain olivine or pyroxene in excess of the amounts that would have crystallized directly from the liquid, and the apparent conflict between

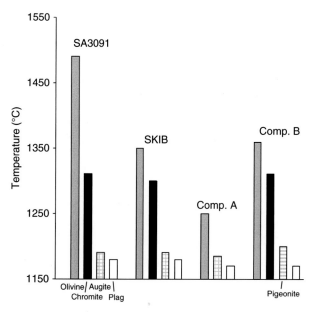

Fig. 7.2. Summary of crystallization sequences at 1 atm pressure in a natural komatiite (SA3091) and in three synthetic komatiitic basalts – SKIB (Arndt, 1977b) and Comp. A and Comp. B (Kinzler and Grove, 1985).

experimental and petrographic data is resolved when it is recognized that pyroxene spinifex lavas may contain 20–30% cumulus pigeonite and augite (Campbell and Arndt, 1982; Barnes, 1983; Kinzler and Grove, 1985).

A useful product of the 1 atm studies was the establishment of a relation between the MgO content and the liquidus temperature of anhydrous komatiite. Nisbet (1982) compiled data available at that time and proposed the following equation:

$$T_{liquidus} = 1000 + 20 * MgO\%.$$

(Nisbet gave the equation as $T_{liquidus} = 1400 + [(MgO - 20)*20]\,°C$ to emphasize that it applies to magmas with liquidus temperatures above 1400 °C.) Figure 7.4 is a compilation of more recent data which are fitted by the equation

$$T_{liquidus} = 1033 + 19.5 * MgO\%.$$

The expression for the 1 atm data is essentially the same as that proposed by Nisbet (1982) and either expression can be used to show that the eruption temperatures of anhydrous komatiites are extremely high.

In a broad-ranging and important paper, Kinzler and Grove (1985) explored the 1 atm phase relationships of komatiitic magmas. They worked

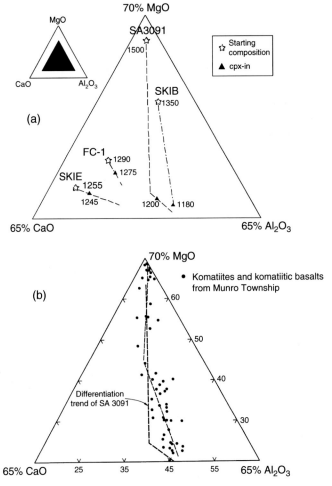

Fig. 7.3. (a) MgO–CaO–Al$_2$O$_3$ diagram showing the compositions of glasses in experiments conducted on a range of natural and synthetic komatiites. In the komatiite sample SA3091, olivine and minor chrome spinel crystallize from the liquidus at about 1500 °C, to 1190 °C. The appearance of clinopyroxene, the second silicate phase, is indicated by the black triangle. In the komatiite, which has CaO/Al$_2$O$_3$ ~1, clinopyroxene starts to crystallize at 1200 °C; in more CaO-rich starting compositions such as FC-1 (a natural komatiitic basalt) and SKIE (a synthetic komatiitic basalt with excess CaO), clinopyroxene appears at higher temperatures. (b) Comparison between the differentiation trend of komatiite SA3091, as determined experimentally, and the compositions of komatiites and komatiitic basalts from Munro Township. In the lavas, the CaO–Al$_2$O$_3$ ratio starts to decrease in relatively MgO-rich liquids, apparently indicating an early onset of clinopyroxene crystallization.

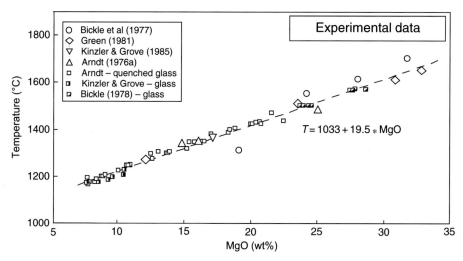

Fig. 7.4. Correlation between liquidus temperatures and MgO contents of experimental liquids at 1 atm.

with two synthetic komatiite compositions. Composition A with 12.6% MgO corresponds to the estimated bulk composition of the upper spinifex layers of Fred's Flow (the differentiated komatiitic flow from Munro Township). Composition B with 17.2% MgO is another estimate of the parental liquid of Fred's Flow, based on the compositions of olivine cumulates in the lower part of the flow. Using these compositions, Kinzler and Grove investigated the equilibrium crystallization sequences of pyroxenes, and the effects that small changes in composition had on these sequences. They also carried out cooling-rate experiments that demonstrated that rapid cooling strongly depressed the temperatures at which pyroxene and plagioclase began to crystallize. This work complements earlier work by Lofgren *et al.* (1974) and Donaldson (1976, 1982) which is described in more detail in Section 7.7.

The results of Kinzler and Grove's study are summarized in Figures 7.2, 7.5 and 7.6. In their equilibrium experiments, the liquidus phase was olivine, joined by chrome spinel at 1320 °C. In the more magnesian synthetic komatiite (composition B), pigeonite appeared at 1205 °C and was joined by plagioclase at 1185 °C. In the less magnesian starting material (composition A, the ana-logue of the upper part of Fred's Flow), the first pyroxene to crystallize was augite, at ∼1195 °C, rather than pigeonite.

Rapid cooling depressed the crystallization temperatures of the two pyrox-enes and plagioclase (Figure 7.6). The effects were more pronounced for pigeonite than augite, and this led to a reversal in the crystallization sequence of pigeonite and augite in composition B: under equilibrium conditions

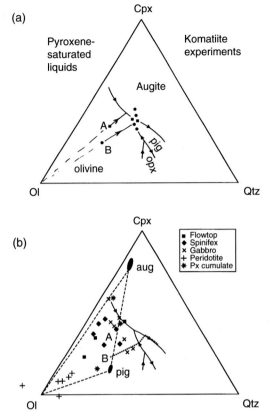

Fig. 7.5. (a) Ol–cpx–qtz diagram showing the 1 atm phase relations of two synthetic komatiite compositions that represent liquids in Fred's Flow. (b) Compositions of samples from Fred's Flow plotted in the ol–cpx–qtz diagram (from Kinzler and Grove (1985)).

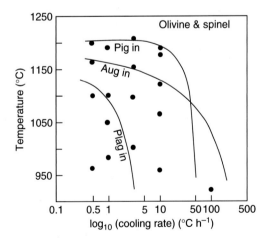

Fig. 7.6. Effect of variation in cooling rate on the crystallization sequence in the synthetic komatiite composition B of Kinzler and Grove (1985).

pigeonite is the first to crystallize whereas during rapid cooling augite crystallizes before pigeonite.

When Kinzler and Grove (1985) compared their experimentally determined crystallization sequences with those inferred from mineralogical relationships in komatiitic lavas, they noticed two discrepancies. (1) From their experiments, they predicted the crystallization path of their composition A liquid, the one corresponding to lavas from the upper part of Fred's Flow, to be olivine + spinel, followed by augite and then pigeonite. In most pyroxene spinifex lavas, a different crystallization is inferred on the basis of petrographic data: in these lavas, pigeonite forms the cores of composite pyroxene needles and this mineral clearly crystallized *before* augite. For composition B, which is slightly more quartz normative, the predicted crystallization sequence is olivine (+ spinel) → pigeonite → pigeonite + augite ± plagioclase, a sequence that corresponds to that inferred from the mineralogy and geochemistry of the flow. (2) At no stage in the crystallization paths predicted from the experimental results does orthopyroxene crystallize, yet orthopyroxene + clinopyroxene cumulates are an important component of the upper portion of the lower cumulate layer of Fred's Flow (Arndt, 1977b).

The first discrepancy was readily explained by the dynamic crystallization experiments (Figure 7.6). Under the conditions of rapid cooling, as can be inferred for the uppermost portions of komatiite flows, the crystallization of pigeonite is suppressed to the extent that it appears close to the solidus, if at all, and augite is the first pyroxene to appear. If the composition of the parental liquid were close to that of composition B, crystallization conditions of relatively slow cooling, as in the interior of the flow, would have followed along the path described above, with pigeonite preceding augite. Closer to the flow top, where cooling rates were higher, pigeonite crystallization would have been suppressed and augite would be the first pyroxene to crystallize.

To account for the second discrepancy is more complicated. As discussed above and emphasized by Kinzler and Grove (1985), the difference between the predicted and observed behaviour may indicate that many spinifex lavas contain excess olivine or pyroxene and do not represent liquid compositions. However, this cannot be the complete explanation, as becomes apparent when the mineralogy and textures of the lavas are considered in more detail. Clasts in the flow-top breccia of Fred's Flow have olivine porphyritic textures and contain little or no pyroxene – these rocks cannot contain any excess pyroxene. The compositions of rocks from the uppermost parts of Fred's Flow plot within the same field as the pyroxene spinifex lavas (Figure 7.5(b)). Likewise, the gabbros of Fred's Flow, most of which have textures and mineralogy inconsistent with accumulation, also plot above and to the left of the primary phase field of orthopyroxene. Although the gabbros contain augite, they are olivine-free, and accumulation of augite alone would cause a displacement towards the cpx apex in the phase diagram (Figure 7.5(a), (b)) and not towards the orthopyroxene field. Many of the pyroxene spinifex lavas contain cumulus pigeonite and augite, but olivine is absent. The accumulation of pigeonite displaces the composition downwards, from a composition like that of A to one closer to B. The effect of pigeonite accumulation is to displace the estimated liquid composition slightly towards the qtz apex. The composition of the liquid that produced the pyroxene spinifex lavas would have been less rich in qtz than the lavas themselves and less likely to crystallize orthopyroxene.

The explanation advanced by Kinzler and Grove for the crystallization of orthopyroxene was mixing between different magmas within Fred's Flow, or contamination of the magma through assimilation of wall or floor rocks. In the text of their paper, they suggested that an evolved liquid had accumulated beneath the upper crust of the flow, and then mixed with an underlying olivine normative liquid. Such mixing would produce a hybrid magma that could have crystallized the orthopyroxene found in the cumulate portion of the flow. In a note added in proof they pointed out that thermal erosion and assimilation of siliceous wall or floor rocks would displace the magma composition first into the pigeonite field, then into the orthopyroxene field, and they advocated this mechanism to explain the puzzling aspects of the crystallization of the flow. The flow-top lavas were interpreted as being uncontaminated, consistent with their pigeonite-free mineralogy, whereas the lavas in the interior were thought to be contaminated. This explanation may well be correct, and there exists independent chemical evidence of progressive contamination of the flow (Arndt and Nesbitt, 1984). Nonetheless, I am not convinced that all aspects of the compositions of spinifex lavas can be explained in this way. The specific problem stems from the pyroxene spinifex lavas that plot well to the left of the pigeonite–augite tie-line. As explained above, their position cannot be attributed to an excess of mafic minerals, because they do not contain olivine; nor to contamination, because inspection of the whole-rock analyses shows that these rocks have relatively low rather than high SiO_2 contents (Table 7.2; Figure 7.7). The aberrant position of these rocks could be caused in part by loss of SiO_2 during hydrothermal alteration or later metamorphism. This is indicated by the abnormally low SiO_2 contents of samples such as C11 and C55, as illustrated by their analyses listed in Table 7.2, and by the data plotted in Figure 7.7. The samples from Fred's Flow fall on a well-defined olivine control line in the MgO vs. Al_2O_3 diagram, but certain samples plot well below the olivine control line in the MgO vs. SiO_2 diagram. The original more SiO_2-rich compositions of these samples probably plotted well to the right of their

Table 7.2. *Compositions of rocks in the upper part of Fred's Flow (data from Arndt and Nesbitt (1982))*

						Sample					
	C11	C55	C67	C54	C53	C7	C72	C66	C73	C6	C4
Rock type	Flow-top breccia	Ol spinifex	Px spinifex	Px spinifex	Px spinifex	Px spinifex	Px spinifex	Gabbro	Gabbro	Gabbro	Gabbro
SiO_2	45.9	46.7	49.2	47.4	49.9	49.4	51.4	50.3	51.4	50.5	52.5
TiO_2	0.49	0.55	0.59	0.54	0.6	0.62	0.59	0.73	0.74	0.82	0.49
Al_2O_3	9.4	10.8	12.0	11.0	11.9	12.8	12.2	13.4	14.2	14.1	13.6
FeO (tot)	11.8	12.3	12.2	11.5	11.7	11.0	11.7	11.8	12.1	12.5	10.5
MnO	0.24	0.2	0.23	0.2	0.19	0.18	0.19	0.23	0.23	0.22	0.2
MgO	19.3	16.4	14.0	15.0	12.1	11.6	11.3	9.2	7.5	7.3	10.5
CaO	10.9	10.2	8.7	11.3	10.4	11.2	9.1	10.5	8.8	9.5	10.7
Na_2O	0.18	1.16	1.42	1.31	1.8	2.06	1.53	2.46	2.72	3.22	1.61
K_2O	0.03	0.06	0.12	0.04	0.15	0.08	0.45	0.35	0.99	0.27	0.01

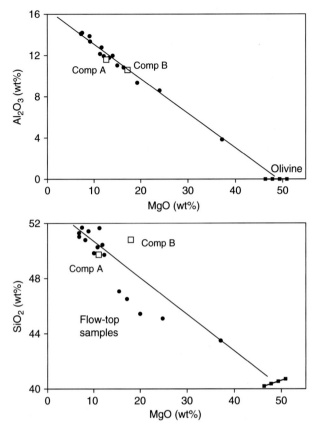

Fig. 7.7. Variation diagrams showing that the Al_2O_3 vs. MgO data of samples from the upper part of Fred's Flow plot on an olivine control line, whereas the SiO_2 vs. MgO data plot on a line at an angle to the control line. The samples from the flow top and upper spinifex zones appear to have lost some SiO_2. Data from Arndt (1977b) and Arndt and Nesbitt (1982).

actual positions in Figure 7.5, closer to that of composition B. This example illustrates the need to bear in mind that no komatiite is entirely fresh and the effects of element mobility during metamorphism must always be reckoned with when attempting to infer how these rocks evolved following their eruption or how they formed within the mantle.

The effect of crystallization in the thermal gradient of the flow top also contributes to the differences between textures and crystallization sequences in experiments and in natural lavas, as discussed in Section 7.7.

Parman *et al.* (1997, 2003) carried out a series of experiments on Barberton komatiites at pressures from 0.1 to 200 MPa, under anhydrous and H_2O-saturated conditions. These experiments were designed to investigate the conditions in which the magmas crystallized and in particular how the

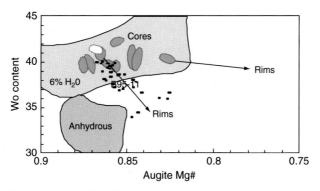

Fig. 7.8. Diagram of Mg# vs. Wo content from Parman *et al.* (1997) showing the fields representing the compositions of pyroxene that crystallized under anhydrous and hydrous conditions in experiments on compositions representing a Barberton komatiite.

compositions of pyroxenes changed as a function of these conditions. A particular goal of the project was to test the idea that the Barberton komatiites were water-rich magmas emplaced as an intrusive complex at depth in the lava pile.

The most important results of the study are plotted in Figure 7.8. In this diagram, the compositions of pyroxenes preserved in Barberton komatiites are compared with pyroxenes that crystallized in experiments (i) at 0.1 MPa pressure from anhydrous magma, and (ii) at pressures of 100 and 200 MPa from H_2O-saturated magmas. The pyroxenes analysed by Parman *et al.* in Barberton komatiites are high-Ca augites that plot in the field occupied by the experimental pyroxenes from hydrous magmas, and low-Ca pyroxene is absent in the komatiites (though present in a komatiitic basalt at a higher stratigraphic level). The coincidence between the compositions of the natural pyroxenes and those that formed in the hydrous 200 MPa experiments appeared to provide strong support for the hypothesis that the Barberton komatiites were hydrous and intrusive.

7.3 Partition coefficients from experimental studies

Mg–Fe partitioning between olivine and komatiitic liquid

Another product of the 1 atm experiments was a series of olivine–liquid partition coefficients, calculated from the major- and trace-element contents of olivines and glasses quenched from the experimental charges. Figure 7.9 is a compilation of Mg–Fe distribution coefficients ($K_D = (\text{MgO/FeO})_{\text{liq}} /(\text{MgO/FeO})_{\text{ol}}$, where FeO is total iron as FeO) determined from electron microprobe analyses of olivines and glasses in 1 atm experimental charges. In the figure, it can be seen that in mafic–ultramafic liquids the average K_D is 0.305 ± 0.02,

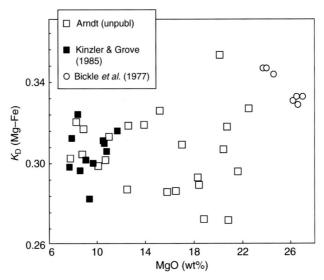

Fig. 7.9. Mg–Fe distribution coefficients measured in experiments on komatiites plotted against the MgO content of the komatiite liquid.

about the same as that measured by Roeder and Emslie (1970) (0.30 ± 0.03). There is some variation that can be related to the oxygen fugacity at which the experiments were conducted, which influences the amount of Fe^{3+} in the glass (see discussion in Bickle (1982)), but there was no indication that higher values of the distribution coefficient are more appropriate for komatiites. Higher values proposed by other authors (Smith *et al.*, 1980; Echeverría, 1982; Barnes, 1983) on the basis of the compositions of olivines and compositions of spinifex-textured samples can be attributed to an excess of olivine in spinifex lavas, or a failure to find the most magnesian olivines in the analysed samples.

At high pressures there does seem to be a trend to higher values. Bickle *et al.* (1977) analysed olivines in high-pressure charges run at conditions very close to the liquidus, and determined K_D values by assuming that these olivines were in equilibrium with liquid with the composition of the starting material. More recent data show a gradual increase in K_D to a value of about 0.36 at a pressure of about 6 GPa (Walter, 1998; Herzberg and O'Hara, 2002; Toplis, 2005).

Ni partitioning between olivine and komatiitic liquids

By measuring the Ni contents of olivines and host liquids in experiments on basalts and komatiites, the Ni partition coefficient (D_{Ni}) was found to be a function of the MgO content of the liquid (MgO_{liq}). This relationship was quantified in an equation similar to that proposed by Hart and Davis (1978) on the basis of their experiments on Fe-free haplobasaltic liquids:

Table 7.3. *Partition coefficients between basalt and komatiite liquids and mantle minerals*

	Olivine		Clinopyroxene		Garnet		Mg-perovskite
	Basalt	Komatiite	Basalt	Komatiite	Basalt	Komatiite	
Th	<0.0001	<0.0001	0.01		0.001		7–100
Nb	<0.0001	<0.0001	0.004–0.06		0.004–0.05		0.01–2.0
Zr	<0.001	0.01–0.02	0.1–0.2		0.2–2.0	0.5	9
Sc	0.1–0.4	0.3	1.0–3.0	0.15–0.2	1.0–3.0	1.0–1.6	0.1–5
V	0.2–0.9	0.12	0.7–5	0.05–0.5	1.5–4.0	1.2–2.6	
Cr	0.7–2.0	0.5	1–4	0.3	2	0.7–2.0	
Ni	10–20	2–5	4–10	3–5	0.1–0.2	0.15–0.20	
Sm	<0.001	0.001–0.01	0.2–0.4	0.01	0.2–1.1	0.16–0.2	2–8
Yb	0.016–0.05	0.01–0.1	0.2–0.6	0.1	4 – 8	0.9–1.1	0.8–1.4

Sources of data: komatiites – Beattie (1993), Bédard (2005), Canil (1997), Li and Lee (2004), Kato *et al.* (1988), Ohtani *et al.* (1989), Lesher and Arndt (1995); basalts – GERM database.

$$D_{Ni} = 124.3/MgO_{liq} - 0.9,$$

where MgO_{liq} is the MgO content of the liquid. As mentioned in Chapters 4 and 5, a similar equation ($D_{Ni} = 124/MgO - 2.2$) can be used to model the Ni contents of natural olivines and host liquids in a komatiite flow from Alexo, Canada. Partition coefficients predicted from both equations are somewhat lower than those obtained using the equation of Hart and Davis (1978), and this led to several years of debate in the late 1970s. The conflict seems now to have been resolved with the publication of results of Kinzler *et al.* (1990), who introduced a term representing the forsterite content of the olivine composition or the more recent work of Beattie (1993) whose thermodynamic approach successfully reproduces the variation of Ni contents of olivine and komatiites from localities such as Belingwe and Alexo (Chapters 4 and 5).

Other partitioning data

Table 7.3 and Figure 7.10 summarize partition coefficients for other major and trace elements obtained from experiments on komatiites. Particularly interesting is the work done by Canil (1997) and Li and Lee (2004) on the partitioning of V and Sc between olivine and liquid. The partitioning of V is redox-sensitive but that of Sc is not, and the V–Sc ratio in basalts and

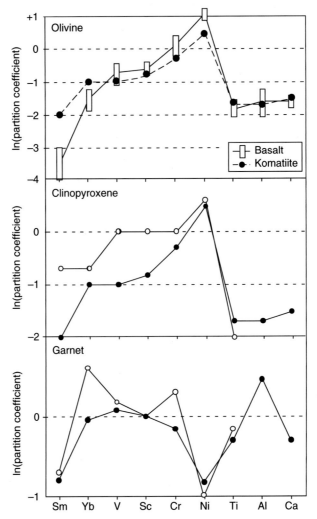

Fig. 7.10. Mineral–liquid partition coefficients of major and trace elements estimated for olivine and clinopyroxene at low pressure and for garnet at high pressure. The data for komatiites are from Bédard (2005) and references listed in this work (Table 7.3), and from Arndt (unpublished results). The data for basalts are taken from the GERM website.

komatiites is a sensitive indicator of the oxygen fugacity in the mantle source of the lavas. Using the V and Sc contents of mafic and ultramafic magmas, Canil (1997) and Li and Lee (2004) showed that both modern and Archean examples were derived from a source with an oxygen fugacity slightly below the fayalite–quartz–magnetite buffer. The implication is that the oxygen fugacity of the mantle has remained relatively constant through geological time.

The partitioning of the PGE between mafic–ultramafic liquids and mantle minerals is relevant to the formation of Ni–Cu–PGE ore deposits. In this case, the concentrations in the liquid of these chalcophile elements is strongly influenced by the behaviour of sulfide during partial melting. Experimental studies (e.g. Ballhaus *et al.* (1994), Peach *et al.* (1994), Crocket (2002)) have shown that sulfide–silicate liquid partition coefficients are $\sim 10^2 - 10^3$ for Ni and Cu and $\sim 10^4 - 10^5$ for the PGE. At oxygen fugacities near the fayalite–magnetite–quartz buffer, sulfide remains in the residue at low to moderate degrees of partial melting, and because high proportions of the Ni, Cu and PGE are retained in this phase, silicate melts formed under these conditions have low concentrations of these elements (Keays, 1982, 1995; Naldrett and Barnes, 1986). Only when the degree of melting is high, greater than 20–30%, does the sulfide totally enter the silicate melt and under these conditions the contents of Ni, Cu and PGE are higher. The elevated chalcophile element content of komatiites is explained in this way. The role of sulfide during the crystallization of komatiite and the formation of sulfide deposits is discussed in Chapter 10.

7.4 Experiments at moderate pressures (1–4 GPa)

Green *et al.* (1974) conducted the first experimental study of a komatiite, and it could be argued that they discovered all we really need to know about the way these magmas formed. They conducted their experiments at low to moderate pressures, from 0.02 to 0.25 GPa, in a piston-cylinder apparatus, and they worked with sample 49J, the now-famous spinifex-textured komatiite from the Barberton greenstone belt. This sample, which was described and analysed by Viljoen and Viljoen (1969c) and Nesbitt and Sun (1976), contains 32% MgO (recalculated anhydrous with $Fe_2O_3 = 1\%$), a value a little higher than that of the probable parental liquid of these komatiites (see Chapter 12). The sample probably contains a small amount of excess olivine that accumulated during growth of the spinifex texture. In subsequent studies, Green (1981) augmented his earlier results with data from three other komatiitic samples, Arndt (1976a) undertook a similar study of the komatiite SA 3091 from Munro Township, and Bickle *et al.* (1977) reported experiments on four komatiite samples from the Belingwe belt in Zimbabwe. The liquidus phase relations of samples 49J and SA 3091 are presented in Figure 7.1.

Green *et al.* (1974) established the phase relations at low to moderate pressures for the komatiite 49J: olivine, the liquidus phase, is joined by clinopyroxene at low pressure (< 1 GPa) and by orthopyroxene at higher pressure. The 1 atm anhydrous liquidus temperature of this komatiite is extremely high, between 1650 and 1700 °C, some 400 °C higher than the extrusion temperatures of modern basalts as Green *et al.* noted. They were aware that Brooks and Hart (1974) had proposed that the liquidus of komatiite could have been lowered by the presence of water, and they addressed the question by conducting several experiments on charges containing 5 or 10% H_2O. The water does indeed depress the liquidus, as shown in Figure 7.1(a), but only at high pressure. Green *et al.* (1974) developed at this early stage of komatiite research several crucial arguments against the wet komatiite hypothesis: since the solubility of water in silicate liquid is negligible at atmospheric pressure, a hydrous komatiite magma should degas as it approaches the surface, a process that would lead to violent vesiculation and/or quench crystallization. If a komatiite can be shown to have erupted and flowed at the surface as a mobile, phenocryst-poor liquid, the parental magma was most probably anhydrous. Green *et al.* (1974) concluded that the eruption temperature of the komatiite 49J was 1650 ± 20 °C and that its water content was less than 0.2%.

For all the more magnesian komatiite samples (MgO > 27%) that had been studied at the time, olivine is the liquidus phase at the highest pressures achieved in the experiments (4 GPa), but in certain less magnesian samples (e.g. SA 3091 and 422/95 from Munro and NG 7638 from Belingwe), pyroxene appeared at the liquidus at pressures above 3–3.5 GPa. Green *et al.* (1974) and Bickle *et al.* (1977) identified orthopyroxene in their experimental charges, but Arndt (1976a) found only a subcalcic augite (probably metastable). In the two samples with still lower MgO contents (NG 7621 from Belingwe and 331/338 from Pilbara), orthopyroxene appeared at still lower pressures. Arndt (1976a) identified garnet slightly above the solidus in the Munro komatiite at 4 GPa pressure, a phase not reported in the other papers.

7.5 High-pressure experiments on komatiites

Two approaches were initially used to study the high-pressure (> 4 GPa) phase relations of komatiites. Shock-wave techniques employed by Tom Ahrens' group at Caltech (Rigden *et al.*, 1984; Miller *et al.*, 1991) produced valuable information about the physical properties of komatiite liquids. Most experiments have been carried out, however, using multi-anvil presses of the type developed in the 1980s by Japanese workers (Ohtani, 1984; Herzberg and Ohtani, 1988; Kato *et al.*, 1988; Ohtani *et al.*, 1988; Wei *et al.*, 1990; Herzberg, 1992; Takahashi *et al.*, 1993; Ohtani *et al.*, 1996; Ohtani *et al.*, 1997; Herzberg and O'Hara, 2002). Two important types of information emerged from these experiments: (1) constraints on the physical properties of high-pressure silicate melts, including the probability that at conditions near the transition zone, komatiitic liquid is as dense or more dense than common mantle minerals such as olivine and pyroxene, and (2) information on the phase relations of mantle peridotite and komatiite at the pressures that prevail in deeper parts of the upper mantle (from 4 to at least 20 GPa).

Densities of komatiite liquids at high pressures

The concept that komatiitic liquids may be denser than mantle minerals such as olivine or pyroxene was introduced by Stolper *et al.* (1981) and Nisbet (1982) on the basis of experiments that had been carried out earlier by Fujii and Kushiro (1977), Olinger (1977) and Hazen (1977). Fujii and Kushiro had shown that the density of basaltic liquid increases significantly with increasing pressure, from $2.7\,g\,cm^{-3}$ at 1 atm to $2.8\,g\,cm^{-3}$ at about 8 GPa, a change that was attributed to the compressibility of the liquid. Stolper *et al.* (1981) recognized that this compressibility was significantly larger than that of olivine, and calculated, by estimating the elastic properties of the silicate liquids, that at a pressure of about 8 GPa, the density of a mantle melt would be the same as that of the olivine and pyroxene in the mantle. Nisbet (1982) independently came to a similar conclusion. Confirmation of the Stolper–Nisbet hypothesis has come from two sources. Rigden *et al.* (1984, 1988) measured the densities of a model basaltic liquid (an anorthite–diopside mixture) at pressures up to 23 GPa using shock-wave techniques, and they proposed that basic to ultrabasic melts become denser than mantle olivine and pyroxene at pressures between 6 and 10 GPa. Miller *et al.* (1991) conducted similar experiments using a natural komatiite sample. The conclusion emerging from these experiments was that the density of komatiite liquid matches that of olivine at pressures above 8 GPa.

This is the same as the value obtained by Agee and Walker (1988a, b), Agee (1998), Ohtani (1984) and Ohtani *et al.* (1995, 1998), who undertook static compression experiments using piston-cylinder techniques. In these experiments the densities of komatiite liquids containing between 28 and 32% MgO were measured as a function of pressure by observing whether spherules of olivine of known composition and calculable density floated or sank when the charges were taken to temperatures near their liquidi. Figure 7.11, from Agee (1998), summarizes the most important results. The diagram shows that the density of the compressible liquid increases more rapidly with increasing pressure than that of the solid phases, olivine and garnet. Olivine with a composition in equilibrium with komatiite liquid (Fo_{93}) has a density equal to that of the liquid at a pressure of about 8 GPa, which corresponds to a mantle depth of 245 km. At greater pressures or depths the olivine floats. For

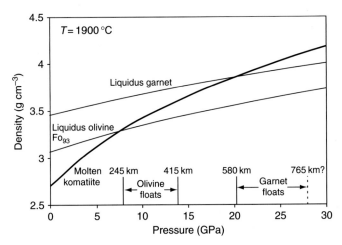

Fig. 7.11. Experimental results showing that at pressures greater than about 8 GPa the density of komatiite liquid is greater than that of olivine, and at pressures greater than about 20 GPa, its density is greater than that of garnet (from Agee (1998)).

the liquidus garnet, the corresponding pressure and depth for the density crossover are 20 GPa and 580 km, respectively.

At pressures between 8 and 20 GPa, therefore, the density of komatiite liquid is greater than that of olivine or clino- or orthopyroxene (whose densities are similar to that of olivine), but still less than that of the bulk peridotite, and significantly less than that of garnet or majorite. These density relationships have been the stimulus for numerous models for the petrogenesis of komatiites and the differentiation of the upper and lower mantle (Ohtani, 1984; Rigden *et al.*, 1988; Miller *et al.*, 1991; Herzberg, 1992; Walter, 1998), as is discussed in Chapter 13.

Phase relations at high pressures

The results of ultrahigh-pressure melting experiments on mantle peridotite and komatiite, which had such a profound influence on our concepts of how these magmas may have formed, were foreshadowed by O'Hara (1968) in his famous paper on basalt genesis, and by Herzberg (1983) in a perceptive paper on the phase relations of mantle peridotite. Most relevant in O'Hara's work is his recognition that as pressure increases, the liquids produced by invariant melting of mantle peridotite become progressively richer in the olivine component. Increasing pressure causes the phase boundaries between olivine and pyroxene or olivine and the aluminous mineral to migrate towards the olivine composition: the primary phase field of olivine shrinks, more olivine enters the liquid formed by low-degree melting, and the liquidus mineral changes from olivine to pyroxene, and, in some basalt liquids at pressures greater than about 3 GPa, to garnet.

Herzberg (1983) drew three important conclusions from his review of experiments carried out on analogues of mantle peridotite, conclusions that have been largely confirmed by later studies (e.g. Herzberg (1992), Walter (1998), Herzberg and O'Hara (2002)). First, he predicted that the liquidus and solidus of mantle peridotite strongly converge at pressures in the range 6.5–15 GPa. Second, he suggested that the liquidus mineral of mantle peridotite changes from olivine at low pressure, to pyroxene at intermediate temperature, and to majorite (a garnet–pyroxene solid solution) at pressures greater than about 10 GPa. Third, he commented on the likelihood that olivine would float in komatiite liquids at high pressures, as discussed in the preceding subsection. The significance of his conclusions is discussed below, after the results of more recent experiments have been presented. However, before doing so, it is useful to explain his reasoning.

The basis of the first conclusion was the work of Davis and England (1964) and Ohtani and Kumazawa (1981), who determined the fusion curve of forsterite at pressures up to 15 GPa. This curve is considerably steeper than the solidus of mantle peridotite, as determined from other experimental investigations. Because olivine is the major component and the liquidus phase of mantle peridotite at pressures up to 15 GPa, Herzberg assumed that the mantle liquidus would be parallel to the forsterite fusion curve, and that the liquidus and solidus would converge. Subsequent experimental studies by Ohtani (1984), Takahashi and Scarfe (1985), Miller *et al.* (1991) and Walter (1998) support Herzberg's contentions. Shown in Figure 7.12 are phase diagrams for mantle peridotite, constructed from the high-pressure experiments of Wei *et al.* (1990), Herzberg (1992) and Walter (1998). Both liquidus and solidus are curved and converge steadily with increasing pressure, reaching a minimum separation between 13 and 16 GPa. Takahashi and Scarfe (1985), for example, determined that the liquidus and solidus of the peridotite sample KLB-1 are less than 100 °C apart at 13 GPa, compared with $\sim$ 500 °C at 1 atm pressure. The solidus and liquidus curve because both solid and liquid are compressible; they converge because the liquid is more compressible than the solid.

Herzberg (1983) investigated changes in the high-pressure liquidus mineralogy of peridotite using existing low-to-moderate-pressure experimental data on basalts and peridotites. Subsequent experiments by Takahashi (1986), Ohtani *et al.* (1986), Takahashi and Scarfe (1985), Walter (1998) and Gudfinnsson and Presnall (2000), as well as those by Herzberg and coworkers (Herzberg and Ohtani, 1988; Herzberg, 1992, 1995, 1999; Herzberg and O'Hara, 2002) demonstrated that in peridotites with compositions like those

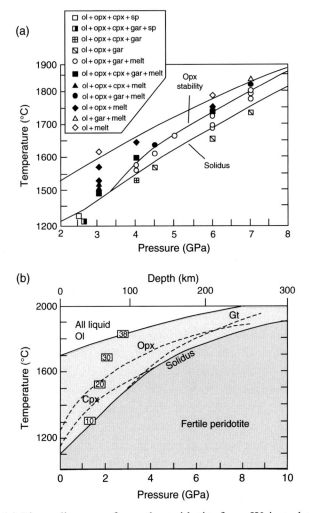

Fig. 7.12. (a) Phase diagram of mantle peridotite from Wei *et al.* (1990); (b) phase diagram of mantle peridotites from Herzberg (1995). The numbers in rectangles indicate the MgO contents in weight percent of partial melts of the peridotites.

inferred for the upper mantle, majorite replaces olivine at the liquidus at pressures between 13 and 16 GPa, and that $MgSiO_3$-perovskite replaces majorite at around 22 GPa, as shown in the phase diagrams in Figures 7.12 and 7.13. Similar diagrams have been produced by Takahashi and coinvestigators (Takahashi and Kushiro, 1983; Takahashi and Scarfe, 1985; Takahashi 1986; Takahashi *et al.*, 1993), who worked on two natural peridotites, and by Gudfinnsson and Presnall (2000), who undertook experiments on the Ca–Mg–Al–Si (CMAS) system. Another series of experiments was carried

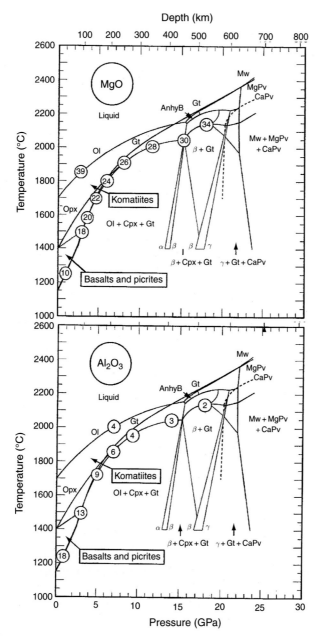

Fig. 7.13. Phase diagrams of mantle peridotite showing the MgO and Al$_2$O$_3$ contents of liquids produced by melting close to the solidus. From Herzberg (1995).

out on hydrous peridotites to establish the phase relations of mantle material containing a small amount of water (Hirose and Kawamoto, 1995; Kawamoto and Holloway, 1997; Inoue *et al.*, 2000; Asahara and Ohtani, 2001; Litasov and Ohtani, 2002).

The phase relations of liquids with the compositions of komatiite were investigated at pressures from 4 to 12 GPa by Herzberg and coworkers (Herzberg and O'Hara, 1985; Herzberg and Ohtani, 1988; Herzberg, 1992), and by Wei *et al.* (1990). In an Al-undepleted komatiite containing 25% MgO (starting material M620) garnet replaced olivine on the liquidus at about 5 GPa pressure whereas in a more magnesian Al-depleted komatiite, HSS-15, which contains 31.5% MgO, garnet appeared on the liquidus only at about 11 GPa.

Takahashi (1986) measured the compositions of liquids produced by melting close to the solidus of mantle peridotite. A summary of these compositions, complemented by more recent results by Herzberg (1992), Herzberg and O'Hara (2002) and Walter (1998) is given in Table 7.4 and Figure 7.13. At low pressure, the liquid is basaltic, containing 8–10% MgO, but with increasing pressure it becomes ultramafic, reaching 21% MgO at 5 GPa and 24% MgO at 7 GPa. For higher degrees of melting corresponding to the point at which orthopyroxene is totally absorbed (the lower curve in Figure 7.14), the liquid produced at 7 GPa and 50% partial melting contains about 30% MgO. This is similar to the MgO contents of komatiitic liquids.

Figures 7.13 and 7.14 show how the MgO content varies as a function of pressure, temperature and degree of melting. In these diagrams, it is seen that the MgO content increases, and the Al_2O_3 content decreases, with increase in the pressure (or depth) or the degree of partial melting. The progression towards more magnesian compositions is due to two factors: first, the decreasing temperature interval between liquidus and solidus, which means that temperatures not far from the solidus correspond to relatively high degrees of melting; second, the changing phase relations, particularly the increase in the proportion of olivine in the liquid, as pressure increases. The latter factor becomes most pronounced at pressures above 13 GPa, where garnet replaces olivine as the liquidus phase. At 20 GPa and 2250 °C, for example, all olivine has entered the melt, and a highly ultramafic liquid with high MgO ($> 35\%$) and low Al_2O_3 ($\sim 1\%$) exists in equilibrium with garnet (Figure 7.13). Such high temperatures are unlikely at such great depths, but even near-solidus liquids produced by melting at pressures between 15 and 20 GPa will have MgO $> 30\%$ and $Al_2O_3 < 2\%$.

From these results, it might appear that the formation of komatiite is straightforward: all that is required is a mantle source that is sufficiently hot

Table 7.4. Liquids produced during Walter's (1998) experiments

Run:	30.12	30.11	40.06	50.01	60.01	70.07	70.08	70.06
Pressure (GPa)	3	3	4	5	6	7	7	7
Temperature (°C)	1580	1630	1590	1680	1710	1790	1850	1950
Percentage melting	14	53	9	10	11	16	66	86
Phase assemblage	ol+opx+cpx+melt	ol+melt	ol+cpx+gar+melt	ol+cpx+gar+melt	ol+cpx+gar+melt	ol+cpx+gar+melt	ol+gar+melt	ol+melt
SiO_2	46.2	48.0	46.4	44.8	45.0	45.2	47.3	45.4
TiO_2	0.91	0.39	1.45	1.26	1.01	1.23	0.28	0.21
Al_2O_3	13.3	9.5	9.8	7.2	6.4	5.1	6.5	5.7
FeO	9.55	9.2	10.7	11.9	12.6	12.5	8.6	8.2
MnO	0.18	0.17	0.20	0.20	0.21	0.23	0.18	0.16
MgO	16.9	23.9	18.6	22.3	23.3	23.9	30.3	34.9
CaO	10.7	7.7	10.3	9.5	9.1	8.6	5.6	4.3
Na_2O	0.96	0.52	0.93	0.86	0.80	1.04	0.41	0.28
K_2O	0.56	0.22	0.83	0.60	0.36	0.91	0.11	0.13

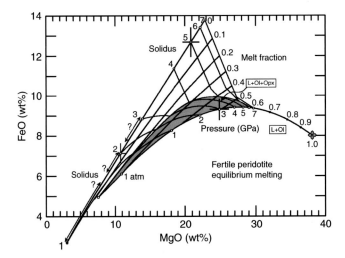

Fig. 7.14. Diagram of MgO vs. FeO illustrating how the compositions of partial melts of mantle peridotites vary as a function of pressure and degree of partial melting. Melts produced at the solidus lie along the upper straight line and those that form in equilibrium with olivine alone plot along the lower curved line. From Herzberg and O'Hara (2002).

that it partially melts at depths exceeding about 200 km. In such a model, liquids with the compositions of Al-depleted komatiites form at great depths under conditions in which garnet is the liquidus phase and liquids with the compositions of Al-undepleted komatiites form at shallower depths, probably by higher degrees of partial melting. As will be evident to anyone following the literature on komatiites, this type of model is not universally accepted. Many authors argue that the temperatures required to form komatiites under these conditions are unreasonably high, and they have appealed to melting of a hydrous source. Another problem has been raised by Walter (1998), who noted that the $CaO-Al_2O_3$ ratio of the experimental melts that he produced by partial melting of peridotite with the chondritic ratio (expected for material from the upper mantle) were lower than those measured in Al-undepleted komatiite. These issues are discussed in Chapters 11 and 13.

7.6 Trace-element partitioning at high pressures

Several authors (Kato *et al.*, 1988; Ohtani *et al.*, 1989) have measured the partitioning of trace elements between silicate liquids and high-pressure mineral phases. Kato *et al.* (1988) showed that trace elements, the REE in particular, partition between majorite and liquid in a manner similar to that for pyrope garnet, in that the HREE are highly compatible and the LREE incompatible. Measured partition coefficients tend, however, to be less extreme for majorite: the incompatible elements are more compatible (D_{La} majorite $> D_{La}$ pyrope); the compatible elements less compatible (D_{Yb} majorite $< D_{Yb}$ pyrope). Majorite coexists with silicate liquids at high pressures and at temperatures far higher than those at which pyrope garnet is stable. The relatively subdued partition coefficients of majorite may be a temperature effect: Draper *et al.* (2003) have shown, for example, that incompatible elements partition far less strongly into silicate liquids under the high-temperatures at which komatiites form.

Kato *et al.* (1988a,b) determined the partitioning of a range of trace elements between silicate liquids and Ca and Mg silicate perovskite. They found that certain elements, such as Nb, Zr and Th, which are incompatible with other silicates, partition strongly into Mg or Ca perovskite. Table 7.3 lists the partition coefficients measured by Kato *et al.* (1988), and the data are summarized in Figure 7.10.

7.7 Dynamic cooling experiments

The discovery of komatiites by Morris and Richard Viljoen coincided with the Apollo missions to the Moon, and the first experimental studies of spinifex textures were undertaken in parallel with studies of textures in lunar basalts. The important papers by Lofgren and Donaldson (1975) and Donaldson (1976) focussed on the morphologies of olivine and pyroxene and how these characteristics change as a function of cooling rate or the degree of super-cooling. These authors ran experiments in which natural olivine basalts or synthetic analogues of these rocks were cooled rapidly at rates varying from $1 \, °C \, h^{-1}$ to at least $1500 \, °C \, h^{-1}$, values that correspond to cooling rates in the upper parts of lava flows. Donaldson's scheme, shown in Figure 4.1, relates the habit of olivine crystals to the cooling rate. This scheme became the basis for the interpretation of textures in the upper parts of komatiite flows, and the notion that skeletal habits are indicative of rapid cooling influenced strongly, and perhaps incorrectly, much of the early thinking about the origin of spinifex textures. Other experimenters (Lofgren *et al.*, 1974; Grove and Bence, 1979; Kirkpatrick *et al.*, 1983) investigated the morphologies of pyroxene and plagioclase as a function of cooling rate. This information has proved useful during the interpretation of the habits of these minerals in komatiitic basalts.

Another outcome of the rapid-cooling experiments was the quantification of the delay in nucleation of olivine, pyroxene and plagioclase as a function of cooling rate or degree of supercooling. In general it was found that elevated cooling rates strongly inhibited the nucleation of plagioclase, had variable effects of high-Ca and low-Ca pyroxenes, but little influence on the nucleation of olivine. Figure 7.6 illustrates this behaviour.

Donaldson (1982) in his important chapter in the first komatiite book (Arndt and Nisbet, 1982a), not only summarized contemporaneous thinking about the mineralogy of komatiites but also discussed the dynamic cooling experiments and how they could be applied to help explain textures in komatiites. In his chapter he formulated what has come to be known as the 'spinifex paradox' – the observation that crystals deep in the crust of komatiite flows, at levels where cooling rates can only have been moderate, have habits like those that grow only at high cooling rates in dynamic cooling experiments. This issue is discussed in Chapters 10 and 11.

Then followed a gap of almost 15 years during which very few dynamic cooling experiments on komatiites were undertaken. Kinzler and Grove (1985) included the results of dynamic cooling experiments in their important study of Fred's Flow komatiites (Figure 7.6) and Ginibre *et al.* (1997) published, but only in abstract form, the results of a study designed to test whether komatiites from Gorgona Island were superheated when they were erupted (the conclusion was that they were not). But the question of how rapid cooling and/or rapid crystallization affected crystallization sequences and mineral compositions in komatiites only re-emerged when de Wit *et al.* (1987) discussed the spinifex paradox in their account of the field characteristics of Barberton komatiites and when Parman *et al.* (1997) included dynamic cooling experiments in the paper in which they investigated the compositions of pyroxenes in these rocks. Parman *et al.'s* work in particular refocussed attention on the origin of spinifex texture, attention that led eventually to new experimental work.

Faure *et al.* (2003) published a reappraisal of Donaldson's (1976, 1982) olivine morphology scheme and showed that certain mineral habits described by Donaldson were merely parts of complex dendritic crystals, sectioned at different orientations. Faure *et al.* (2006) then used a novel experimental setup, illustrated in Figure 7.15, that allowed them to investigate crystallization during rapid cooling in a thermal gradient, mimicking the conditions that exist in the upper crust of a komatiite flow. Under these conditions, olivine and pyroxene crystals acquire a preferred orientation parallel to the temperature gradient, and even at relatively low cooling rates their habits were markedly skeletal. In other words, they were able to reproduce spinifex textures. Finally Bouquain *et al.* (2008) used the same experimental procedure to investigate how the compositions and crystallization sequences of pyroxenes are related to the conditions in which they form. Bouquain *et al.* showed that the peculiar compositions of augite in Barberton komatiites are a consequence

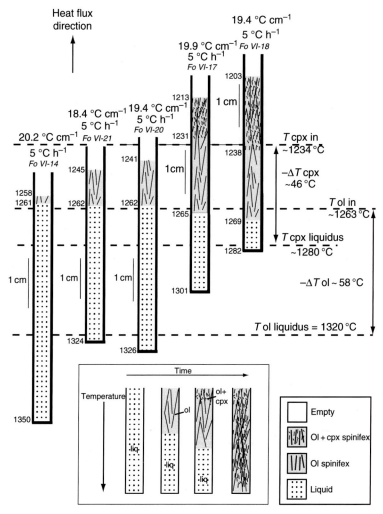

Fig. 7.15. Diagram summarizing the experimental results of Faure *et al.* (2006), who showed that spinifex textures are reproduced by rapid cooling in a thermal gradient.

of the unusual high CaO/Al_2O_3 compositions of these Al-depleted komatiites and not an indication of crystallization in water-rich intrusions.

7.8 Huppert and Sparks' experiments

Huppert and Sparks (1985b) and Turner *et al.* (1986) described a series of experiments conducted to help understand how komatiite behaves during eruption and how it crystallizes following emplacement. In one type of experiment they allowed hot water to flow over a wax substrate to simulate thermal erosion of floor rocks by turbulently flowing komatiite (Huppert *et al.*, 1984). They developed their ideas using numerical modelling and concluded that the most magnesian komatiites would flow at such a high rate that they would rapidly assimilate the rocks over which they flow. They proposed that the assimilation of thermally eroded sulfur-bearing wall or floor rocks triggered the formation of the Kambalda Ni-sulfide deposits. Their papers elicited considerable comment and discussion, as explained in Chapters 9 and 10.

In a second series of experiments, Turner *et al.* (1986) observed the growth of crystals downwards into a cooling salt solution, an experimental setup designed to reproduce conditions in the growing crust of a komatiite flow. They produced crystals that superficially resembled the olivine and pyroxene in spinifex textures and they developed a new model for the growth of spinifex in which they attributed the form and distribution of the constituent crystals to constrained growth downwards into a layer of expelled fluid. These ideas formed the basis of all subsequent models for the formation of spinifex texture, most notably those of Arndt (1986b), Renner *et al.* (1994) and Faure *et al.* (2006).

7.9 Summary and implications for komatiite petrogenesis

The relevance of these findings to broader questions of komatiite petrogenesis will be discussed more fully in following chapters, and here only general conclusions are listed.

(1) At low pressures the temperature interval between the liquidus and solidus of komatiite is very large, up to 500 °C in the case of the most magnesian magmas. Throughout most of this interval only olivine and minor chrome spinel crystallize. Other minerals – pigeonite, augite and plagioclase – appear at temperatures within 100 °C of the solidus. Because of this, komatiite liquids contain only sparse concentrations of crystals for much of their cooling intervals. A komatiite cooled by 200 °C below its liquidus is still highly mobile; a basalt 200 °C below its liquidus is almost totally solid.

(2) Discrepancies exist between the equilibrium phase relations determined by experiment and crystallization histories inferred from mineralogy and textures of komatiitic lavas These are explained by a combination of factors: pyroxene accumulation during the growth of spinifex texture, depression of pigeonite and plagioclase crystallization by rapid cooling, selective contamination of komatiitic liquids, and post-emplacement alteration.

(3) The eruption temperatures of komatiites may have been as high as 1600–1700 °C, some 400–500 °C higher than those of basalts.

(4) At pressures greater than about 8 GPa, corresponding to depths of about 240 km in the mantle, the density of komatiite liquid is similar to that of olivine and

pyroxene, although still less than that of garnet. Under these conditions, the komatiite liquid cannot escape upward from an olivine-rich matrix, although the denser garnet may segregate downwards. If the matrix + liquid continues to move upward (as in an upwelling limb of a convection cell or in a rising plume), then eventually the neutral density threshold will be broached and the magma will escape towards the surface.

(5) With increasing pressure, liquids produced by melting close to the peridotite solidus become increasingly magnesian. At pressures around 3 GPa, the liquid has a highly mafic composition with about 15% MgO; at 7 GPa the liquid is komatiitic and contains 23% MgO and at 15 GPa it is ultramafic with about 30% MgO.

(6) At pressures above 7 GPa, majoritic garnet replaces olivine as the liquidus phase of mantle peridotite. Melts forming under such conditions coexist with garnet and acquire very high MgO, high CaO/Al_2O_3 and low Al_2O_3/TiO_2. Barberton-type, Al-depleted komatiite forms under these conditions.

(7) Munro-type, Al-undepleted komatiite probably forms through higher degrees of melting at shallower depths.

(8) Dynamic cooling experiments have successfully reproduced spinifex and other textures characteristic of komatiite.

Part II

Interpretation – the manner of emplacement, the origin and the tectonic setting of komatiites

8

Physical properties of komatiites

8.1 Introduction

Komatiitic liquids have extreme chemical compositions, being very rich in MgO and poor in SiO_2, Al_2O_3 and alkalis. Given these compositions, their physical properties must have been very different from those of basalts and other more familiar types of magma. Although it should be possible to measure these properties in the laboratory on molten natural komatiites, there will always remain some uncertainty as to whether the sample chosen accurately represents a real komatiite liquid: the familiar problems of volatile content, excess olivine component and the effects of post-emplacement alteration will always introduce an element of doubt. To my knowledge little direct work of this type has been done. As discussed in Chapter 7, a series of experiments has been conducted to measure the density of komatiite at high pressure in order to investigate whether the density of ultramafic liquid exceeds that of common mantle minerals. Liquidus temperatures at atmospheric pressure, and the relationship between temperature and crystal content, have also been established experimentally, but no systematic measurements of the density and viscosity of komatiitic liquids at low pressure appear to have been made.

Certain rheological properties can be inferred from the structures and textures of natural komatiites, but, as will become clear in Chapter 9, the interpretation of such textures is often ambiguous, and in many cases the interpretation hinges on assumptions about the nature of the komatiite liquids themselves. Most of our ideas about the physical properties of these liquids therefore come from calculations based on models of the properties of silicate liquids (Shaw, 1969; Bottinga and Weill, 1970, 1972; Shaw, 1972; Mo *et al.*, 1982; Ryerson *et al.*, 1988), augmented by experience gained from experimental and field studies of basaltic and less mafic lavas.

The physical properties of komatiite were discussed by Viljoen and Viljoen (1969b) and Nesbitt (1971), who realized that the unusually high temperatures and very low viscosities of komatiite would strongly influence the way these magmas erupted and crystallized. Lajoie and Gélinas (1978) estimated the physical and rheological properties of komatiite liquids and used them to explain structures in spinifex-textured flows, and Usselmann *et al.* (1979) developed a quantitative model of the cooling and crystallization of a Ni sulfide deposit at the base of a komatiite flow. The need to explain the formation of magmatic ores prompted further discussion of the physical properties of komatiites by, e.g., Lesher *et al.* (1984), Groves *et al.* (1986), Lesher and Groves (1986), Evans *et al.* (1989) and Frost and Groves (1989a,b) and more general aspects of komatiite eruption were considered by Barnes *et al.* (1988) and Hill *et al.* (1987, 1995).

The most important papers in the field were written by Hubert Huppert, Steve Sparks and coworkers (Huppert *et al.*, 1984; Huppert and Sparks, 1985a,b; Turner *et al.*, 1986; Huppert and Sparks, 1989) in the course of their investigations of the eruption and crystallization of komatiite. These papers provide a comprehensive account of the probable physical properties of ultramafic liquids and the fluid dynamics of flowing and crystallizing komatiite lavas, and, together with the work of Williams *et al.* (2002), who refined the modelling of Huppert and Sparks, they are the source of most of the information in this chapter. Further ideas have been drawn from Hill *et al.*'s (1995) account of the physical volcanology of komatiites, and by the detailed consideration of the subject by Rebecca Renner in her Ph.D. thesis (Renner, 1989).

Huppert and Sparks' ideas were developed before it was realized that many or most komatiite flows were emplaced through the inflation process. In their modelling Huppert and Sparks assumed that the komatiite lava flowed in open channels and that the solid crust at the top of the flow was thin and transitory, being destroyed by internal flowage or convection as soon as it formed. Under these conditions the flowage of lava would have been rapid and turbulent, and when the flow ponded internal convection would have been vigorous. If the lava is emplaced by injection beneath a thick, insulating crust, however, the internal flow regime is quite different from that in an open channel, as discussed in Chapter 9.

8.2 Temperature

The probable eruption temperatures of komatiite lavas can be predicted from liquidus temperatures determined in low-pressure experimental studies. Most important is the work of David Green (1974, 1981), who predicted that the eruption temperature of the Barberton komatiite 49J was slightly above 1650 °C. As described in Chapter 7, the relationship between the 1 atm liquidus temperature (T_{liq}) and the MgO content of anhydrous komatiitic liquid (MgO_{liq}, in weight per cent) is given by the expression $T_{liq} = MgO_{liq} * 20 + 1000$ (Nisbet, 1982).

To estimate the temperature of erupting komatiite lavas requires some knowledge of the composition of the liquid, its volatile content and of the kinetics of degassing. These issues are discussed at length in Chapters 11 and 12, but can be summarized as follows. There is good evidence for the existence of komatiite liquids that had an MgO content of around 30% and for these liquids the 1 atm dry liquidus temperature would be about 1600 °C. In addition there is some evidence for even more magnesian liquids with up to 34% MgO and a 1 atm dry liquidus temperature of close to 1700 °C. These lavas erupted under submarine conditions and, if the water depth were 1–2 km, the liquids could have retained a small quantity of water. In the case of the sample 49J, Green *et al.* (1974) assumed that eruption took place at a water pressure of 0.1 GPa (on the floor of an ~1 km deep ocean) and that the komatiite contained about 0.5% H_2O; the presence of this small amount of water depressed the liquidus by about 50 °C.

There is abundant evidence that these ultramafic magmas flowed out onto the sea floor as mobile, phenocryst-poor lava. Had the magma contained higher water contents, the water would have degassed as the pressure decreased as the magma approached the surface. Loss of water would then have caused the magma to crystallize olivine in a proportion related to its original water content (Green *et al.*, 1974; Arndt *et al.*, 1998a). Under equilibrium conditions, an originally water-rich magma, once degassed, would crystallize partially to become highly porphyritic lava with no resemblance to the phenocryst-poor compositions preserved in the chilled margins of komatiite flows. Parman *et al.* (2004) and Dann (2001) have proposed that kinetic factors may have prevented degassing, but this seems improbable for low-viscosity magmas like komatiites. It seems likely, therefore, that komatiites erupted at a range of high temperatures, up to a maximum near 1700 °C.

At higher pressures the liquidus temperatures are greater. At 3 GPa, for example, temperatures are ~300 °C greater than at atmospheric pressure, corresponding to a slope of 0.3 °C per GPa on the liquidus, as shown in Figure 7.4. At high pressures, volatiles are soluble in the silicate liquid and their presence depresses the liquidus by roughly 10 °C for each per cent of dissolved H_2O or CO_2 (Shaw, 1972). The experimental results of Asahara *et al.*

(1998) show that in a liquid produced by 40–60% partial melting of mantle peridotites at 6.5 GPa, the presence of 1% water in the source decreases the liquidus temperature by about 100 °C; and the presence of 5% water decreases it by 250–300 °C. Such a high water content in the source (not the partial melt) is most unlikely, however.

8.3 Crystallization interval

An important characteristic of all komatiites is the large temperature interval between the liquidus, which almost always corresponds to the beginning of olivine crystallization, and the temperature at which the next silicate mineral starts to crystallize. As discussed in Chapter 3 and illustrated in Figure 3.1, the liquidus for a komatiite containing 25% MgO is about 1500 °C and the second silicate mineral to crystallize is clinopyroxene, which appears at about 1180 °C. (Chromite crystallizes earlier but in very minor quantities.) There is thus an interval of close to 400 °C in which olivine is the only important solid phase. The rate at which this phase crystallizes as temperature decreases is remarkably low: each 100 °C decrease in temperature results in the crystallization of only 20% additional olivine. Compare this with a basalt with 8% MgO for which the interval between liquidus and solidus is only about 100 °C. Komatiite that has cooled through 100 °C remains a highly mobile magma containing only 20% crystals, capable of flowing rapidly and differentiating readily; basalt that has cooled the same amount has frozen into immobility.

8.4 Viscosity

Huppert and Sparks (1985b) used the experimental data of Urbain *et al.* (1982) to modify Shaw's (1972) empirical method for calculating the viscosities of komatiitic liquids. They derived two empirical formulas that relate the viscosity of a Barberton-type komatiite with 30% MgO to its temperature. The first applies to liquid without crystals:

$$\mu_1 = \exp(15.095 - 0.0104T)$$

where μ_1 is the dynamic viscosity of the liquid in pascal seconds and T the temperature in degrees centigrade. The second gives the viscosity of lava that crystallizes olivine as temperature decreases:

$$\mu_2 = \exp(34.551 - 0.0226T)$$

where μ_2 is the viscosity of the crystal–liquid mixture. The variations in viscosity calculated with these equations are illustrated in Figure 8.1 and

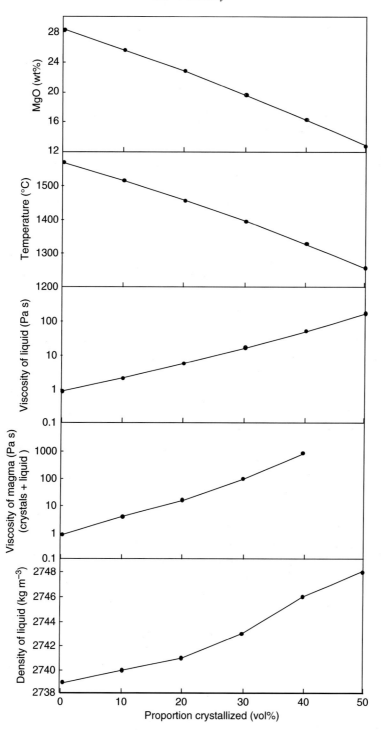

Fig. 8.1. Variations in the physical properties of a crystallizing komatiite liquid. The procedure used to calculate the plotted values is given in Table 8.1.

Table 8.1. *Physical properties of selected komatiites and basalts*

	Huppert & Sparks (1985b) model komatiite	Kambalda model komatiite (Williams *et al.*, 2002)	Cape Smith komatiitic basalt (Williams *et al.*, 2002)	Tholeiitic basalt (Williams *et al.*, 2002)
SiO_2	47.2	46.0	46.9	52.0
TiO_2	0.22	0.33	0.6	0.90
Al_2O_3	4.1	7.2	9.8	16.6
Fe_2O_3		1.2		
FeO	10.8	10.9	14.4	9.8
MnO		0.18	0.3	
MgO	30.3	27.1	18.9	6.3
CaO	5.8	7.5	8.6	9.1
Na_2O	0.56	0.06	0.3	4.7
K_2O	0.21	0.01	0.05	0.23
Liquidus temperature (°C)	1600	1555	1419	1160
Solidus temperature (°C)	1200	1170	1150	1080
Density at liquidus ($kg\,m^{-3}$)	2800	2780	2800	2750
Specific heat $J\,kg^{-1}\,°C^{-1}$	730	1740	1640	1480
Dynamic viscosity at liquidus (Pa s)	0.13	0.17	0.74	39

Tables 8.1 and 8.2. The viscosity of the 30% MgO komatiite liquid is 0.13 Pa s, almost two orders of magnitude less than that of 6% MgO basalt. If the olivine crystals that form during cooling remain suspended in the liquid, by the time that the temperature has dropped to 1400 °C and the MgO content to 20%, about 50% of olivine has crystallized and the viscosity of the crystal–liquid mixture has increased by many orders of magnitude.

Kushiro (1986) and Scarfe (1987) investigated the effect of pressure on the viscosity of silicate liquids. They conducted their experiments using a piston-cylinder apparatus at pressures up to 3 GPa and they monitored the rate at which round crystals of olivine settled in charges of molten basalt. Using this method they showed that, at constant temperature, viscosity increases with pressure. However, with increasing depth in the mantle, temperature increases as pressure increases, and the combined effect is to cause the viscosity of the melt in equilibrium with mantle peridotites to decrease as the depth in the mantle increases. These results have been confirmed by Liebske *et al.* (2005),

Table 8.2. *Variations in physical characteristics of a komatiite undergoing olivine crystallization*

SiO_2	46.1	46.6	47.2	47.8	48.6	49.5
TiO_2	0.36	0.39	0.44	0.49	0.54	0.60
Al_2O_3	7.11	7.90	8.77	9.75	10.8	12.0
FeO	11.0	11.5	12.0	12.4	12.8	12.9
MgO	28.3	25.6	22.8	19.6	16.3	12.8
CaO	6.60	7.33	8.15	9.05	10.1	11.2
Na_2O	0.29	0.33	0.36	0.40	0.45	0.50
K_2O	0.10	0.11	0.13	0.14	0.15	0.17
Fraction crystallized (wt%)	0	10	20	30	40	50
Temperature (°C)	1567	1514	1456	1394	1327	1256
Viscosity of liquid (Pa s)	0.9	2.2	5.9	17	52	171
Viscosity of lava (liquid + crystals) (Pa s)	0.9	4	16	96	820	15482
Density of liquid (kg m^{-3})	2739	2740	2741	2743	2746	2748

who developed a new technique to establish how the viscosity varies in the pressure range 2–13 GPa. Using a multi-anvil apparatus and an X-ray radiography technique, they showed that the viscosity of molten peridotite changes from 0.13 Pa s at 1800 °C and 2.5 GPa, to 0.02 Pa s at 1950 °C and 3.4 GPa. The viscosities of partial melts with komatiitic compositions will be a factor of 2–3 higher because of the difference in magma composition and temperature, but even these viscosities are low enough to facilitate rapid crystal–liquid segregation in situations where the phases have contrasting densities.

8.5 Density

The density of a silicate liquid depends on its composition, and particularly on the amount of dissolved volatiles, as well as on temperature and pressure. Again, no direct measurements appear to have been made on komatiite liquids at atmospheric pressure and the densities of these liquids must be calculated using indirect methods such as those developed by Bottinga and Weill (1970). Such methods give a density of about 2738 kg m^{-3} for a komatiite liquid containing 30% MgO.

During crystallization, olivine with high forsterite content is removed from the liquid. This causes the content of MgO, which has a low molar volume and contributes positively to the density, to decrease and the contents of most other major elements to increase (Table 8.2). The contents of FeO and MgO, the components that has the greatest influence on density, increases only slowly, and passes through a maximum when the liquid contains about 20% MgO. These changes in composition cause the liquid density first to increase, then to decrease. When the effect of falling temperature is taken into account, the combined effect is to cause a steady increase in liquid density, as shown in Figure 8.1 and Table 8.2. The effect of suspended olivine crystals, which have a density $\sim30\%$ greater than that of the liquid, is to increase the density of the lava; for example, the presence of 30 vol% suspended crystals increases the density from about $2700\,kg\,m^{-3}$ to about $2800\,kg\,m^{-3}$. These values are slightly greater than that of typical Fe-poor MORB with 8% MgO and 12% FeO, but are comparable with those of many Fe-rich tholeiites.

Measurement of the density of ultramafic liquid at high pressure was described in Chapter 7. Silicate liquids are compressible and their density increases significantly with increasing pressure. At 8 GPa, for example, the density of komatiite liquid has increased to about $3300\,kg\,m^{-3}$, essentially the same as that of olivine at that pressure. The important consequences for komatiite petrogenesis are discussed in Chapters 13.

8.6 Other thermal properties

The thermal properties of komatiite, such as specific heat, thermal conductivity, and latent heat of crystallization, were estimated by Usselman *et al.* (1979) Huppert and Sparks (1985b) Renner (1989), Silva *et al.* (1997) and Williams *et al.* (2002) in the course of their treatments of the cooling and crystallization of komatiite flows. These values do not differ dramatically from those of less mafic magmas, and will not be discussed further. The values adopted by Williams *et al.* (2002) are listed in Table 8.1.

Plate captions

Plate 1. (a) General view of outcrop in the Komati Formation of the Barberton greenstone belt, South Africa, the type area of komatiite. The high-standing, tree-covered ridges mark the positions of thick units of massive olivine cumulates, either thick flows or sills. Good outcrop is limited to the beds of small streams, such as Spinifex Stream, which is just out of view in the foreground of the photo.

(b) A differentiated spinifex-textured komatiite flow on Pyke Hill, Ontario. This rare photograph, taken in 1971, illustrates the most photogenic example of a komatiite flow. The surface of the flow has since been destroyed by waves of visitors. The flow dips at about 80° to the north (to the left of the photograph). From top to base we see the polyhedrally jointed flow top (A_1), the spinifex layer (A_2), and layers B_1–B_4 of the cumulate layer (see descriptions in Chapters 2 and 3).

(c) A differentiated, inflated spinifex-textured flow on Gilmour Island, Hudson Bay, Canada. The lighter coloured spinifex layer, to the left, is clearly distinguished from the chocolate-brown (80% cocao) cumulate layer, and the flow top is marked by the conspicuous curved fracture. Peter, my guide, is standing on pillow basalts that overlie the komatiite. Along strike, towards the foreground of the photo, the flow grades through massive olivine porphyry to pillow lava, and it is about 2 m thinner than along the spinifex-textured portion. Within the curved termination of the spinifex lens, the orientation of columnar joints and of spinifex crystals rotates to remain parallel to the flow surface. These structures are taken as evidence that the spinifex portion of the flow was inflated through the input of magma into the flow (see descriptions and discussion in Chapters 2, 3 and 9).

Plate 2. (a) The upper part of a komatiite on Pyke Hill. Clearly illustrated are the polyhedrally jointed flow top (A_1), the spinifex layer (A_2), the B_1 layer and the upper part of the cumulate layer. Note the progression from fine random olivine spinifex, shown in detail in (b), to platy olivine spinifex, shown in (c). In each photograph the coin is 1.8 cm in diameter.

Plate 3. (a) A polished slab illustrating the base of the spinifex layer and the upper part of the B_1 layer of a small komatiite sill from Dundonald Township in Ontario. The olivine crystals are completely serpentinized and have a dark green colour; they lie in a paler matrix of fine augite crystals and altered glass. The preferred orientation of small platy olivine crystals is evident in the B_1 layer. Some of the large dendritic olivine crystals of the spinifex layer have nucleated on, and grown upward from, the B_1 layer; others have grown downwards from the top of the sill. The polished slab is 8 cm across.

(b) The upper contact of the spinifex sill. The grey-brown material in the upper part of the photograph is olivine cumulate of the overlying unit. The upper contact of the sill is sharp and large skeletal hopper olivine crystals appear in the komatiite sill within 1 cm of the contact. This rapid transition distinguishes the sill from komatiite flows which have a thick chilled and jointed upper layer in which all crystals are microscopic (see Plate 1(a)).

(c) Spinifex-textured vein in the lower olivine cumulate layer of a complex komatiitic unit from 'Serpentine Mountain' in Sothman Township, south of Timmins in the Abitibi belt. Large platy olivine crystals have grown inwards from both margins of the vein. A photograph and sketch illustrating the broader-scale features of the unit are presented in Figure 11.3 and a complete description of the outcrop is provided by Houlé et al. (2008a). (photograph from Mike Lesher).

(d) Olivine orthocumulate from Murrin Murrin in the Yilgarn craton of Western Australia. The olivine grains are serpentinized and have weathered to a cream colour so that they stand out clearly in the darker grey-brown matrix (photograph from Steve Barnes).

In each photograph the coin shown is 1.8 cm in diameter.

Plate 4. (a) Photomicrograph of remarkably fresh olivine adcumulate from a dunitic unit in the Wildara area, Yilgarn craton, Western Australia (photograph from Steve Barnes).

(b) Highly skeletal hopper olivine grains from the lower cumulate part of the spinifex flow from Gorgona Island, Colombia.

(c) Spinifex-textured komatiite from remarkably fresh komatiite from the Zvishavane locality in the Belingwe Belt of Zimbabwe.

(d) The same as (c) in crossed nicols.

(e) Zoned pynoxene crystals with pigeonite cores and augite margins, in the interior of a spinifex-textured komatiite from Gilmour Island.

(f) Detailed view of curved augite crystals in spinifex lava of the Alexo komatiite flow. A platy olivine crystal (completely serpentinized) lines the lower part of the photo. Dendritic chromite crystals have nucleated upon and grown outward from this crystal. Other, less detailed, views of the same sample are shown in Figure 3.11.

Plate 1. (See end of section for plate captions.)

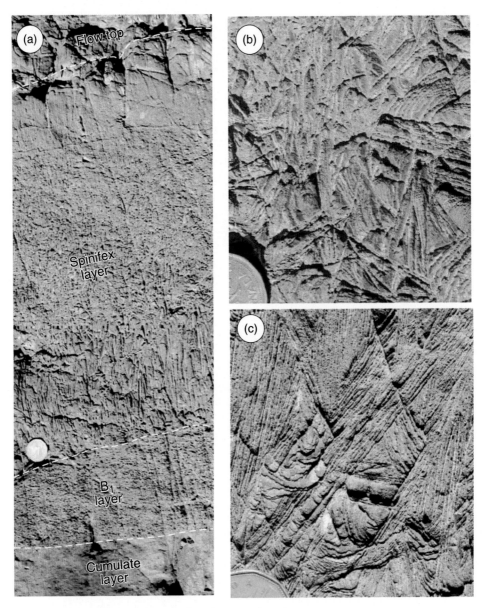

Plate 2. (See end of section for plate captions.)

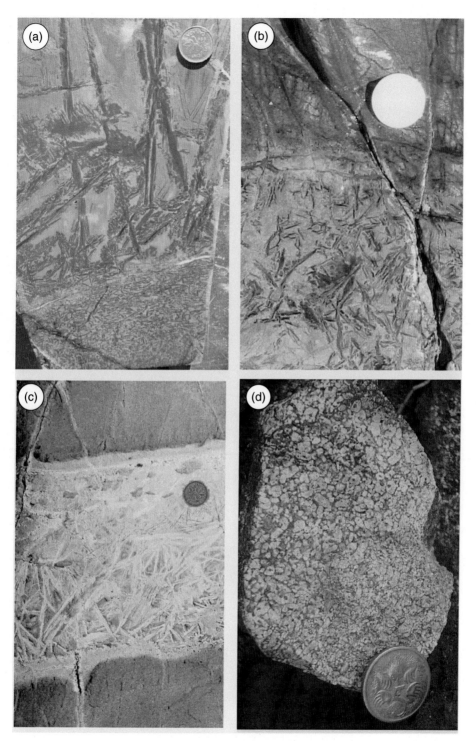

Plate 3. (See end of section for plate captions.)

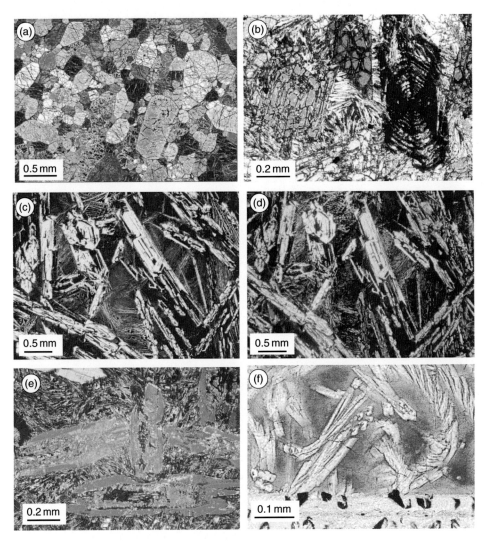

Plate 4. (See end of section for plate captions.)

9

Physical volcanology

S. J. BARNES AND C. M. LESHER

9.1 Introduction

Most komatiites have been metamorphosed and deformed, and most occur in Archean greenstone belts where exposure may be locally spectacular but normally limited. Archean geology is sometimes like peering through a keyhole at tiny areas of exquisitely preserved detail, and at other times like looking down from the air through a thick ground fog. Only rarely do we have enough detailed knowledge over a large enough area to be able to obtain a clear image of komatiites at the scale of a single eruption. Understanding of komatiite volcanology must therefore be built up from a fragmentary base, by combining detailed local information with broad regional syntheses and comparisons with modern basalt flow fields. Our current state of knowledge is derived largely from areas described elsewhere in this volume, particularly the well-preserved flows of the Abitibi, Barberton and Belingwe belts (Chapters 2, 3), and the well-mineralized, markedly more olivine-rich, but generally less well-preserved flows of the eastern Yilgarn Craton (Chapter 2).

One of the challenges of komatiite research has been to relate the diversity we observe to the volcanic processes that produced it, and to construct a broadly applicable volcanological model for komatiite flow fields. The conclusions have turned out to be of great importance in understanding the origin of komatiite-associated Ni–Cu–PGE deposits (Chapter 10). They are also of great importance in unravelling the tectonic history of greenstone belts.

We have seen in Chapter 8 that komatiite liquids had remarkable physical properties that distinguished them from the nearest modern counterparts, basalts. They had extremely low viscosities and high densities, which lead us to predict that (with all else equal) they would have formed laterally more extensive flow fields (or sill systems) with lower aspect ratios, which may have been emplaced, in some cases, by turbulent flow. They crystallized over a very

wide temperature interval, with much of the interval occupied by crystallization of olivine as the only silicate phase. The combination of the low viscosity
and the tendency of olivine to display a wide range of crystal forms depending
on the conditions of growth, accounts for much of the observed diversity.

This chapter will focus on two fundamental questions: how komatiites were
erupted and emplaced, and how komatiites crystallized. Along the way we will
encounter two important paradoxes: komatiites commonly form unusually
thick flows despite the low viscosity of the liquids; and komatiite flows contain
spectacularly dendritic textures, which imply high cooling rates but which are
developed within the slowly-cooled interiors of thick flows.

9.2 Nomenclature and terminology

It is necessary to define some terms at the outset. We adopt much of the
nomenclature used by volcanologists who have studied subaerial and submarine basalts in the Hawaiian islands (e.g., Walker (1970)). A *lava flow* is the
product of a single, uninterrupted eruption from the same source, and a *lava
flow field* forms by multiple sequential eruptions from the same source. *Cooling
units* are bodies of igneous rock bounded by distinct continuous cooling
surfaces. *Simple flows* are composed of single cooling units, whereas *compound
flows* are composed of multiple overlapping cooling units formed during the
same eruptive event.

Simple flows may be *massive*, or texturally and/or geochemically *differentiated*. The terms *differentiation* and *fractionation* have been used synony-
mously in igneous petrology to describe the formation of a variety of rock
compositions from a single parental composition. In this chapter *differentiation* is used to indicate macroscopic textural and/or compositional variations
within a single cooling unit due to the physical–chemical segregation of crystalline and liquid components, whether by gravity settling, fractional crystallization or other processes such as flowage differentiation.

Compound flows may be composed of many small rounded tubular units
(*pillows*), multiple thin overlapping sheets (*flow lobes*), or thicker *sheet flows*,
comprising combinations of volcaniclastic, pillow and/or massive lava (see
Dimroth *et al.* (1978)). Many flows are fed by *lava pathways*. *Lava channels*
(also called *lava trenches*) are uncovered lava pathways, whereas *lava conduits*
(larger) and *lava tubes* (smaller) are covered lava pathways. Lava channels
crust over to form lava conduits.

By analogy with sedimentary and metamorphic lithofacies, we define *volcanic lithofacies* as rocks with similar textural (e.g., aphyric, volcaniclastic,
vesicular, spinifex, porphyritic, crescumulate, orthocumulate, mesocumulate,

adcumulate) or compositional (e.g., basalt, gabbro, komatiitic basalt, komatiite, peridotite, dunite) characteristics. Cumulate rocks were originally interpreted to reflect gravitational accumulation of suspended phenocrysts or primocrysts, but it has subsequently been shown that many crystallize *in situ* (e.g. Campbell (1978)). In this chapter *olivine accumulation* is used to indicate that the bulk composition of a rock or an entire cooling unit is enriched in olivine via settling of olivine phenocrysts and/or via *in situ* growth and accumulation of olivine, relative to the presumed composition of the parent silicate liquid.

A *flow facies* has structural, textural and/or compositional features (e.g., massive vs differentiated, pillowed vs lobate vs sheet) that distinguish it from other sequences containing similar lithofacies. *Facies assemblages* or *facies associations* are groups of facies that commonly occur together and provide a basis for defining broader, interpretive facies based on the environment of emplacement (e.g., vent facies, proximal facies, distal facies) or mode of emplacement (e.g., pathway vs sheet vs channelized sheet) for the purpose of volcanic flow field reconstructions.

9.3 Komatiite facies

As described in Chapter 3, komatiite flows and sills vary widely in thickness, structure, texture, degree of internal differentiation, degree of olivine accumulation and composition. Much of this diversity is manifested in a spectacular range of olivine textures, which reflects a wide spectrum of crystallization conditions. Lesher *et al.* (1984), Lesher (1989), Hill *et al.* (1989, 1995), Prendergast (2001) and Barnes (2006) have applied facies analysis to komatiites. Expanding their work and using terms analogous to those employed by sedimentologists, we can define a range of komatiite *lithofacies, flow facies, facies assemblages*, and therefore *facies environments*. Lithofacies and flow facies are designed to be descriptive and usable in the field where limited exposure often hampers interpretations. Facies assemblages are much more interpretive: some are characterized by distinctive lithofacies and/or flow facies, but many are not. Similarly, some lithofacies and/or flow facies are characteristic of particular facies environments, but many are not.

Komatiite lithofacies

The lithofacies classification of komatiitic rock types is introduced in Chapter 1 (Table 1.1). The main subdivisions are into aphyric and phyric quenched flow-top material, spinifex-textured rocks and conventionally cumulus-textured rocks, ranging from orthocumulates through to adcumulates and showing a wide range of olivine morphologies and grain sizes. The relative distributions and proportions of spinifex and cumulus rock types vary widely, and this is one of the defining features of different volcanic environments.

Other lithofacies are much less abundant, but are significant as indicators of environment. Crescumulate rocks exhibit dendritic (commonly called 'harrisitic') crystal morphologies, contain a high cumulus component and are normally enclosed within polyhedral olivine cumulates. They are distinct from platy and acicular spinifex which occur in A zones of differentiated flows and have therefore grown downward. Spinifex rocks in some cases contain a component of cumulus crystals, and therefore do not necessarily reflect liquid compositions. They are distinct from crescumulates in containing a much lower cumulus component, and in having strongly dendritic fractal crystal shapes.

As noted in Chapter 3, most komatiite flow tops typically contain few (<1%) vesicles (Pyke *et al.*, 1973; Arndt *et al.*, 1998a) but some komatiite flows enclose strongly vesicular zones or layers, defining a separate mappable lithofacies. Vesicles occur in flow tops (e.g., Dundonald: Eckstrand and Williamson (1985); Barberton: Dann (2001)), in porphyritic lavas (e.g., Lewis and Williams 1973), and in some cumulate rocks (Keele and Nickel, 1974; Stolz and Nesbitt, 1981; Beresford *et al.*, 2000; Hill *et al.*, 2004). Beresford *et al.* (2000) distinguished *amygdales* (vesicles filled with late-stage minerals) from *segregation vesicles* (partially filled with interstitial silicate or sulfide melt). *Sulfide-filled segregation vesicles* are described and illustrated in Chapter 10.

Volcaniclastic lithofacies are produced by consolidation of volcanic fragments formed by ejection from a vent (*pyroclastic*), by quenching with water (*hydroclastic*), by flow fragmentation (*autoclastic*), or by erosion (*epiclastic*). Volcaniclastic komatiites are rare, but have been described at Scotia, Western Australia (Page and Schmulian, 1981), in Sattasvaara, Finland (Saverikko, 1985), on Gorgona Island (Echeverría and Aitken, 1986), in Karasjok, Norway (Barnes and Often, 1990), in the Steep Rock and Lumby Lake greenstone belts, Ontario (Schaefer and Morton, 1991; Tomlinson *et al.*, 1999), at Dachine, French Guiana (Capdevila *et al.*, 1999), at Gabanintha in the Murchison Province, Western Australia (Barley *et al.*, 2000) and in the Wallace greenstone belt, Ontario (Sasseville and Tomlinson, 2000). Some have been interpreted to be *pyroclastic*, but contain fragments of rock types that do not form pyroclastically (e.g., spinifex-textured or cumulate lava) and are therefore more likely epiclastic. Others contain only fine-grained glassy material and are limited in stratigraphic extent, so they are more likely *hydroclastic*. We are not aware of any komatiites *sensu strico* that contain unequivocal spatter and bombs, but komatiitic basalts with such features have been described in the Möykkelmä area in Finland (Räsänen *et al.*, 1989) and ferropicrites with such features have been described at Kotselvaara in the Pechenga belt (Green and Melezhik, 1999). *Autoclastic* lavas have been reported at Kambalda by Beresford *et al.* (2002) at Bannockburn (Houlé *et al.* 2005), and are also observed at Sattasvaara in Lapland (Thordarson, pers. comm.), but these are blocky rather than a'a lavas. Komatiitic a'a flows are unknown, and are very unlikely to have formed from such low-viscosity lavas. *Epiclastic* komatiites have been described at Spinifex Ridge by Gélinas *et al.* (1977b), at Karasjok by Barnes and Often (1990), and probably occur locally in most localities.

Komatiite flow facies

Komatiites form many of the *flow facies* recognized in basaltic flow fields, particularly submarine basaltic flow fields, including simple and compound volcaniclastic units, compound pillowed and lobate flows, simple differentiated and undifferentiated sheet flows, and simple and compound differentiated and undifferentiated lava conduits (Table 9.1).

Sheets and conduit facies exhibiting many of the same lithofacies as flows may also form in invasive (downward burrowing) flows, deeply erosive flows, intrusive sills, or feeder conduits. For example, the differentiated sills at Dundonald Beach (Arndt *et al.*, 2004b) and Boston Creek (Houlé *et al.*, 2001) exhibit many of the compositional and textural characteristics of differentiated flows, and were originally interpreted as flows (see Muir and Comba (1979), Stone *et al.* (1987)). Because the contact relationships that are required to distinguish between extrusive, invasive and intrusive modes of emplacement are often not exposed, we have grouped them all under flow facies.

Very-low-viscosity high-Mg komatiites normally form compound flow lobes rather than compound pillowed flows, but pillows occasionally form in low-Mg komatiites (e.g., Belingwe: Nisbet *et al.* (1977)) and commonly form in komatiitic basalts.

Archean geologists normally encounter komatiites in small outcrops and drill cores, so one-dimensional profiles through individual cooling units are commonly used as a basis for their description and interpretation. Many of the *flow facies* in Table 9.1 can be portrayed on a simple matrix (Figure 9.1) of two variables: (1) the amount of excess olivine (vertical axis) and (2) the degree of *in situ* textural and compositional differentiation (horizontal axis).

The first variable is the amount of accumulated olivine, calculated by comparing the weighted average bulk composition of the unit with the aphyric

Table 9.1. *Komatiite flow facies*

Flow facies	Most common lithofacies[a]	Examples[b]
Pyroclastic		
Proximal	Pyroclastic (cg–fg, heterogeneous, often heterolithic)	Koselvaara (ferropicrite), Möykkelmä (komatiitic basalt), Dismal Ashrock (Steep Rock)?
Distal	Pyroclastic (fg tuffaceous)	Upper Onverwacht (Barberton)
Pillow/lobe		
Pillow	Thin massive aphyric, massive porphyritic	*High-Mg komatiite*: not known *Low-Mg komatiite*: Belingwe, eastern Munro Twp. *Komatiitic basalt*: Cape Smith Belt (Chukotat Group), Pontiac Group (Baby Group)
Undifferentiated lobe	Thin massive aphyric, massive porphyritic	Pyke Hill, Belingwe, Kambalda (Tripod Hill member), parts of most komatiite sequences
Differentiated lobe	Spinifex-orthocumulate	
Sheet		
Undifferentiated	Massive aphyric, massive porphyritic, massive ortho-meso-adcumulate, spinifex greatly subordinate or absent	*Extrusive*: Kambalda (Silver Lake member), Scotia, Spinifex Ridge, Silver Swan
Differentiated	Spinifex-orthocumulate, spinifex-mesocumulate, with or without gabbro	*Extrusive*: Kambalda (Silver Lake member) *Intrusive*: Boston Creek, Damba-Silwane *Unknown*: Kurrajong (Walter Williams)
Channel/conduit/chonolith		
Undifferentiated	Massive mesocumulate or adcumulate	*Extrusive*: Scotia *Invasive*: Katinniq, Perseverance? *Intrusive*: Thompson, Mt Keith, Perseverance?
Differentiated	Spinifex-adcumulate, spinifex-mesocumulate or gabbro-adcumulate	*Extrusive*: Kambalda (basal host units), Alexo–Dundonald, Shaw Dome, Silver Swan *Intrusive*: Dumont, Thompson

[a] Lithofacies in Table 1.2.
[b] Locations in Chapter 1.

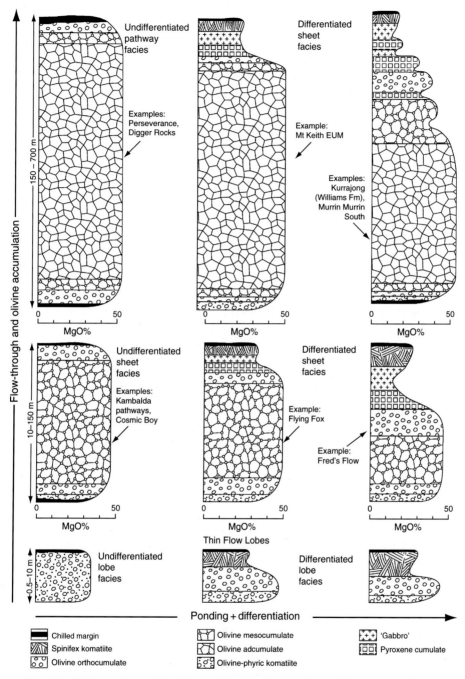

Fig. 9.1. Schematic profiles through komatiite cooling units, representing a matrix of variability in degree of flow-through and olivine accumulation (vertical) and degree of static internal differentiation (horizontal). The width of each profile is proportional to MgO. Modified from Lesher *et al.* (1984), Lesher and Keays (2002), and Barnes (2006).

chilled margins (see below). The second variable represents the degree of internal differentiation due to a combination of fractional crystallization in the upper parts of the flows and accumulation of olivine (or less commonly pyroxene) in the lower parts. The narrow D-shaped profiles on the lower left represent units that have not accumulated much olivine and have not differentiated significantly, including massive, pillowed/lobate and pyroclastic facies. The wide D-shaped profiles on the upper left represent units with bulk compositions that are enriched in olivine relative to their chilled margins and therefore contain a large component of accumulated olivine. The S-shaped profiles on the lower right represent units that have differentiated *in situ* under closed system conditions to produce olivine cumulates in the lower parts (B zones) and less magnesian lavas in the upper parts (A3 zones), but they have bulk compositions similar to the chilled margins and therefore contain little excess olivine. The transitional profiles in the centre and on the upper right represent units with complex histories of early flow-through and accumulation of olivine, followed by late ponding and differentiation. The interplay of the two variables generates the observed spectrum of flow facies between end-members represented by massive undifferentiated non-cumulate units (lower left), differentiated non-cumulate units (lower right), undifferentiated adcumulate units (upper left) and differentiated cumulate units (middle and upper right).

Komatiite facies assemblages and facies environments

Some komatiite flow facies occur together consistently enough to define *facies assemblages* (Table 9.2) and some facies and facies assemblages can be used to define *facies environments* (Table 9.3). Of the facies assemblages represented in Table 9.2, 'compound flow lobes' and 'compound sheet flows' are by far the most common, but as discussed in Chapter 10, channel/conduit facies assemblages are less common but economically very important. We consider the hypothetical geometry of komatiite flow fields and the relationship of these facies assemblages to one another at the conclusion of this chapter.

9.4 Komatiite flows and flow fields: size, structure and emplacement processes

Size and scale

As discussed in Chapter 3, the dimensions of komatiite flows vary considerably. Their lengths range from tens of metres in the case of individual lobes within compound flows to correlatable cooling units at least kilometres, and

Table 9.2. *Komatiite facies assemblages*

Facies assemblage	Characteristic flow facies[a]	Examples[b]
Compound flow lobes	Multiple thin lava lobes with little or no interflow sediment; very common	Alexo–Dundonald, Belingwe, Kambalda (Tripod Hill member), Mt Clifford, Pyke Hill, Shaw Dome
Multiple/ compound sheet flows	Multiple thick, laterally extensive, variably differentiated sheet facies lacking conduits	Spinifex Ridge
Thick composite sheets	Single or multiple thick differentiated or undifferentiated mesocumulate or adcumulate sheet facies	Shaw Dome (Lower Komatiite Horizon), Walter Williams Formation, Murrin Murrin, Boston Creek (ferropicrite)
Channelized sheet flows	Thick poorly- to strongly- differentiated mesocumulate conduit facies flanked by differentiated sheet facies with variable degrees of differentiation	Kambalda (Silver Lake Member), Raglan (Cross Lake Member)
Channel/ conduit complex	Multiple, thick, poorly-differentiated meso- to adcumulate conduit facies, in some cases flanked by peperites in sediment-dominated environments	*Extrusive*: Alexo *Deeply erosive*: Raglan (Katinniq Member) *Invasive* or *intrusive*: Mt Keith, Perseverance *Intrusive*: Damba-Silwane, Dumont, Thompson
Lava lake	Difficult to unequivocally identify, but should be topographically-controlled and areally restricted, relatively thick, and composed of single or multiple, predominantly porphyritic or differentiated spinifex-cumulate facies	Munro Twp lava lake, Lion Hills, Lake Harris?

[a] Flow facies in Table 9.1.
[b] Locations in Chapter 1.

probably tens of kilometres long. The largest individually correlatable cooling unit, the Walter William Formation in the East Yilgarn, is at least 130 km long, but may well be a sill rather then a flow. The sizes of flow fields are difficult to estimate because of limits of outcrop and the deformation that has affected most greenstone belts, but have been well established on the scale of at least 80 km in Zimbabwe (Prendergast, 2003) and Abitibi, and at least 60 km in the East Yilgarn. There is good reason to suppose that they could have been ten

Table 9.3. *Komatiite facies environments*

Facies/assemblage[a]	Vent	Proximal and/or high flow rates	Mid-distal and/or moderate flow rates	Distal and/or low flow rates
Volcanic/invasive				
Proximal pyroclastic	X	(X)		
Lava lake	X	X	X	
Composite sheet		X		
Lava channel/conduit		X		
Channelized sheet			X	
Lava lobe				X
Distal pyroclastic				X
Intrusive				
Intrusive conduit		X		
Intrusive sheet			X	

[a] Facies and facies assemblages in Tables 9.1 and 9.2.

times that size in the latter two belts. The thicknesses of komatiite flows are better constrained and here we can say confidently that while most individual cooling units in compound flow facies environments are only a few metres thick, there is compelling evidence in a number of localities (particularly in the East Yilgarn) for extrusive or very shallow intrusive cooling units several hundred metres thick.

Here we reach the first of the paradoxes that run through this chapter. How did such extremely fluid lavas form such thick flows? As Cas *et al.* (1999) have pointed out, highly fluid lavas should form extensive but very thin flow units. To explain the paradox, we look at some of the advances that have been made in recent years in understanding the emplacement of large basaltic flows.

Emplacement of thick basalt flows: the inflation hypothesis

The vast lava flows of continental flood basalt provinces had universally been thought to be the result of cataclysmic eruptions (e.g. Swanson *et al.* (1975)) until detailed studies of flow emplacement processes of Hawaiian and Columbia River basalt flow fields led to the development of the inflation hypothesis (Hon *et al.*, 1994, 1995; Kauahikaua *et al.*, 1998; Keszthelyi *et al.*, 1999; Self *et al.*, 1996; Thordarson and Self, 1998). According to this model, based in part on direct observation of emplacement of present-day flow fields on Kilauea, lava flows large and small are emplaced through processes of endogenous growth of flow lobes beneath an inflating visco-elastic skin.

A new flow lobe forms when hot lava is erupted from a point or vent source, and it propagates initially with low aspect ratio, often only a few centimetres thick, beneath a developing quenched crust. As the crust thickens it develops sufficient strength to retard lobe advance and retain incoming lava. The lava lobe continues to cool through its upper surface and accretes cool viscous lava beneath the upper cooling surface, forming a composite skin of visco-elastic lava overlain by a brittle crust which, paradoxically, is the oldest part of the flow. This progressively thickening composite crust is moved upwards by lava injection into the interior, or core, of the lobe, driven by the hydrostatic head of lava up slope. This process of endogenous growth is referred to as *inflation* (Figure 9.2).

At high flow rates, early-formed lobes coalesce and the flow evolves into a single thick sheet, consisting of an extensive sheet-like core capped by the merged original lobe crusts. Large continental flood basalt sheet flows are thought to develop in this way (Thordarson and Self, 1998) at eruption rates of the order of $10^4 \, m^3 \, s^{-1}$, compared to rates of 10^2–$10^3 \, m^3 \, s^{-1}$ observed for tube-fed compound Hawaiian flows (Keszthelyi *et al.* 2006). Large sheet flows are fed by high-flux zones or 'pathways' within a continuously recharged liquid core. As flow rates decline, through continued expansion of the flow field and/or through declining eruption rate, the sheets probably evolve into multiple internal pathways.

At lower flow rates, multiple small lobes undergo radial cooling and progressive inward solidification. This forces the flux of lava to concentrate in the centres of the lobes, forming preferred lava 'pathways', which may evolve into lava tubes as radial solidification advances. The well-insulated nature of these tubes provides for continuous efficient transport of lava to the advancing flow front with minimal temperature loss (e.g., as low as $\sim$0.1 °C km^{-1} (Thordarson and Self, 1998)).

Continued uplift and fracture of the roofs and edges of established lobes at low flow rates results in budding of new lobes as 'breakouts'. These newly-formed lobes themselves go through the life cycle of propagation, advance-ment, inflation and further breakout. The resulting compound flow is a fractal network of parent lava pathways flanked by episodically emplaced, inflated, subsidiary lobes and flows, fed by their own network of tubes. This is the structure of the low-volume flow field of the present-day eruption of the Kilauea East Rift (Hon *et al.*, 1994; Kauahikaua *et al.*, 1998).

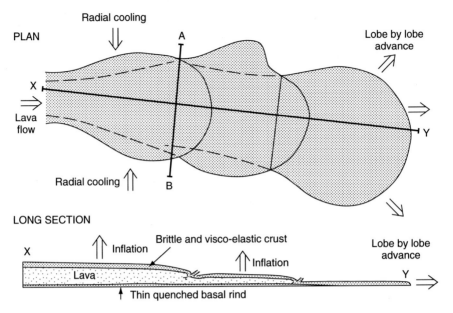

PLAN

Radial cooling ⇓

A

Lobe by lobe advance ↗

X ⇒

Lava flow

Radial cooling ⇑   B

Y

⇒

↘

LONG SECTION

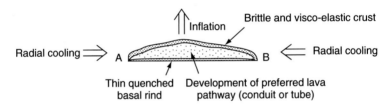

X

Inflation ⇑   Brittle and visco-elastic crust

Lobe by lobe advance

Lava

Inflation ⇑

Y ⇒

↑ Thin quenched basal rind

CROSS-SECTION THROUGH PATHWAY – TIME 1

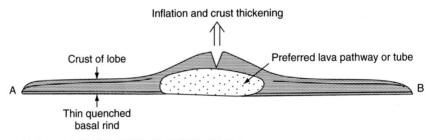

Inflation ⇑   Brittle and visco-elastic crust

Radial cooling ⇒   A   B   ⇐ Radial cooling

Thin quenched   Development of preferred lava
basal rind   pathway (conduit or tube)

CROSS-SECTION THROUGH PATHWAY – TIME 2

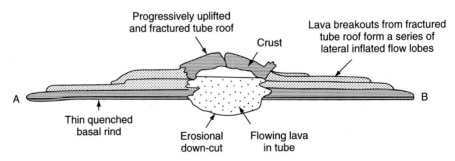

Inflation and crust thickening ⇑

Crust of lobe   Preferred lava pathway or tube

A   B

Thin quenched
basal rind

CROSS-SECTION THROUGH PATHWAY – TIME 3

Progressively uplifted
and fractured tube roof   Crust

Lava breakouts from fractured
tube roof form a series of
lateral inflated flow lobes

A   B

Thin quenched
basal rind   Erosional   Flowing lava
                  down-cut   in tube

Fig. 9.2. Endogenous growth model for propagation and development of basaltic flow lobes (modified from Hill (2001)).

The notion that much of the lava flux takes place beneath an insulated crust is crucial when considering the eruption of lava flows under water. A common misconception is that submarine lava flows are characteristically pillowed; in fact, pillowed lavas are largely a consequence of initial lobe propagation, particularly on steep slopes, at relatively low eruption rates (Moore, 1975). Steady prolonged eruptions give rise to extensive inflated sheet flows in much the same way as on land; once the lava is flowing beneath a crust, the influence of seawater cooling is greatly reduced, to the extent that individual tube-fed flows over 100 km long have been identified on the sea floor off Hawaii (Clague *et al.*, 2002). This is clearly of crucial importance in understanding komatiites.

The inflation hypothesis gives us a way of explaining the low-viscosity, thick-flow paradox, although we still have a disparity between the thicknesses of the more portly komatiites and typical flood basalt flows. We can now consider the evidence for inflation processes in komatiite flows.

Inflation in komatiite flows

The suggestion that komatiite flows were inflated by internal injection of lava was made by Arndt (1982), Barnes *et al.* (1983) and Lesher *et al.* (1984), based on elongate lava rises or tumuli in komatiite flows on Gilmour Island (see Chapter 2), mass balance constraints in thick flows at Alexo and the convex geometries of thick flows in the lower part of the komatiite sequence at Kambalda. Hill and Perring (1996) applied the inflation paradigm to komatiites and concluded that the spectrum of komatiite flow profiles could be interpreted as inflationary compound flow fields. This conclusion was born out by observation of characteristic inflationary features such as inflationary tumuli (upward projections in the roofs of lava tubes), breakouts and rotated crusts in some Barberton komatiites (Dann, 2001).

Williams *et al.* (1998) pointed out that at constant eruption rates, volume conservation requires flow thickness to increase as flow rates decrease. Cas *et al.* (1999) estimated the thickness of the propagating fronts of komatiite flows as a function of flow velocity, viscosity, density and slope of the terrain, based on the fluid dynamic Jeffreys equation. They concluded that most flows would have propagated at thicknesses of at most a few metres regardless of eruption rates, implying that inflation was essential to generate thick flows. They further suggested that Phanerozoic continental flood basalts were the best modern analogue to thick, cumulate rich komatiite flows.

Hill (2001) interpreted the characteristic relationships within the compound flow facies as a direct analogue of the inflationary basaltic flow model, with thin spinifex lobes forming as breakouts at the propagating edges of the flow field, or subsequently by lateral breakouts and overflow from lava pathways or tubes. In this model, the thin compound flow lobe facies assemblage is equivalent to modern compound pahoehoe flows, and consists of a network of interconnected tubes and inflated lobes. The thick cumulate-rich flow units represent extensive sheet flows formed by coalescence of lobes during continuous, high-volume eruptions.

Direct field evidence for inflation in komatiites is rare, but the tumuli (upward projections in the roofs of lava tubes), breakouts and rotated crusts described in the Gilmour Island flows and the Komati River outcrops are unequivocal. Beresford *et al.* (2002, 2005) describe rare multiple stratiform vesicle-rich layers within single cooling units, features which are taken as evidence of inflation in modern flows (Thordarson and Self, 1998). Fred's Flow in Munro Township (Chapter 2: Arndt *et al.*, 1979), the basal host unit at Langmuir (Green and Naldrett, 1981), and the basal host unit at Durkin Deeps at Kambalda (Lesher, 1983) all contain spinifex-textured flow tops now stranded in olivine cumulate rock near the bases of the flows.

Houlé *et al.* (2008b) describe spinifex-textured 'sills' or thick veins, typically a few centimetres thick but in rare cases up to 2 m thick, injected as subhorizontal sheets into B zones of differentiated flows at Pyke Hill and the Serpentine Mountain locality, both in the Abitibi belt. Textural evidence indicates that the host cumulates were incompletely solidified at the time of injection, and these features are interpreted as a form of late-stage inflation. This is a plausible explanation for the thicker sill-like bodies, but the thin, discordant spinifex veins which are widespread in komatiite flows are more likely to be due to injection of segregated, fractionated intercumulus liquid, analogous to segregation veins in modern basalts.

Supporting evidence from the study of thick olivine-rich komatiite units is illustrated in the idealized profiles in Figure 9.1. The bulk composition is considerably more olivine-rich than that of the parent magma, not only for the dunite bodies and the thick cumulate-rich basal flows at Kambalda, but also for some thin differentiated spinifex-textured flows. Thick piles of olivine cumulates with variable compositions could never have been emplaced all at once over large areas. The inflation hypothesis provides a mechanism to explain upward accretion of thick cumulate zones from bodies of flowing magma much thinner than the eventual total thickness of cumulates. According to this model, which we explore in much greater detail below, a continuous flow of lava took place over a bed of accumulating/growing crystals while simultaneously inflating the overlying visco-elastic crust. Thus, the inflation process potentially explains not only the paradox of how thick flows formed from such highly fluid lavas, but also how thick sequences of cumulates formed in lava flows.

The case for inflationary processes in the eruption of komatiite flow fields is strong, but not proven. A number of objections have been raised, some of which will be discussed below in the context of the crystallization of komatiites and the origin of the distinctive textures and characteristic patterns of internal differentiation.

One important objection is the relative rarity of features such as tumuli, inflation clefts, lava rises, tilted crusts and other prominent structures of modern inflationary pahoehoe flow fields. These features have been identified by Dann (2000) in the Barberton type section, but are apparently rare elsewhere. In part, this may simply be due to limitations of exposure, but such features have not been observed even where extremely well exposed, for example in the spectacular glacially-polished outcrops of Pyke Hill or Spinifex Ridge in the Abitibi belt of Ontario and Quebec. This may also be a function of local environment: tumuli, extension cracks and related features are the result of pulsed lava emplacement and sudden surges of lava. However, it is also possible that such features did not form as frequently in komatiites, because the crusts were thinner, more elastic and more flexible. In any case, it is necessary to consider alternative mechanisms.

Another objection to the inflation mechanism for the largest flood basalt flows has been raised by Anderson *et al.* (1999) and relates to the 'magma-static' pressures needed to sustain the process over the hundreds of kilometres lengths of these flows. This is a complex argument that is still unresolved (Self *et al.*, 2000; Anderson *et al.*, 2000) and is beyond the scope of this chapter, but which will have great impact on interpretations of inflation in komatiites.

Alternatives to the inflation mechanism

Nobody would argue that komatiite flows advanced along the sea floor like a vertically-fronted breaking wave, so it is inescapable that thick flows started out as thin flows and grew thicker. However, there are several alternatives to inflation, none of which excludes another:

(1) Local ponding: lava channels or conduits may have been localized in topographic depressions in the substrate (Lesher *et al.*, 1984), which may have become enlarged by thermomechanical erosion (Huppert *et al.*, 1984; Williams *et al.*, 1998, Lesher, 2007). For example, the thickest units at Kambalda (Figure 2.18) and the Shaw Dome are localized in topographic depressions in the footwall rocks and the thickest units at the Alexo deposit (Figure 10.3) and Raglan (Figures 10.5, 10.8) appear to have generated their own embayments by thermomechanical erosion. However,

some thick flows are not obviously confined by embayments, so local ponding provides at best a partial explanation.

(2) Regional ponding: komatiite flows will have been ponded in places within basins or sub-basins, calderas or other features providing topographic relief on a regional scale. For example, Davis (1997, 1999) and Houlé *et al.* (2008a) have argued that komatiites in the Kidd–Munro assemblage of the Abitibi greenstone belt were localized by topography in the underlying felsic volcanic sequences. Indeed, if komatiites were capable of flowing such long distances, it is inevitable that many would become ponded. However, ponding should produce thick differentiated flows, not thick cumulate flows, so thick cumulate units must have accumulated olivine prior to ponding, which is consistent with the geochemical variations in many thick differentiated cumulate units (see below).

(3) Flood flow: komatiites may have erupted at extremely high eruption rates. The flows must still have started thin and become thicker, but if eruption rates were high enough, this could have happened very quickly, before much crystallization had taken place. Flows formed in this manner would have been turbulent (Huppert and Sparks, 1985b). They may have been bounded by very thin and probably ephemeral viscoelastic skins or they may have flowed as open channels with very thin, transient crusts, rather like the 'Rivers of Fire' of the 1984 Mauna Loa eruption commonly depicted in volcanology texts. This scenario was the basis for numerical modelling of komatiite flow emplacement by Huppert and Sparks (1985b) and Williams *et al.* (1998, 2002).

'Flood flow' and 'inflation-accretion' mechanisms are end-member models, and it is probable that real flows fell somewhere between the two extremes. How can we decide which was dominant? Much of this discussion depends on details of crystallization processes and constraints on thermal erosion, both of which are covered below, but some purely physical considerations apply.

Submarine komatiite flows have an important physical distinction from subaerial basalts: the crusts were only mildly vesicular, and in many cases non-vesicular (especially in deep water). Massive lava is denser than the liquid from which it crystallizes and massive crust will sink, in contrast to vesicular subaerial basalt crust, which happily floats. Submarine komatiite crusts will fragment and sink, unless supported by the sides of the channels. Because flood flows were turbulent (Huppert and Sparks, 1985b), they will contain abundant autoliths of their own foundered crusts, a feature that is only rarely observed in komatiites. In the inflationary process, the crust above the slowly-flowing lava is supported by its own elastic strength, and by the hydrostatic head of lava upstream.

In summary, although it seems likely that some komatiite flows were ponded in topographic depressions, the inflation process was almost certainly operative on some scale in most komatiite flows.

Rates of eruption

Rates of eruption of komatiite flows cannot be determined directly. High-precision, sensitive high-resolution ion microprobe (SHRIMP) zircon dates bracket the age range of komatiite sequences in some cases, but this gives what are very much upper-end estimates. For example, the Kambalda komatiite sequence has been bracketed between zircon dates of 2692 ± 4 Ma for the overlying Kapai Slate, and an interflow metasediment near the base of the Kambalda Komatiite Formation at 2702 ± 4 Ma, which means that the Kambalda komatiites erupted over an interval as short as 2 My or as long as 18 My (Nelson, 1998). The age of the komatiites in the Kidd–Munro assemblage of the Abitibi belt has been more precisely bracketed between 2717 and 2714 Ma using single zircons extracted from overlying and underlying felsic volcanic horizons (Bleeker, 1999; Ayer *et al.*, 2002, 2005a,b; Sproule *et al.*, 2002). These ranges are within the durations of modern plume-related *volcanism* (1–10 My (Coffin and Eldholm, 1993)). However, these estimates give no idea of the duration of individual eruptive events, which were certainly much shorter than the precision of even the most precise U–Pb zircon ages.

Gole *et al.* (1990) estimated eruption rates from thermal relaxation profiles within stacks of thick flows. A very rapidly emplaced stack of flows develops a sinusoidal thermal profile because residual heat in the interior of the lobes is not dissipated before the next lobe arrives. Gole *et al.* (1990) describe an example at Honeymoon Well in Western Australia where spinifex textures in thin flows sandwiched between two thick dunite bodies were thermally metamorphosed and partially remelted during conductive relaxation of such a profile, implying that a stack of hundreds of metres of olivine-rich komatiite had been erupted on a time scale of tens–hundreds of years.

Studies have derived velocities from possible magma discharge rates and ascent rates, based on assumed fissure-vent dimensions, and buoyancy and density differences driving magma uprise to the Earth's surface. Assuming ascent through dykes, Huppert and Sparks (1985b) estimated two-dimensional discharge rates (i.e. per length of linear vent) of between 0.5 and $100 \, m^2 \, s^{-1}$ at the vent, or between about 0.5 and $100 \, km^3$ per day per kilometre width of flow front. The lower end of the range is comparable with the largest Hawaiian eruptions; the higher end is comparable with continental flood basalts.

Did komatiites flow turbulently?

This question is closely bound to the last one. At the lower end of the range of discharge rates estimated by Huppert and Sparks, 0.3 m thick lava streams flowed at velocities around $2\,\mathrm{m\ s^{-1}}$, yielding Reynolds numbers of around 5000 at the vent. At the higher end of the range, the figures are $10\,\mathrm{m\ s^{-1}}$, $10\,\mathrm{m}$, and 10^6. On this basis, Huppert and Sparks concluded that most komatiite flows would have flowed turbulently. Williams *et al.* (1998) reached similar conclusions, but because of several improvements to the model predicted that the flows would have been turbulent over a shorter distance. On the other hand, calculations of flow thicknesses and Reynolds number using the Jeffreys equation by Cas *et al.* (1999) suggested that most komatiites should have flowed laminarly, at least at the propagating fronts of flows.

The presence of thin horizons of interflow sediments between the flanking sheet flow facies at Kambalda suggests that these units were not emplaced turbulently (see discussion by Lesher *et al.* (1984), Lesher (1989), Lesher and Arndt (1995), Cowden and Roberts (1990) and Cas *et al.* (1999)) and the same probably applies to thin compound lava lobes such as those at Pyke Hill, Barberton, Belingwe, or in the Tripod Hill Member at Kambalda. However, the great thicknesses, large amounts of olivine accumulation and evidence for thermomechanical erosion beneath the lava conduit facies at localities like Kambalda, Alexo and Katinniq strongly suggest that they flowed turbulently during at least part of their development (see below).

There is a discrepancy between the field relationships, which suggest that many komatiites did not flow turbulently, and the high Reynolds number predictions of Huppert and Sparks (1985b) and Williams *et al.* (1998), which suggest that most should flow turbulently. This arises from two factors: (1) the one-dimensional nature of the models, which does not allow for lateral and temporal variations (and therefore implies flood flow); and (2) the assumption in the models that komatiite flows were covered by only very thin and transient crusts that did not constrain the advancement of the flow. Where such crusts are present, as in inflated flows, the rate of advance of the front of the flow is very much less than the rate of flow at the vent. According to Cas *et al.* (1999), ratios of near-vent channel flow velocities to flow-front advance velocities vary from 3:1 to 5:1 in the first few hours, increasing to >100:1 for flows lasting a few days, to >10 000:1 for pahoehoe flows where flow has become confined to insulated tubes. Cas *et al.* conservatively estimated that the ratio would have been about 10:1 for large komatiite flows. Depending on the internal structure of the flow field, initial turbulent emplacement would be expected to pass with time and distance into laminar flow within inflated lobes, which is consistent with the Williams *et al.* models.

The Cas *et al.* model predicts low Reynolds numbers, but considers only the flow front, and not the subsequent evolution of the flow regime within the liquid core of inflated flows. Turbulent flow may well take place within the pathways or tubes after the flow has thickened by endogenous growth. If we consider the conduit facies at Kambalda, we can envisage flow pathways having dimensions at least 10 m thick. A flow rate of 5 m s^{-1}, in the middle of the Huppert and Sparks range, implies a discharge rate of 50 m^2 s^{-1} and a Reynolds number of around 10^5, well within the turbulent range. The Kambalda scenario could therefore be explained by a sequence of events beginning with near-vent turbulent flood flow followed by separation into contrasting flow regimes – turbulent flow in channels/conduits near the vent and in narrow pathways further afield, and laminar flow in flanking areas of sheet flow and at the flow front. This situation would be enhanced if lava were able to drain freely from the ends of the tubes or pathways. In modern Hawaiian flows, this happens where the tubes drain into the ocean. In a sea-floor volcanic setting, it may happen where the flow field encounters steep topographic features such fault scarps or calderas.

Turbulent flow within established pathways can also explain the setting of thick olivine adcumulate bodies, which we attribute to turbulent lava flow over a growing bed of crystals. Thus, adcumulate komatiites may be a hallmark of turbulent flow.

Thermomechanical erosion beneath komatiite flows

The idea that komatiite lavas melted the substrates over which they flowed was initially proposed by Nisbet (1982) and modelled by Huppert, Sparks, and co-workers (Huppert *et al.*, 1984; Huppert and Sparks, 1985b). The major impetus was the recognition that the linear host units at Kambalda represented lava conduits (Lesher *et al.*, 1984) and that the ore-localizing embayments and adjacent sediment-free zones might reflect thermomechanical erosion. Details of the application of the thermal erosion concept to ore formation are dis-cussed in Chapter 10.

The premise for the initial Huppert *et al.* model, subsequently refined by Williams *et al.* (1998, 2002), was that the lava conduits at Kambalda were emplaced rapidly as high-volume, rapidly-flowing, turbulent flows, and that erosion and flow emplacement happened essentially instantaneously and simultaneously. Cas *et al.* (1999) and Cas and Beresford (2001) have criticized this model on the grounds that these assumptions are unrealistic, and specifi-cally, as discussed above, that turbulent flow is likely to be found only close to the vent. Rice and Moore (2001) argued that thermal erosion beneath lava flows is impossible, based on numerical analysis and analogue wax models. However, this contradicts all of the increasingly wide range of geological, geochemical and isotopic evidence for thermomechanical erosion beneath mid to distal komatiite lava conduits (see Chapter 10 and reviews by Lesher (1989), Lesher *et al.* (2001), Lesher and Keays (2002), Barnes (2006)) and the evidence that has been accumulating for many years for thermal erosion beneath basaltic lava flows. The critical evidence came when the process was observed in action.

Evidence for substrate erosion by tholeiitic basalts, which are much cooler and much more viscous than komatiites, has been established for decades (see the review by Greeley *et al.* (1998)). Field observations on the Cave basalt on Mount St Helens, for example, indicate minimum amounts of down-cutting of 5–15 m, with the maximum extent of erosion corresponding to the maximum palaeoslope. The erosion process was both thermal and mechanical. Other localities include the Ainahu Ranch and Earthquake lava tubes on Kilauea, where lava tubes cut down through underlying palaeosols into earlier flows (Swanson, 1973). The existence of the phenomenon was put beyond doubt by observations of active lava tubes on Kilauea by Kauahikaua *et al.* (1998), who determined average down-cutting rates of up to 10 cm per day, by direct measurement of the distance between the upper crust of a partially-filled tube and the base of the flowing lava over a period of months.

Based on these observations, substrate erosion in basalts involves several stages: (1) the establishment of a well-insulated lava tube by the processes of inflation and radial cooling discussed above; (2) prolonged lava flow through the tube, which raises the temperature of the floor basalt to its liquidus; and (3) mechanical plucking and entrainment of blobs and fragments of plastic, partially-molten floor material into the eroding lava. The process is therefore most accurately referred to as thermomechanical erosion. Erosion is strictly limited to major long-lived lava pathways, post-dates the inflation phase of initial flow emplacement, and pre-dates the olivine accumulation phase of emplacement, precisely as inferred for the lava conduits at Kambalda (Lesher *et al.*, 1984; Lesher, 1989). No erosion whatsoever occurs at the base of newly-propagating pahoehoe lobes, as demonstrated dramatically by the lack of erosion at the base of Kilauea pahoehoe lobes emplaced on asphalt roads. This lack of erosion is consistent with the calculations of Cas *et al.* (1999) and completely negates the criticism of Rice and Moore (2001).

The degree of erosion depends on the melting temperature of the substrate, which is greater for felsic substrates than for mafic substrates, greater for glassy or aphanitic substrates than for phaneritic substrates, and greater for unconsolidated substrates than for consolidated substrates (Williams *et al.*, 1998, 2002). Erosion rates also vary with distance from the discharge site, although the lava in well-insulated tubes or conduits may remain near the eruption temperature over great distances (Heliker *et al.*, 1998). The presence of significant amounts of erosion beneath Kilauea basalt lava tubes (Kauahikaua *et al.*, 1998) and only minor amounts beneath Kambalda komatiite lava channels (Lesher *et al.*, 1984; Lesher, 1989), both of which overlie basaltic substrates, may reflect differences in the surface properties of the basalts (more fractured on Hawaii) that facilitated mechanical erosion or differences in the duration of eruption (longer on Hawaii).

Direct evidence for thermomechanical erosion by komatiites comes mainly from detailed studies of Fe–Ni–Cu sulfide deposits, with particularly good evidence coming from the Alexo locality in the Abitibi belt (see Chapter 10). In the present context, the critical question is how large thermal erosion channels could have been, and in many Archean locations the evidence is greatly complicated by poor exposure, structural complexity and metamorphism. The lenticular morphology of the Perseverance dunite body and the linear nature of the Kambalda troughs have been the examples most commonly cited, but in both cases the primary geometry has been substantially modified by deformation, and except at Kambalda where the depths have been determined by mapping the footwall rocks in least-deformed areas (Lesher, 1989), the depth of the erosional troughs cannot be established unequivocally (Trofimovs *et al.* 2003).

New evidence has emerged from better preserved, less deformed localities. Clear evidence of thermomechanical erosion is recorded in continuous, glacially-polished exposures at the old Alexo nickel deposit, where a lava conduit clearly transgresses underlying pillowed andesites (Figure 10.5). Another example is seen in the almost continuous vegetation-free outcrop at the Katinniq deposit in the Raglan belt of northern Quebec, where the lava conduit cuts progressively downwards through bedding in underlying semipelites and through igneous layering in a thick gabbro sill, producing a very large embayment that completely contains the ultramafic complex (Figure 10.8).

This brings us back to the question of whether komatiites flowed turbulently. The flow regime of lava in the eroding Kilauea lava tube was laminar throughout, which means that turbulent flow is not necessary, consistent with theoretical modelling (Kerr, 2001). The key appears to be the development of long-lived lava channels or conduits, which allows prolonged pre-heating of the floor beneath the conduits. However, the deep embayments implied by Katinniq and the geometry of the Yilgarn dunite lenses require a substantial energy input, as addressed by the various fluid dynamic models which have been attempted.

The original Huppert and Sparks model envisaged very-large-volume flood flow eruptions, with flows bounded by very thin, ephemeral crusts. Jarvis (1995) made several improvements to the Huppert and Sparks model, including a two-dimensional analysis that simulated the undercutting of the embayments at Kambalda mapped by Lesher (1983, 1989). Williams *et al.*

(1998, 1999, 2002) made several additional refinements, making allowance for convective heat loss in overlying seawater, compositional changes in the lava due to the crystallization of olivine and assimilation of substrate, and consequent changes in rheological properties, among others. These modifications resulted in a reduction in the erosion rate by a factor of 4–5 times the Huppert and Sparks estimates.

The field data and fluid dynamic models therefore reach the same conclusion that very extensive erosion is possible beneath komatiite flows if the flow is channelized. With all else equal (e.g. discharge rate, slope, magma physical properties), the degree of erosion will be greater for more felsic substrates, for unconsolidated substrates and especially for water-saturated unconsolidated substrates. For example, Williams *et al.* (1998) concluded that a 10 m thick flow could have fluidized and eroded wet sediment at a rate of 20 cm to 5 m per day at distances of 10–100 km from the source. Embayments tens–hundreds of metres deep could have formed over periods of prolonged flow of the order of weeks.

The high erosion rates seemingly permitted by the model raise interesting possibilities regarding the Perseverance and Mt-Keith-style dunite lenses. As the modelling of Jarvis (1995) suggests, deep erosion channels in low-melting, felsic substrates could gradually transform into deeply undercut trenches. In bimodal komatiite–dacite volcanic settings at Mt Keith and Black Swan, these trenches appear to have become so erosive on their margins that they became essentially intrusive, as proposed for Katinniq by Lesher (2007). This may well be the explanation for the field relationships observed by Hill *et al.* (2004) at Black Swan and conceivably even those of Rosengren *et al.* (2005) at Mt Keith. Because of thermomechanical erosion and the effects of superimposed deformation and metamorphism, distinguishing between extrusive flows, deeply erosive flows, invasive (downward burrowing) flows, and intrusions becomes very difficult.

One further aspect of thermal erosion needs to be considered, and that is the relative timing of erosion and cumulate formation. Contamination accelerates olivine crystallization, but the formation of a blanket of accumulated olivine crystals at the base of the flow inhibits thermomechanical erosion. A delicate balance would have to exist between the two processes. We consider this balance within the broader context of the crystallization and solidification of komatiite flows.

Timing of crust development

An underlying theme throughout this discussion is that komatiite flows were dynamic and need to be understood in dynamic terms. Not all of the components of a komatiite flow, as observed at one place, necessarily formed at the same time. A case in point is the relative timing of crust formation in and erosion beneath the channelized sheet flows at Kambalda, which is discussed in detail in Chapter 10.

There is evidence in the flows at Victor and Lunnon that the A_1 upper crust and A_2 spinifex zone of the channel facies formed after the A_1 upper crust of the flanking sheet facies, and that the A_3 spinifex zone and uppermost part of the B zone of the channel facies formed after the majority of the cumulates in the B zone (Lesher, 1989). This means that the models of Barnes *et al.* (1983) and Hill (2001), which have the A_3 spinifex zone forming simultaneously with the cumulate zone, are not applicable to this system. The applicability of the Lesher or Barnes–Hill models to other channel facies is limited by the availability of data. Evaluating the timing of spinifex and cumulate zones requires very detailed and complete geochemical and mineralogical profiles through entire flows, and such data are currently scarce, at least in the thicker flows. Further work is needed here, and the general issue of the timing and stability of crusts on dynamic thick flows remains an open question

9.5 Crystallization of komatiite liquids and the origin of distinctive komatiite textures

Mechanism of growth of spinifex texture

As we have seen in Chapter 3, coarse-grained platy olivine and acicular pyroxene spinifex textures are some of the most extreme examples of dendritic crystal growth observed in any igneous rocks. The existence of such coarse-grained spinifex textures within rapidly-cooled submarine lava flows represents another of the major paradoxes in komatiite volcanology.

Although early workers (Viljoen and Viljoen, 1969b; Nesbitt, 1971; Green, 1975) referred to these textures as 'quench' textures, Donaldson (1976) demonstrated in a series of dynamic cooling experiments that dendritic olivines similar (but not identical) to those in A_3 zones formed under conditions of strong undercooling ($\sim 50\,°C\,h^{-1}$), not necessarily under conditions of rapid cooling. Donaldson (1982) noted that those rates are much higher than the $\sim 1\,°C\,h^{-1}$ rate calculated for the position of the A_3 layer in a relatively thick 5–10 m komatiite flow, assuming that the flow was cooling conductively through an upper crust. Olivines crystallizing at these cooling rates should have relatively equant, hollow polyhedral morphologies like those in cumulate B zones.

Many hypotheses have been advanced to explain the discrepancy. Turner *et al.* (1986) suggested on the basis of fluid dynamic calculations and some

ingenious analogue tank experiments that spinifex growth was aided by convection within the flow. Their experiments indicated that ponding of buoyant,
olivine-depleted rejected solute would be expected between the growing crystals at the top of the flow. This developing boundary layer would inhibit the
growth of all but those favourably oriented crystals which penetrated the
boundary layer into the underlying convecting lava core (Figure 9.3).

Noting that komatiites could have erupted superheated (Lesher and Groves,
1986; Echeverría, 1980), Arndt (1994) suggested that delayed nucleation triggered dendritic growth at lower than normal cooling rates – a process that
would be analogous to the formation of dendritic ice crystals on the surface of
a pond. Shore and Fowler (1999) suggested that anisotropic conductive and
radiative heat transfer along olivine plates, enhanced by convection of seawater through cracks in the flow crust, increased cooling rates deep within the
flow. Furthermore, olivines are anisotropic with regard to their thermal conductivity and capacity to transmit radiant heat, and crystals oriented with their
c-axes normal to the cooling front (the right orientation for platy spinifex)
conduct heat fastest. Williams *et al.* (1998) have shown that convection of the
overlying water column also has a major effect on the rate of cooling. All of
these effects may well contribute to the process, but no one hypothesis completely solved the problem.

An important breakthrough came with a series of experiments by Faure
et al. (2006), who were the first to successfully grow A_3 platy olivine spinifex
textures under controlled laboratory conditions (Figure 9.3). The key factor
turned out to be the thermal gradient. Platy olivine spinifex, with oriented
plates identical to those in komatiite flows, developed at appropriate cooling
rates of 2–5 °C h^{-1} in thermal gradients of 7–35 °C cm^{-1}. Such conditions
would prevail approximately 60 cm below the top of the flow crust, which is
exactly where the transition from A_2 (random) to A_3 (platy) spinifex is typically observed (Figure 9.3).

This observation turned out to be the key to understanding the whole
process of formation of A zones of komatiite flows. The upper crust of the
flow forms under a combination of very high thermal gradients and very high
cooling rates which, as the Faure experiments show, produces fine-grained
random spinifex typical of the A_1 and A_2 zones. At the critical point at the base
of the A_2 zone, high thermal gradients coincide with declining cooling rates,
and dendritic olivine plates begin to form at degrees of supercooling of the
order of 40–50 °C. Plates that nucleate in the ideal orientation, i.e. with their
c-axes oriented vertical to the isotherms, grow rapidly and are able to penetrate
the boundary layer of olivine-depleted liquid which forms around them. At
this point the preferential heat transfer along the olivine c-axis (Shore and

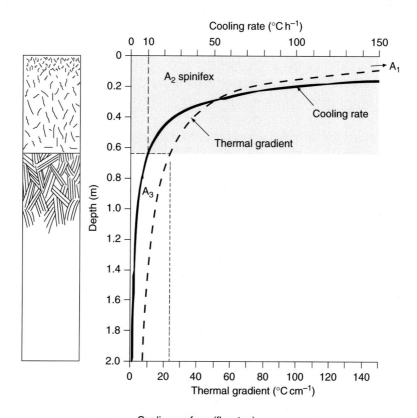

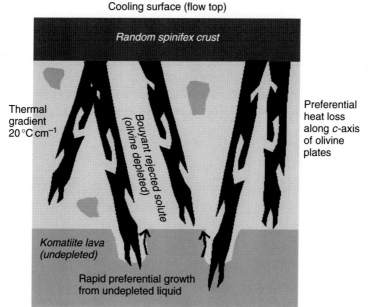

Fig. 9.3. Cooling rate vs thermal gradient plot for a hypothetical komatiite flow (after Faure *et al.* (2006)), and cartoon showing the various processes which have been postulated to operate during the development of spinifex textures.

Fowler, 1999) and ponding of buoyant rejected solute both come into play, helping to maintain a steep thermal gradient around the tips of the growing dendrites and establishing a zone of depletion between the plates. Olivines in the other orientations grow much more slowly, intersect the more rapidly-growing vertical plates and stop growing. The parallel A_3 plates therefore develop via a kind of crystallographic Darwinism – survival of the longest and best oriented.

This raises the very important question of why spinifex textures are confined to komatiites, if the thermal gradient is the dominant control. Steep thermal gradients might also be expected at the margins of submarine olivine-saturated tholeiite flows. A number of factors are probably involved. The main one is the unique property of komatiites of having such a very wide temperature interval (up to $460\,^{\circ}C$) between liquidus and solidus (Chapter 6), for a large part of which olivine (with minor chromite) is the only liquidus phase. Komatiite flow lobes therefore have the potential to develop very high thermal gradients, sustained during long periods of olivine growth, and for crystallization to proceed in conditions in which sparse crystals are aligned perpendicular to the flow top in abundant silicate liquid.

This potential is enhanced by the very low viscosity, which facilitates rapid settling of any suspended olivine crystals. This has two effects: (1) it clears the melt in the upper part of the flow of nuclei, helping to delay crystallization and promote supercooling; and (2) it forms a blanket of olivine cumulates on the floor of the flow, which insulates the flowing lava from the cooler footwall rock. This means that very high thermal gradients can be maintained for long periods across the base of the growing A_3 zone, which is sandwiched between the top of the B zone (maintained close to the liquidus temperature) and the top of the crust (maintained close to ambient temperatures by convecting seawater).

Crystallization and accumulation of polyhedral olivine

As is clear from the idealized profiles in Figure 9.1, many komatiite units contain large excesses of olivine. The bulk composition of the unit is commonly much higher in olivine components (MgO being the most obvious chemical manifestation) than the composition of the chilled margin. This olivine component is 'cumulus' in the strict petrological sense (Irvine, 1982): it represents a concentration of crystals formed at the liquidus of the parent magma. This point is emphasized by the 'real life' profiles in Figure 9.4, which are simplified versions of well-documented flows (or sills) in a number of localities, showing the major zonal subdivisions, and the MgO profile and (where available e.g. Figure 2.20) data on olivine compositions.

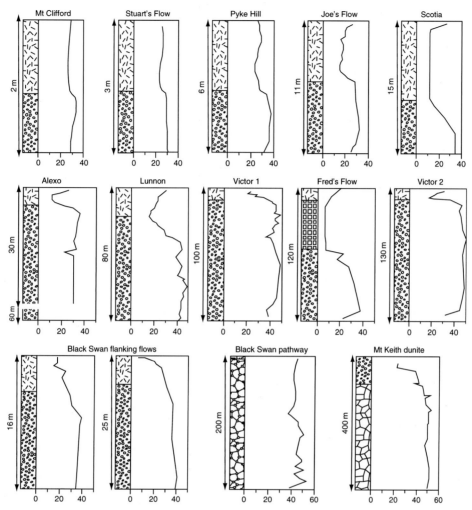

Fig. 9.4. Summary of published MgO profiles through a variety of komatiite units, from thin to very thick. Data sources, *Pyke Hill*, Arndt, unpublished data; *Scotia*, Page and Schmulian (1981); *Alexo* (Arndt, 1986b); *Mt Clifford*, Barnes *et al.* (1973); *Stuart's Flow* and *Fred's Flow*, Huppert *et al.* (1984); *Joe's Flow*, Renner *et al.* (1994); *Victor* and *Lunnon*, Lesher (1989); *Black Swan*, Hill *et al.* (2004); *Mt Keith*, Barnes (1998). Patterns as in Figure 9.11.

The proportion of excess olivine can be estimated from the equation:

$$\% \text{ excess olivine} = 100 \times \frac{\text{MgO}_{\text{flow}} - \text{MgO}_{\text{lava}}}{\text{MgO}_{\text{olivine}} - \text{MgO}_{\text{lava}}} \qquad (9.1)$$

where MgO_{flow} is the weighted bulk weight per cent of MgO of the flow, MgO_{lava} is the weight per cent of MgO of the chilled margin (including

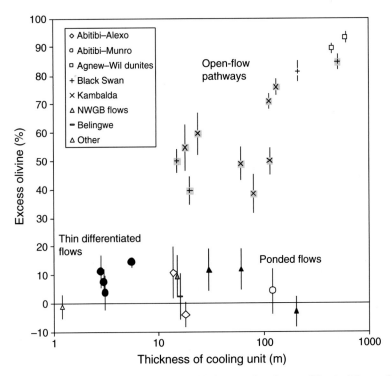

Fig. 9.5. Plot of excess olivine vs flow thickness, for the profiles in Figure 9.4, calculated from equation (9.1): Agnew–Wil dunites = Perseverance, Mt Keith; NWGB = Norseman Wiluna greenstone belt; 'others' includes profiles through Ultramafic Hill, Buldania, and Hayes Hill flows cited by Barnes *et al.* (1983).

entrained phenocrysts, if any), and $MgO_{olivine}$ is the average weight per cent MgO of the excess olivine. Calculations of this type yield excess olivine contents ranging from 0% for many thin differentiated lobes, through 35% for the thick basal unit at Lunnon Shoot, Kambalda and 65% for the thick basal unit at Victor Shoot, Kambalda (Lesher, 1989), to 85–90% for the Perseverance and Mt Keith dunites (Barnes, 2006 and unpublished data) (Figure 9.5).

There are two broadly defined trends on a plot of excess olivine against unit thickness (Figure 9.5): a 'no excess' trend of less than 20% excess olivine within flows ranging from 3 to 200 m thick; and one of increasing excess olivine content with increasing unit thickness, which forms the basis for the vertical axis on Figure 9.1. The Lunnon Shoot (Kambalda) and Katinniq (Raglan, not shown) profiles fall between the two trends.

The weighted bulk MgO contents of the flows along the 'no excess' trend are similar to their chilled margins, suggesting that they differentiated as closed systems, and that the olivines in the lower cumulate B zones formed by a combination of sedimentation of a small component of transported olivine (represented in the flow-top composition as frozen-in phenocrysts) and fractional crystallization *in situ*. This is the simplest case, although the issue of how the B zones actually accumulated is still a point of debate. Whether or not ponded komatiite lava would convect is a determining factor, as we will see. Regardless of this, understanding the thick, cumulate-enriched flow facies is a difficult problem.

How did all that extra olivine get there? A variety of alternative views have been aired over the years, falling broadly into three categories: (1) one-off emplacement of olivine rich crystal mushes, (2) sedimentation of transported olivine from flowing magma, and (3) *in-situ* accretion of olivine cumulates by nucleation and growth from magma flowing over the top of the crystal pile.

We can immediately dismiss the first alternative. The extreme olivine enrichments observed in thick dunite bodies, or indeed in most of the thick olivine-enriched flows, could never have been emplaced as crystal mushes, because their effective viscosities would be impossibly high. Marsh (1981) concluded that crystallinity of around 36% increases viscosity by an order of magnitude and that 50% crystallinity is the practical limit for flowing lavas. Even allowing for complexities introduced by uncertainties in the degree of clustering of crystals or the influence of crystal shape, it is clear that adcumulates and probably mesocumulates could not have been emplaced as mushes. Flowage differentiation has been proposed as a way around this problem, whereby suspended solids are forced into the centre of a flowing suspension by a mechanism called the Bagnold effect. However, it was shown many years ago (Barriere, 1976) that while this process might operate within flow channels a metre or two wide, it is a very weak effect and is completely inadequate at the scales we need.

The remaining two mechanisms both share an essential feature, which is that the flow channels involved must have been open systems to account for the presence of excess olivine (beyond the 10–25% or so which might be able to be emplaced in crystal mush form). Much, in fact most, of the liquid which either carried or crystallized the olivine has flowed on out of the part of the system we observe. The excess olivine in the lower parts of komatiite units records the passage of many times their own volume of liquid. Just how many times is a matter of speculation, and is virtually impossible to determine. In the cases illustrated in Figure 9.4, the composition of the accumulated olivine becomes more magnesian further up the section (see Fig. 2.20), implying the continuous arrival of hotter magmas with time. This implies that the amount of fractional crystallization represented by each increment of olivine is too small to have had an appreciable effect on the composition of the liquid; the liquid–olivine ratio must have been at least 10:1, but as far as we can tell from the chemical profiles, it may have been orders of magnitude more. We have to consider where the missing went: presumably to the margins of extensive flow fields.

Distinguishing between nucleation and growth *in situ* on one hand and mechanical sedimentation from phenocryst-bearing lava on the other requires us to think about the detailed crystallization mechanisms of the various types of olivine cumulates: orthocumulates (coarse- and fine-grained, laminated and not), mesocumulates, adcumulates and dendritic crescumulates. We will start with the most common type of olivine cumulate, characteristic of most thin flows: the fine- to medium-grained orthocumulate-textured B zones composed of sub- to euhedral polyhedral olivines, often showing mildly skeletal 'hopper' morphologies. Patterns of grain size and chemical variation in the cumulus olivine population vary between flows.

Some flows, such as the exceptionally well-preserved 'Tony's Flow' in the Belingwe belt described by Renner *et al.* (1994), have constant olivine core compositions throughout, with some superimposed variability and zoning due to postcumulus overgrowth and re-equilibration. 'Tony's Flow' displays reverse size grading of the cumulus crystals, which Renner *et al.* ascribed to postcumulus overgrowth following an initial phase of gravitational settling of clusters of olivine crystals. In an elegant application of quantitative textural analysis to these rocks, Jerram *et al.* (2003) showed that the spatial distribution of clustered glomerocrysts in cumulates is identical to that quenched into the chilled margin of the flow; differences in modal olivine contents result simply from mechanical accumulation and overgrowth. This similarity provides strong evidence for purely mechanical accumulation of transported crystal clusters.

Other flows, such as the Victor unit at Kambalda described by Lesher (1989), show 'reversed fractionation', where olivine becomes more magnesian with height above the base. This pattern is widespread in the thick dunite lenses and sheets, notably Perseverance (Barnes *et al.*, 1988). This observation clearly requires crystallization in an open system with an influx of more magnesian magmas with time. For example, the thick B zones in the Lunnon and Victor profiles shown in Figures 2.20 and 9.4 commonly comprise coarse, closely-packed olivine mesocumulates, which could be attributed to postcumulus enlargement and filter pressing of trapped intercumulus liquid from original mechanically sedimented orthocumulates. Alternatively, they may have formed by *in situ* nucleation and growth of olivine crystals at the top of the crystal pile. A combination of mechanical sedimentation followed by *in situ* overgrowth (in contact with the lava core) is also an option. We need some criteria to choose between these alternatives.

In the Victor flow, the basal part of the B zone is a crescumulate and typical mesocumulates are intimately interlayered with branching crescumulate olivines which clearly have an origin by *in-situ* nucleation and growth (Figures 3.8(a) and 10.4). The implication is that at least some of the olivine in the B zone of this unit nucleated and grew *in-situ* from flowing magma. Systematic variations in Fo content with stratigraphic height (Figure 2.20), high Ca, Mn and Cr contents of olivine, and intercumulus microspinifex textures are also consistent with *in situ* growth (Lesher, 1989). We have good reason to think this is true of the adcumulate dunites also, as we will see.

The preferred orientation of olivine grains may be an indicator of mechanical sedimentation, as opposed to *in situ* nucleation and growth. Laminated cumulate textures, with elongate or platy olivines aligned parallel to the original horizontal, are observed in some areas but are not common. The extremely coarse-grained orthocumulates of the Black Swan deposit have a strong lamination, but this unit is unusual among orthocumulates in having exceptionally coarse olivines, up to 4 cm in length. The Black Swan unit may be an atypical case of mechanical sedimentation of transported intratelluric phenocrysts.

This brings us to another question. If the accumulated olivine in some komatiites was mechanically deposited, where did it originally nucleate? Either it nucleated and grew after eruption during flow of the lava, or the lavas were erupted carrying 'intratelluric' phenocrysts derived from the sub-volcanic plumbing system. There is no definitive way to resolve this question, but we can make some inferences based on theoretical considerations, the populations of phenocrysts within chilled margins of flows and also the degree of equilibration between olivines and transporting liquids.

The abundance of phenocrysts in chilled margins varies widely. Phenocrysts are relatively rare in the chilled margins of many East Yilgarn komatiites, which along with other arguments regarding low viscosities and rapid ascent rates, led Lesher and Groves (1986) to suggest that most komatiites were superheated when they erupted and contained no intratelluric phenocrysts.

Renner *et al.* (1994) reported 16–23% polyhedral olivine phenocrysts within chilled margins of the particularly fresh, well-preserved SASKMAR flows. Olivine cores have roughly constant compositions at around Fo_{92} throughout the B zones, with a small population of larger, more forsteritic grains up to nearly Fo_{94}. Olivines were transported, and accumulated, as glomerophyric clusters, and the average core olivine composition appears to be consistent with equilibrium with the liquid component of the chilled margin (although regrettably Renner *et al.* did not provide full whole-rock analyses to allow this to be tested). A plausible interpretation is that the large, more magnesian, phenocrysts represent a minor component of intra-telluric (pre-eruption) phenocrysts, whereas the bulk of the B zones is made up of clustered phenocrysts that grew in the lobe cores upstream within the flow field.

Clustered phenocrysts, or glomerocrysts, with restricted compositions in equilibrium with the quenched liquids, are exactly what we would expect from such 'upstream growth'. Clusters of crystals typically form at mild degrees of supercooling, as a result of the well-known kinetic process of heterogeneous nucleation. It is much easier for crystals to nucleate on existing crystals of their own type than to nucleate on their own from a homogeneous body of solute (Campbell *et al.*, 1978). This is the origin of the common clustering of like crystals in igneous rocks (Jerram *et al.*, 2003; Jerram and Cheadle, 2000). Clustering, as revealed by the type of quantitative textural analysis practised by Jerram and coworkers, combined with evidence for well-equilibrated compositions, may be the key to revealing mechanically sedimented, endogenously-derived crystal populations. Nevertheless, the abundance of quenched-in phenocrysts in the SASKMAR flows is anomalous, and most komatiites contain many fewer. Other examples of chilled margins containing olivine phenocrysts are described in Chapter 3.

There are several explanations for the variations in phenocryst contents, the most obvious being that even if the komatiites erupted without any intratel-luric phenocrysts, they may have crystallized as primocrysts during emplacement. This is supported by the presence of two populations of olivine (a smaller population of higher Fo olivines representing less-evolved lava and a larger population of lower Fo representing more-evolved lava) in the SASKMAR flows. The high-Fo olivines in conduit facies at Alexo, Kambalda and Perseverance must have crystallized from relatively unevolved magmas, suggesting that they did not contain abundant phenocrysts. There may also be a sampling bias at work. With the exception of the lava channels discussed above, chilled flow tops represent the earliest-formed crust inherited from the original breakout of the lobe; in a dynamic compound flow system, transported phenocrysts may lag behind in lobe cores and not be represented in the initial breakouts. This has been observed in Hawaiian picritic flows (S. J. Barnes and R. E. T. Hill, unpublished data). Some care is therefore needed in making broad conclusions from chilled flow tops, especially in the Western Australian examples, where the degree of alteration is often intense.

Relative degrees of equilibration between cumulus olivine and 'host' liquids are also a potential criterion for assessing the relative amount of transport of olivine crystals. In most cases fresh olivine is unavailable for analysis, so we rely on trends on whole-rock variation diagrams (Chapter 5), such as FeO vs

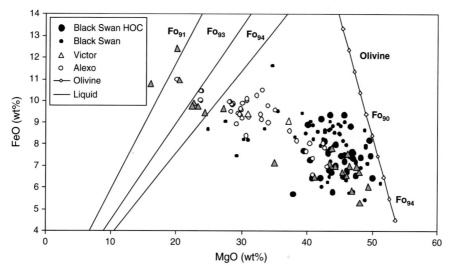

Fig. 9.6. Plot of FeO (corrected for Fe_2O_3 in the liquid) vs MgO (both elements recalculated volatile free), showing data for Alexo (Abitibi), Victor (Kambalda) and Black Swan; the trend for pure olivine compositions is labelled with forsterite content, and the lines for liquid compositions in equilibrium with the specified olivine composition. (HOC – 'Hopper' (hollow) olivine orthocumulate.)

MgO (Figure 9.6). On this plot and other variations (see Chapter 12 and Lesher (1989)), flows containing mixtures of komatiite liquid and well-equilibrated cumulus olivine of roughly constant composition should define straight-line trends, intercepting the pure olivine line at the average olivine composition, as discussed in Chapter 5. Although there is some variation reflecting systematic variations in olivine and liquid compositions in the Alexo and the Victor flows (e.g. Fo_{94-91} and 31–16% MgO at Victor), the trends are consistent with flow-through under relatively steady-state conditions with *in situ* growth of *most* of the olivine from a liquid with a narrower compositional range. A very different and much more erratic pattern is observed in the Black Swan cumulate-rich flow, characterized by a strong preferred orientation of elongate and extremely coarse olivine grains, implying transport and mechanical deposition. Here there is no linear trend, and the accumulated olivine evidently had a widely variable composition bearing no systematic relation to position within the flow. This is another reason to identify Black Swan as a good candidate for a large accumulation of poorly-equilibrated intratelluric phenocrysts, inherited from pre-eruption processes. This diagram does not prove the point, but it does provide supportive evidence for variability in process and product.

Crystallization of adcumulates

A number of features of the coarse-grained adcumulates in thick dunite units suggest *in situ* growth. As noted above, earlier suggestions that these bodies formed by flowage differentiation of a sulfide-liquid-bearing olivine-rich crystal mush (Binns *et al.*, 1977; Burt and Sheppy, 1975; Marston, 1984) are unfeasible on rheological grounds. Coarse-grained, entirely monomineralic adcumulates are virtually unknown outside major layered intrusions, and the komatiite bodies are unique in containing adcumulates within cooling units as thin as 200 m. Clearly the 'filter pressing' models for porosity elimination in thick crystal piles, commonly applied in layered intrusions, are inapplicable here, for a number of reasons discussed in detail by Hill *et al.* (1995). They are probably wrong in layered intrusions too, for reasons discussed by Campbell (1978, 1987).

Processes involving convection of the interstitial melt can reduce the porosity from the orthocumulate to mesocumulate range (Kerr and Tait, 1985), but below 10% porosity the permeability seals up, locking in the remainder of the pore fluid (Wilson, 1992). The dunite bodies typically show interlayering of mesocumulates and adcumulates on a scale of tens of metres, and olivine shows fine-scale fluctuations in grain size related to presence of interstitial sulfide, a feature incompatible with pore fluid migration at a scale greater than a few tens of metres. Dunites at Mt Keith and Mt Clifford contain a highly unusual poikilitic chromite texture (Barnes and Hill, 1995) in a pure biminera-lic olivine–chromite rock. The chromite could not have crystallized from trapped liquid, which would only have contained 0.5% or so of dissolved chromite, and is interpreted instead as developing from bursts of growth of isolated chromite grains from Cr-supersaturated magma at the top of the crystal pile, locally overtaking slower-growing neighbouring olivines. Like the mesocumulates at Victor, adcumulates are in a number of cases inter-layered with coarse-grained crescumulates.

All of these lines of evidence imply that the adcumulates grew, at least in part, by *in-situ* accretionary growth within the top few centimetres of the crystal pile, under conditions of minimal supercooling. Turbulent flow provides a mechanism for breaking down depleted boundary layers around growing crystals, favouring the growth of existing crystals over nucleation of new ones (Figure 9.7). This allows the development of coarse-grained adcumulate textures despite shallow depths of emplacement.

Barnes *et al.* (1988) and Hill *et al.* (1995) concluded that adcumulates form in major feeder conduits, by rapid flow-through of magma crystallizing very close to the liquidus temperature at very low degrees of supercooling. The upward increase in Fo content of olivine observed at Perseverance and

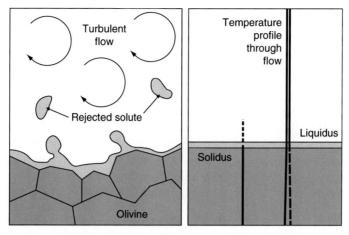

Fig. 9.7. Cartoon illustrating crystallization of olivine adcumulates.

elsewhere is most likely a consequence of thermal maturation of the upstream flow pathway: prolonged flow of magma progressively lines and heats the conduits, possibly all the way from the mantle source. Consequently successive magmas arrive hotter with time at any given point along the flow trajectory, rather like the water from the bathroom tap, having lost less heat along the way.

Origin of mesocumulates

Converting an orthocumulate to a mesocumulate involves physical removal of trapped liquid to reduce the porosity, and can be achieved either through compaction or through crystallization convection of liquid through the inter-cumulus pore space (e.g. Kerr and Tait, 1985, 1986). In thin komatiite flows, the time available for strain-related compaction is too short, leaving only two possibilities: mesocumulates form either in a similar way to adcumulates, by *in situ* crystal growth down to low porosities within tens of centimetres of the top of the cumulate pile, or through extensive convective circulation through the crystal pile. Both processes would produce similar bulk rock compositions, zoning profiles and other features and there is no obvious way of distinguishing between them. Renner (1989) concluded on physical grounds that inter-cumulus convection was unlikely in 10 m thick flows, but the question remains open in thicker flows.

9.6 Models for the emplacement and solidification of komatiite flows

As we have seen, there is a spectrum of komatiite flow profiles that broadly falls on a matrix of two variables: internal differentiation and accumulation of excess olivine (Figure 9.1). These variables define two trends on a plot of excess olivine content vs unit thickness (Figure 9.5). One of the two trends is of increased thickness without extra accumulated olivine, and reflects some form of ponding and *in situ* closed system differentiation. The other is of increased olivine accumulation with increased thickness, and reflects prolonged flow-through and olivine accumulation. Simultaneous inflation and olivine accretion, by either *in situ* growth or mechanical sedimentation, provide a good explanation for the correlative trend: the longer the flow channel is active, the more olivine accumulates and the thicker the final package. This mechanism provides a solution to the paradox of thick flows and low-viscosity lavas. Some flow units, like the Lunnon examples (Figure 9.4), go through an initial inflation–accretion stage and then pond to form thick upper differentiated spinifex zones and complementary low-Mg cumulates. In this concluding section, we present a series of models to account for the range of natural variability.

Solidification of thin differentiated flow lobes

We consider first the simplest one-dimensional case of a ponded komatiite flow lobe, losing heat to seawater, and quenched against a cold floor, which may well be an earlier lobe formed within the same compound flow.

This model, and those that follow, starts from the premise that the thermal profile through a cooling sheet of lava has the general conductive form illustrated in Figure 9.8, after Renner (1989) and the empirically-calibrated conductive model of Peck *et al.* (1977) based on the Alae Lava Lake in Hawaii. Heat loss is faster through the top of the flow than the base, as the top of the flow is maintained at ambient temperature by convecting seawater, while the temperature of the floor rocks initially increases due to conductive heat transfer from the flow. Consequently, the last part of thin, non-cumulate flows to solidify is below the centre of the flow. The exact shape of this profile will vary widely according to the thickness, the presence or absence of convection in the liquid core and other factors, but this overall geometry is fixed.

Thin flow lobes are emplaced as components of compound tube-fed flows analogous to modern pahoehoe flows formed at low eruption rates, and cool primarily by conduction (Figure 9.9, top). Generally low phenocryst contents in chilled margins imply that they were emplaced in most cases as crystal-poor lava breakouts, which form a viscoelastic crust almost instantaneously. The upper A_1 and A_2 zones form as part of this crust by simple progressive quenching. This establishes a conductive thermal gradient into the upper parts of the flow. Once the crust is about 60 cm thick, the thermal gradient

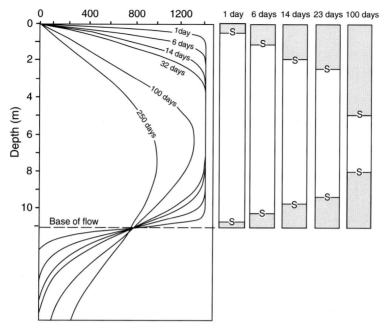

Fig. 9.8. Temperature profile through an 11 m thick komatiite flow based on modelling of Renner (1989), combined with models and observations on the comparably thick Alae lava lake, Kilauea (after Peck *et al.* (1977)). Temperatures are computed using a purely conductive model, assuming a thermal conductivity of liquid and solid komatiites of $3.0 \, \mathrm{J \, m^{-1} \, K^{-1} \, s^{-1}}$, and assuming uniform release of the latent heat of crystallization, following the method of Carslaw and Jaeger (1959). The columns on the right show the position of the solidus isotherm (S) at different times.

immediately beneath it (Figure 9.3) is enough to trigger the growth of coarse plate dendrites, or pyroxene needles in more evolved liquids. These dendrites preferentially grow down into the interior of the flow, their parallel growth being driven by two major factors: (1) accumulation of buoyant, olivine-depleted fractionated liquid in the space between the plates, and (2) anisotropic radiation and conduction of heat along the long axis of the appropriately oriented plates. The spinifex dendrites continue to grow while the composition of the liquid in the molten core of the lobe gradually evolves due to olivine crystallization. This accounts for the gradual downward decrease in the MgO content of the A zone observed in several of the profiles in Figure 9.4.

Simultaneously, polyhedral olivines are growing in the slowly-cooled molten core of the lobe. In relatively thin flows, in the absence of transported polyhedral olivine phenocrysts, euhedral and hopper olivines probably grow

Thin ponded flow

(a)

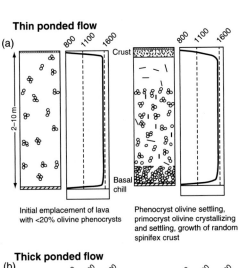
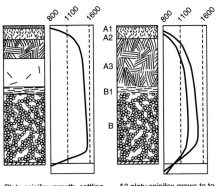

Initial emplacement of lava with <20% olivine phenocrysts

Phenocryst olivine settling, primocryst olivine crystallizing and settling, growth of random spinifex crust

Platy spinifex growth, settling of platy dendrites to B1 Zone

A3 platy spinifex grows to top of B1 zone, slow cooling towards final solidification

Thick ponded flow

(b)

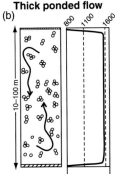
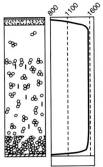
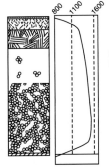
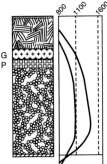

Initial emplacement, convection inhibits settling

Convection wanes as crust thickens, polyhedral olivine clusters crystallising and settling

A_3 spinifex growth from fractionating liquid, polyhedral olivines grow in interior and settle

Pyroxene crystallisation to form spinifex and upper B zone pyroxenite (P), final crystallization of eutectic gabbroic melt (G)

Dynamic pathway

(c)

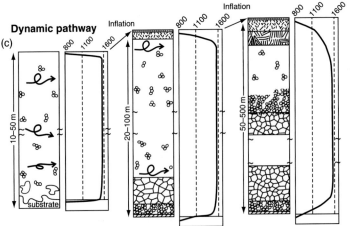
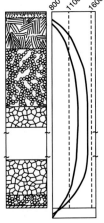

Turbulent flow through open channel/tube, thermal erosion of substrate

Continuous flow, transient thin crust, *in situ* adcumulate and mesocumulate accretion at low cooling rates at base, rapid heat loss through thin crust

Waning flow rate, transition to laminar flow, orthocumulate and harrisite (H) growth at top of thick cumulate pile, thick crust stabilised, onset of plate spinifex growth

Ponding due to blockage of pathway, A_3 spinifex and upper cumulates crystallise from fractionated lava as heterogeneously nucleated clusters (Jerram *et al.*, 2003). These crystals continue to increase in abundance in the lower part of the flow. Exactly how they accumulate depends on whether or not the liquid core is convecting.

Arndt (1986a) and Turner *et al.* (1986) considered that convection was likely in the liquid core, even for thin flows. If so, the lava core continues to convect while the overlying spinifex zone forms, gradually becoming enriched in polyhedral olivine crystal clusters. The effective viscosity of the suspension increases as the crystallinity increases and the temperature drops, until the rheology evolves to the point where convection is no longer possible at around 40–50% crystals. From this point the lower part of the flow is transformed into the B zone, which then solidifies via conductive heat loss to the floor and crystallization of the trapped pore liquid (we discuss the perplexing B_1 zones below). At this point the cumulus crystals are modified by postcumulus overgrowth, becoming reverse zoned and in some cases slightly compacted (Jerram *et al.*, 2003). Renner *et al.* (1994) came to the opposite conclusion from that of Turner *et al.*: they concluded that convection within the liquid core of the flow was weak, and that formation of the B zone took place largely by gravitational settling of transported poly-hedral olivine clusters. In both models, however, the low-Mg, fractionated part of the A_3 spinifex zone is the last part of the flow to crystallize (Figure 9.9).

Solidification of flow lobes without spinifex texture

There are two possible origins for thin 'undifferentiated' flow lobes, those consisting entirely of fine-grained olivine porphyritic rocks without spinifex. These commonly occur intercalated with thin flow lobes containing spinifex in Pyke-Hill-type sequences.

The first possibility is that they may have contained a higher proportion of transported olivine phenocrysts, which is consistent with their more irregular morphology (Arndt *et al.* 1977). These units may represent olivine-choked tubes, in which the content of phenocrysts had built up to the point where the effective viscosity was very high and flow ceased. The high viscosity, arising in part from the clustered nature of the phenocryst distribution, further inhibited

Caption for Fig. 9.9.
Series of qualitative theoretical profiles showing thermal and textural development of komatiite flows of varying thickness with time. (a) Thin ponded flow; assumed to contain approximately 15% transported olivine phenocrysts; (b) thick ponded flow or lava lake, also with 15% initial phenocryst load, undergoing convection of liquid core in early stages, followed by ponding and *in situ* differentiation; (c) dynamic lava channel, tube, or pathway with minor phenocryst content, undergoing progressive inflation and olivine accumulation during flow.

the ability of the olivine clusters to settle. However, if we examine the bulk compositions of these units, we see that they contain no excess olivine (Figure 9.5). This means that most of the olivine crystallized *in situ*.

A second possibility, suggested by Turner *et al.* (1986), arises from the constraints on spinifex growth. Spinifex development requires that the thermal gradient be maintained long enough to allow spinifex blades to grow; this takes at least hours or possibly days. If another flow lobe, perhaps formed as a breakout from the same master tube, arrives on top of the crystallizing lobe within this time, the thermal gradient will be reduced or eliminated and spinifex growth will be impeded. A very rapidly emplaced stack of thin lobes will develop a sinusoidal thermal profile as a result of residual heat within the interior of the lobes, which is not dissipated before the next one arrives (Gole *et al.*, 1990). In extreme cases of rapidly emplaced very thick flows, Gole *et al.* (1990) showed that previously formed spinifex texture could be thermally metamorphosed and even remelted. In the simpler and more common case of rapidly emplaced thin lobes, sudden pulses of lava through the tube system could produce rapidly emplaced stacks of thin lobes whose spinifex content will be a function of exactly which lobes arrived when.

Both mechanisms may operate in different places, but there is good evidence for the first mechanism. In several examples at Pyke Hill, as described by Arndt *et al.* (1977) and shown in Figure 3.4, the same lobe changes along strike from spinifex-textured to undifferentiated, but field relationships preclude a 'blanketing' effect from succeeding flows. The interplay of crystallization during flow, crystallinity, crystal clustering and viscosity provides the best explanation for lateral variability within lobes.

Origin of B_1 zones

The origin of the B_1 zone in komatiite flows has also been a long-standing problem. The reader will recall from Chapter 3 that B_1 zones are commonly, but not always, developed between the top of the typical orthocumulate-textured B_2 zone and the bottom of the coarse dendritic spinifex A_3 zone in fully differentiated flows. The B_1 zone is typically defined by the presence of moderately elongate, hopper olivine crystals, which in many (but not all) cases define a strong lamination parallel to the flow top. Detailed studies of B_1 zones in the Belingwe belt (Silva *et al.*, 1997), at Alexo (Arndt, 1986), and at Kambalda (Thomson, 1989) showed that they contain relatively Mg-rich olivines, consistent with forming early during the crystallization history of the flow, and often display reverse grain-size grading. Some B_1 zones are also enriched in chromite (Hanski, 1980).

One commonly proposed (but probably incorrect) explanation is that the B_1 layer is an accumulation of suspended broken spinifex crystals, with the characteristic layer-parallel lamination being produced during late stage flow-through (Arndt, 1986b; Turner *et al.*, 1986). A version of this model was favoured by Thomson (1989) and Hill (2001), with the additional suggestion that the B_1 zone was triggered by a breakout downstream in the tube system, resulting in flux of new lava through the tube quenching against the cooler A and B zones above and below.

None of these mechanisms fully explains the most puzzling feature of B_1 zones, which is the primitive character of the olivines (Silva *et al.*, 1997) and their skeletal habit, or the concentration of chromite in some units. These two features do not usually coincide because chromite normally crystallizes from komatiitic basalt magmas, not komatiite (*sensu stricto*) magmas (Murck and Campbell, 1986). B_1 olivine evidently began to crystallize early in the history of the flow, before significant amounts of fractionation and (as discussed above) before growth of the spinifex zones, and while cooling rates were high, but the crystallization persisted or accumulation was retarded as the magma fractionated into the chromite stability field.

The explanation for this may lie in the preferential concentration in residual liquid in the flow of suspended solid phases which settle slowly: skeletal magnesian olivines and fine chromite grains. Silva *et al.* (1997) argued that platy skeletal low-density hopper olivine grains settle more slowly than the clumped polyhedral olivines that accumulate early to form the B_2–B_4 zones, consistent with the observation that platy olivine crystals settle much more slowly than equant ones in picritic basalt flows (S. J. Barnes and R. E. T. Hill, unpublished data). Furthermore, owing to their lower density, more magnesian olivines settle more slowly than more iron-rich grains of similar size and shape. Hence, platy magnesian olivines (with or without fine-grained chromite) may be preferentially retained among the population of suspended crystals that finally accumulates as the B_1 layer. If this model is correct, olivine crystallization, fractionation of the lava into the chromite stability field, and settling of olivine and chromite must have occurred rapidly enough (within a few days) for the olivine not to reequilibrate.

Another part of the story of B_1 zones may be that lava flows can deflate as well as inflate, when lava drains out of the tube or sheet flow downstream. This could lead to compression of a liquid-rich mush and preferential accumulation of crystals, flattened by the sudden collapse of the liquid core and preferential squeezing out of liquid. This could account for the fact that B_1 zones are typically laminated but not lineated, as they should be if the alignment was due to flow.

Thick differentiated flows

Strongly differentiated flows with minimal excess accumulated olivine, such as
Fred's Flow (Chapters 2, 3 and 4) must have formed from thick, ponded sheets
of lava, with at most a small percentage of transported olivine phenocrysts.
These could have formed by flood-flow emplacement, by inflation with minimum crystallization, or more likely by topographic ponding, but regardless of
the exact mechanism the main distinction between these and thin flows is the
likelihood of convection within the ponded lava.

Rayleigh number calculations (e.g. Turner *et al.* (1986)) suggest that thick
flows with liquid layers more than 10 m thick would have convected, at least
initially. This tendency would have been reduced rapidly by the growth of the
upper crust, which would have diminished the temperature difference between
the top and base of the liquid interior that drives convection.

Growth of upper spinifex zones would be expected to follow the path
described above for thin flows, although at higher cooling rates the random
spinifex crust might be expected to be thinner. A_3 platy (or needle) spinifex
zones are characteristically thick in ponded flows and sills, in some extreme
cases extending over tens of metres (Stone *et al.*, 1995; Wilson *et al.*, 2003). The
implication here is that a thermal gradient was maintained at the propagating
base of the A_3 zone even when it was several metres thick. Efficient segregation
of liquid from growing spinifex would contribute to the cumulate component
of the spinifex crystals.

Deeper in the flow, growth of suspended polyhedral olivine crystals would
be expected along the lines proposed by Turner *et al.* (1986) and Arndt (1986a),
and a proportion of these crystals would be expected to be retained within the
convecting core. Accumulation would be expected in a basal B zone owing to
'leakage' of crystals from the convecting suspension (Martin and Nokes,
1988). The presence of layered olivine and pyroxene cumulate layers in the
upper part of the cumulate sequence implies that convection had largely ceased
by this stage, such that crystals were able to accumulate gravitationally as they
formed. In the case of Fred's Flow and other strongly differentiated units,
crystallization culminated in the formation of a 'sandwich layer' of gabbro
representing a near-eutectic residual liquid.

Thick cumulate-rich flows

As we have seen, there are two trends on a plot of 'excess olivine' vs thickness
of komatiite cooling units: one of variable thickness with little or no excess
olivine, and one of increasing excess olivine with increasing thickness. Some

units fall between the two trends, but none fall above the positive correlation trend. Thick flows exist with no excess olivine, but thin flows with more than a small percentage of excess olivine, over and above that preserved as pheno-crysts within chilled margins, are not observed (Figure 9.5). From this, we conclude that the processes of bulk olivine accumulation and inflation go hand in hand in large open-system conduits (Figure 9.9(c)).

The model illustrated in Figures 9.9 and 9.10 is one plausible sequence of events, beginning with the spread of a broad thin lobe through laminar flow of initially crystal-poor lava under a thin flexible crust. As the flow begins to cool and crystallize, lava flux gradually becomes concentrated within preferred pathways. Heat loss through the crust is initially rapid, resulting in the forma-tion of a transient A_2 spinifex zone and a suspended load of polyhedral olivine crystals which accumulate gradually at the floor. As this basal cumulate zone develops, it forms a favourable site for nucleation of new crystals, and the cumulate zone accretes through a combination of mechanical sedimentation and *in situ* growth. There is a continuous flow of lava through the preferred pathways beneath a gradually inflating crust, the thickness of the liquid core being a function of the rate of flow and the degree of constriction downstream within the pathway. Particularly large pulses may result in local breaching of the crust and formation of new lateral breakouts.

Most thick flows have relatively constant olivine compositions up through the cumulate zones, and in some cases olivines become more magnesian up-section, implying that the flowing lava at any given point along the flow path either had a steady-state composition or actually became hotter with time. This implies that the floor, walls and roof of the pathway became gradually better insulated with time. In part this would be due to accretion of the crystal pile and the establishment of a thick bed of cumulus olivine crystals close to the liquidus temperature of the lava, losing heat only very slowly to the floor. However, most of the heat loss would have been through the crust, so the implication of the cumulus composition trends is that the crust either attained a steady state or thickened progressively with time.

The timing of crust growth is clearly a critical factor in controlling many aspects of komatiite flow crystallization, and remains probably the least well-understood aspect of the whole process. A critical aspect in the thick flows is the timing of growth of the A_3 platy spinifex zone. In a number of cases, such as the Victor and Lunnon flow profiles shown in Figures 2.20 and 9.4, the A_3 zone shows clear evidence of having formed from progressively fractio-nating lava, implying that it grew in a closed system after cessation of flow. The two Black Swan flow profiles in the same figure show a trend of increasing MgO from A_2 downward into A_3, indicating that in these units at least the

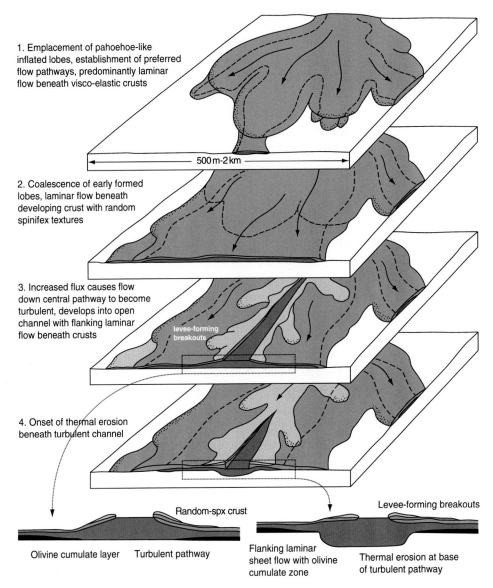

1. Emplacement of pahoehoe-like
inflated lobes, establishment of preferred
flow pathways, predominantly laminar
flow beneath visco-elastic crusts

500 m-2 km

2. Coalescence of early formed
lobes, laminar flow beneath
developing crust with random
spinifex textures

3. Increased flux causes flow
down central pathway to become
turbulent, develops into open
channel with flanking laminar
flow beneath crusts

levee-forming
breakouts

4. Onset of thermal erosion
beneath turbulent channel

Random-spx crust

Levee-forming breakouts

Olivine cumulate layer   Turbulent pathway

Flanking laminar
sheet flow with olivine
cumulate zone

Thermal erosion at base
of turbulent pathway

Fig. 9.10. Schematic block diagram showing emplacement of a thick cumulate-
rich sheet flow with an early episode of thermal erosion.

upper part of the A_3 zone could have been crystallizing while the cumulate pile
was accreting. The detailed data available to evaluate this question are in most
cases lacking, and more work is clearly necessary.

Figure 9.10 shows a cross-section through a thick cumulate-rich flow, with
the additional complication of an episode of thermal erosion beneath the base
of a pathway.

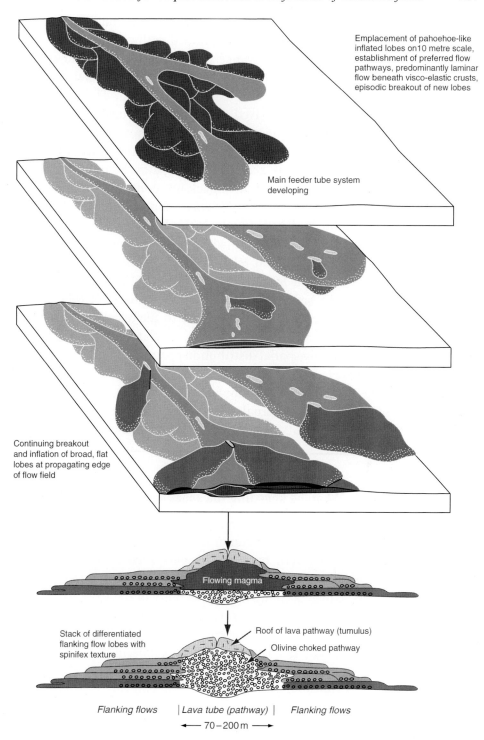

Emplacement of pahoehoe-like inflated lobes on10 metre scale, establishment of preferred flow pathways, predominantly laminar flow beneath visco-elastic crusts, episodic breakout of new lobes

Main feeder tube system developing

Continuing breakout and inflation of broad, flat lobes at propagating edge of flow field

Flowing magma

Stack of differentiated flanking flow lobes with spinifex texture

Roof of lava pathway (tumulus)

Olivine choked pathway

Flanking flows | *Lava tube (pathway)* | *Flanking flows*

◀— 70 – 200 m —▶

Fig. 9.11. Schematic block diagram showing emplacement of a tube-fed compound flow as a series of inflationary lobes.

The relative timing of thermal erosion, inflation and cumulate accretion is critical. An initial phase of prolonged, constrained pathway flow is necessary to heat the substrate to its melting range and to instigate thermal erosion. However, thermal erosion cannot happen once the floor is blanketed with an insulating cumulate layer, so a delicate balance is implied. The sequence of events at Kambalda evidently involved some initial channelling within pre-existing topographic features, followed by a period of erosion and ore formation (covered in more detail in Chapter 10), followed by the onset of the main phase of inflation and cumulus accretion. The pathway must then have become blocked, causing lava within it to back up, pond and differentiate *in situ* giving rise to differentiated A$_3$ spinifex zones and a corresponding upper layer of low-Mg chromite-enriched cumulates (Lesher, 1989).

The final scenario, based on the flow-field emplacement model of Hill (2001) (Figure 9.11), shows a hypothetical relationship between a thick cumulate-rich unit developed within a feeder tube, and thin differentiated flanking flows formed by episodic breakouts from a central pathway or tube system. This sequence of events might be expected close to the distal end of a major distributory tube system, and illustrates the processes at the propagating edge of a flow field. The resulting spatial distribution of flow facies is a microcosm of what might be expected in the flow field as a whole.

To conclude this rather long and complicated discussion, there is no single model which accounts for the internal and external structure of all komatiite units, but much of the diversity can be explained through combinations of a few controlling variables.

9.7 What did komatiite volcanoes look like?

Nobody has ever seen a komatiite volcano erupt, so we can only rely on indirect evidence and inference based on modern analogues. The most readily accessible of these are modern basaltic volcanos such as Kilauea, where volcanologists have been making detailed observations of eruptions for decades. The critical issue then becomes the applicability of these observations to komatiites, given the differences in viscosity and temperature, and the fact that almost all of the komatiite sequences we know about evidently erupted under water. We can go some way to answering this by comparing the attributes of komatiite flows with those of their modern basaltic counterparts.

Without modern exposures we have no way of knowing what komatiite volcanoes looked like, but we can make some inferences based on regional-scale studies in Ontario and Quebec (e.g. Jensen (1985), Chown *et al.* (1992), Ayer *et al.* (2002, 2005a,b), Sproule *et al.* (2002)), Western Australia (e.g.

Lesher (1989), Hill *et al.*, (1995), Swager *et al.* (1990, 1997), Perring *et al.* (1995)), and Zimbabwe (e.g. Prendergast (2003)). Below we combine information derived from regional-scale studies with the information on emplacement processes discussed above.

Structure of komatiite flow fields

Lesher (1989) noted that the proximal parts of lava systems must process more lava than distal parts and that the observed variations in facies probably result from variations in flow rate (decreasing with time and the distance from the vent), the degree of channelization (decreasing with time and the distance from the vent, but variable across strike), and the rate of cooling and crystallization (increasing with the distance from the vent). He suggested that lava facies should grade from komatiitic dunite through komatiitic peridotite and komatiite to komatiitic basalt, along the length of the flow, and identified volcanic facies assemblages characteristic of central, proximal, mid-distal and distal volcanic environments (Figure 9.12(a)).

Lesher (1989) suggested that the common progression from thicker, more magnesian, more cumulate sheet and channelized sheet flows separated by interflow sediments to thinner, less magnesian, non-cumulate flow lobes in many komatiite sequences (e.g., Kambalda, Alexo, Langmuir, Raglan) reflected an evolutionary change from more voluminous, more episodic eruptions to smaller, more continuous eruptions. Lesher and Arndt (1995) interpreted this as a *regression* in lava facies: more 'proximal' facies in the lower parts and more 'distal' facies in the upper parts.

Cas *et al.* (1999) concurred with the previous interpretations that komatiite flow fields developed basal channels near their vents and transformed into sheets in medial to distal settings, but suggested that they may have also transformed upwards into sheets near their vents due to in-channel sedimentation of olivine (and/or sulfides) or through increased magma discharge. Distinguishing between these two possibilities will be very difficult to test, but it is clear that one or the other occurred in areas like Alexo, where there are multiple cycles of komatiites, each comprising thicker lower channelized sheet flows overlain by thinner lobate flows (Houlé *et al.*, 2008a).

Prendergast (2003) interpreted the discontinuous Reliance unit in Zimbabwe as a series of essentially discrete komatiitic flow fields 33–85 km in diameter with centres up to 2.5 km thick that were submarine in most regions, but partly emergent at Belingwe (Nisbet *et al.*, 1993b). As is typical of many komatiite sequences, each displays a simple and coherent sequence from komatiite or komatiitic basalt sheet or channelized sheet flows and lava channel complexes

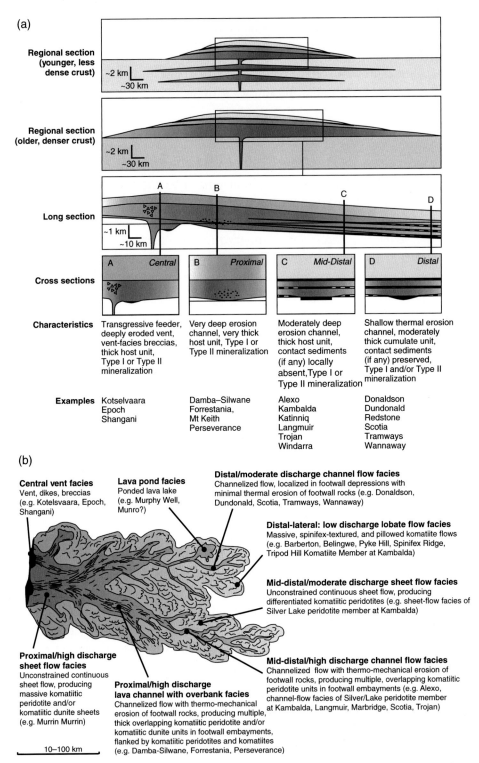

Fig. 9.12. Modified versions of the komatiite facies and flow-field models of (a) Lesher (1989) and (b) Hill *et al.* (1995).

in lower and proximal positions to komatiitic basalt marginal lava lobes in upper and distal positions. Interflow sediments are a minor component. However, unlike many other komatiite sequences, each flow field is spatially associated with a locally composite, subjacent sill of ultramafic, or locally mafic, composition up to several tens of kilometres in lateral extent.

A similar association of flows with geochemically related subvolcanic intrusions was also inferred in the Shaw Dome region of the Abitibi belt (Stone and Stone, 2000), and in the Mt Keith region of the Norseman–Wiluna belt (Rosengren *et al.*, 2005). This association may be the defining characteristic of komatiite shield volcanoes,

Houlé *et al.* (2002) and Sproule *et al.* (2005) were not able to define any systematic regional-scale variations in individual komatiite-bearing assemblages in the Abitibi greenstone belt, but noted important differences in the physical volcanology of the komatiites and komatiitic basalts in each of the different komatiite-bearing assemblages:

(1) The 2750–2735 Ma *Pacaud assemblage* contains undifferentiated non-cumulate pillowed komatiitic basalts and lesser differentiated non-cumulate komatiitic basalts. Cumulate lava conduits and channelized sheet flows were not identified in this assemblage.

(2) The 2723–2720 Ma *Stoughton–Roquemaure assemblage* contains undifferentiated non-cumulate pillowed komatiites and komatiitic basalts and lesser differentiated non-cumulate komatiites and komatiitic basalts. Cumulate lava conduits or channelized sheet flows were not identified in this assemblage.

(3) The 2719–2710 Ma *Kidd–Munro assemblage* contains differentiated non-cumulate komatiite and komatiitic basalt lobes, undifferentiated non-cumulate lava lobes and sheet flows, differentiated cumulate komatiites and komatiitic basalt lava conduits and channelized sheet flows, rare undifferentiated non-cumulate pillowed komatiites and komatiitic basalts, and very rare volcaniclastic komatiitic basalts.

(4) The 2710–2703 Ma *Tisdale assemblage* also contains differentiated non-cumulate komatiite and komatiitic basalt lobes, undifferentiated non-cumulate lava lobes and sheet flows, differentiated cumulate komatiite and komatiitic basalt lava conduits and channelized sheet flows, rare undifferentiated non-cumulate pillowed komatiites and komatiitic basalts, and very rare volcaniclastic komatiitic basalts.

The komatiites in the Tisdale assemblage are quite similar to those in the Kidd–Munro assemblage, but the Tisdale assemblage contains more andesitic rocks and sulfide facies iron formations, and the Kidd–Munro assemblage contains more basalt. Importantly, the younger assemblages (Kidd–Munro and Tisdale) contain larger volumes of komatiitic rocks, in particular larger volumes of thick, highly magnesian cumulate lava channels and channelized sheet flows than the older assemblages (Pacaud and Stoughton–Roquemaure). This suggests that the magma discharge rates were higher and/or that they formed more proximally to the eruptive site (Sproule *et al.*, 2002, 2005).

Following similar reasoning to Lesher (1989), Hill *et al.* (1995) proposed a model for the spatial relationship of facies assemblages in a large komatiite flow field (Figure 9.12(b)), loosely based on the extensive but poorly exposed komatiite sequence of the Widgiemooltha–Wiluna corridor of the Eastern Goldfields Superterrane. This model was not intended to be a literal representation of field relationships in any particular belt, but it is a useful working hypothesis for the likely form, structure and evolution of komatiite flow fields.

The model illustrates the spatial relationship of the main facies assemblages. According to this scheme (Figure 9.12(b)), thin spinifex lobes of the compound flow facies occupy the distal portions, forming at the propagating edges of the field. Closer to the vent, flow rates were higher, and the thick cumulate-rich facies making up 'compound sheet flows' were predominant. In the original Hill *et al.* model (originally sketched on a table napkin in the back bar of the Central Hotel in the Western Australian metropolis of Leonora), laterally extensive sheet flows dominated by dunites (olivine adcumulates and mesocumulates) formed closest to the vent representing the maximum effusion rates, while the channelized, lenticular dunite bodies represented the feeder channels to the compound sheet flows. As we have seen, aspects of this model have been challenged by new observations on the origin of the dunitic facies, but the general relationship between thick proximal sheet flows and thin distal compound flows probably holds. In this modified version, the proximal units may be represented by thick extrusive sheeted bodies such as the Murrin Murrin complex of Western Australia (Barnes, 2006), and the conduits between these and the distal compound flows correspond to thick meso- to orthocumulate dominated channel complexes such as Black Swan.

The form of komatiite volcanoes

Owing to the complicating effects of structural deformation (which may increase or decrease thicknesses) and erosion (which reduces thicknesses), we may never know the true form of an Archean komatiite volcano. Komatiite volcanoes probably had a variety of morphologies, as modern basaltic edifices do, depending on the interplay of pre-existing topography, eruption rate and other external factors.

If permitted to flow unimpeded, the higher densities and much lower viscosities of komatiite lavas should have permitted them to flow further than basaltic lavas. Indeed, most komatiite lava piles are generally less than 2–3 km thick, commonly less than 1 km thick and normally less than associated basalts. This, combined with the great lateral continuity of komatiite assemblages in Ontario–Québec, Western Australia and Zimbabwe, suggests that they formed broad, flat shield volcanoes with shallow slopes.

Modern subaerial basaltic shield volcanoes have positively-skewed slope distributions with median slopes of $\sim8°$ during constructional stages and $\sim20°$ during inactive stages (Bleacher and Greeley, 2003). Because of their intimate associations with pillow basalts and/or fine-grained sulfidic sediment, most komatiites appear to have erupted under water. Submarine lava should crust over more quickly, insulating the lava and allowing it to flow further from the vent, but lava flows should spread more slowly in submarine environments, because of the lower density contrast between komatiite and seawater, explaining why lavas on subaerial parts of the Kilauea East Rift Zone have lower relief than those on the submarine extension (Puna Ridge) (Zhu *et al.*, 1999). Regardless, it seems likely that the greater density and much lower viscosity of komatiite magma should have resulted in very low slopes for submarine komatiite volcanoes.

Komatiite sequences may also have been thicker if erupted onto a more rugged topography, such as that created by multiple spaced felsic volcanic centres (e.g. Abitibi greenstone belt (Davis, 1997, 1999)), and in settings that may have restricted the rates of eruption (see above), both of which would have resulted in the accumulation of lavas closer to the vent.

Komatiite volcanoes most likely formed very broad, relatively flat, submarine lava shields built from a combination of large inflated sheet flows, channel-fed sheet flows and tube-fed compound flows. In places where the bases of thick komatiite piles are exposed, it appears that the first eruptions were the largest and that eruption rates diminished with time (Lesher, 1989; Lesher and Arndt, 1995). The first eruptions appear to have been laterally-extensive sheet flows, which probably inflated during emplacement (as discussed above), but in some cases may have been huge flood flows fed in the early stages by broad open channels. These would have been truly spectacular sights: submarine rivers of lava at up to 1600 °C, glowing blue-white where the crusts were thinnest, and flowing possibly turbulently at velocities up 40 km h^{-1}.

Where these flows were emplaced onto 'mafic plains' they should have spread laterally as sheets on either side of early rifts, which by analogy with modern Hawaiian eruptions would have quickly evolved into point sources feeding channelized pathway systems. Subsequently, much of the transfer of lava to the propagating margins of the shield would have been in conduits and tubes beneath insulating crusts. As eruptions waned, the lava facies would have regressed, leaving pillow lavas and flow lobes both as the only facies in the most distal parts of the system and sheet and channelized facies covered by flow lobes and pillow lavas in the proximal parts of the system.

9.8 Concluding comments

It is evident from the lengthy discussions in this chapter that many aspects of komatiite volcanology are unresolved, and considerable differences of opinion exist. The following is not an exhaustive list, but highlights some of the major questions.

(1) Intrusions and flows. How common is the association described in the greenstone belts of Zimbabwe by Prendergast (2003) in which flow fields are underlain by sill complexes of similar extent? Is the stratigraphy seen at Mt Keith, where laterally restricted dunitic bodies underlie extensive flows, or is the Mt Keith 'intrusion' (and others like it) a deeply-entrenched rille? Do sill/flow complexes imply proximal eruption, or can sills 'flow' as far as flows?

(2) What was the original phenocryst content that most komatiites were erupted with, and what proportion of komatiites was erupted superheated?

(3) Was turbulent flow in komatiites widespread, or was it restricted to large feeder conduits and channels?

(4) When did crusts form on thick channelized flows? Did crusts form early in some cases and late in others? And did large lava conduits flow as open channels without crusts?

(5) What is the balance of *in-situ* nucleation and growth of new crystals, *vis à vis* mechanical deposition of transported ones, in cumulate zones, thick and thin? What is the relationship between crystal load and viscosity, and what proportion of crystals can lavas transport before their effective viscosity becomes too great? This depends critically upon the size and geometry of olivine crystal clusters, and more quantitative textural data are urgently needed to address this.

Notwithstanding these questions, considerable progress has been made, and we hope to have identified some testable hypotheses that can be addressed by further observation. We need more detailed petrochemical profiles through well-preserved flows, particularly based on systematic sampling on the scale of a flow field, and combined with quantitative textural data such as crystal size distributions and crystal clustering measurements. Sophisticated numerical modelling codes are becoming available that, provided they are well constrained by realistic assumptions, have great potential to advance our understanding of komatiite flow and thermomechanical erosion. Billions of dollars worth of undiscovered Ni–Cu–PGE deposits may lie just beyond the next breakthrough.

10

Komatiite-associated Ni–Cu–PGE deposits

C. M. LESHER AND S. J. BARNES

10.1 Introduction

There are two main types of ore deposits associated with komatiites: primary
magmatic sulfide deposits, mined chiefly for their nickel content with minor
by-product Cu and PGEs, and secondary laterite deposits enriched in nickel
and cobalt, developed over thick olivine cumulate units, the latter known thus
far only in the Yilgarn Craton of Western Australia (Elias, 2006). The chro-
mitite deposits of the Sherugwe district in Zimbabwe are probably also associated with komatiites (Tsomondo, 1995; Cotterill, 1969), but this chapter is
concerned only with the sulfide deposits, which currently account for about
10% of the world's annual nickel production and about 18% of global sulfide
nickel reserves (Hronsky and Schodde, 2006).

The first Ni–Cu–PGE deposit found in what are now recognized to be
komatiites was the Alexo deposit in northern Ontario (Naldrett, 1966), which
was discovered in 1907 and which has been mined intermittently since.
However, not until 1966, after much greater amounts of mineralization were
discovered by the Western Mining Corporation in the Kambalda region of
Western Australia (Woodall and Travis, 1969), was it appreciated that a new
class of Ni–Cu–PGE deposit in ultramafic rocks had been discovered. The
association with komatiite lava flows was recognized soon thereafter (Ross
and Hopkins, 1975).

Ni–Cu–PGE deposits associated with komatiitic magmas have since been
discovered in other parts of the Abitibi greenstone belt in Canada (Naldrett
and Gasparrini, 1971; Barnes and Naldrett, 1987; Coad, 1979), in other parts
of the Yilgarn Craton of Western Australia (Barnes, 2006), in several areas in
the Bulawayan Province of Zimbabwe (Prendergast, 2003), in the Sao
Francisco Craton of Brazil (Costa et al., 1997), in the Karelian Craton of
Finland and Russia (Dragon Mining NL, unpublished data), and in several
Proterozoic terrains, including the Raglan area of the Proterozoic Cape Smith

belt in northern Quebec (Lesher, 2007) and the Thompson nickel belt in Manitoba (Layton-Matthews *et al.*, 2007). The Proterozoic Ni–Cu–PGE deposits in the Pechenga area of the Kola Peninsula are associated with high-Mg ferropicrites, but are fundamentally very similar (Green and Melezhik, 1999; Barnes *et al.*, 2001). Ironically, there are no significant komatiite-associated Ni–Cu–PGE deposits in the original type locality of Barberton Mountain Land.

Numerous reviews already exist of the geology of komatiite-associated Ni–Cu–PGE deposits (Marston *et al.*, 1981; Marston, 1984; Lesher, 1989; Lesher and Keays, 2002; Naldrett, 2004; Barnes, 2006; Hoatson *et al.*, 2006). The purpose of this chapter is to provide brief descriptions of selected deposits that illustrate the range of variability, to point out the particular characteristics that distinguish komatiite-associated ores from other styles of magmatic sulfide mineralization and to show how the distinctive properties of komatiites influence ore-forming processes.

10.2 General characteristics

Komatiite-associated Ni–Cu–PGE deposits are associated with thick cumulate-dominated lava flows and with associated subvolcanic intrusions. A variety of classification schemes have been proposed over the years. A bimodal classification into 'volcanic peridotites-associated' and 'intrusive dunite-associated' (Marston, 1984) was used until the late 1980s, but fell out of favour with the interpretation of many of the mineralized dunite bodies as extrusive. Subsequently, the pendulum has swung back the other way and the Mt Keith body, host of the largest dunite-associated orebody, has been re-reinterpreted as intrusive (Rosengren *et al.*, 2005). In the interest of having a classification scheme that does not depend on the often very uncertain determination of the intrusive or extrusive status of the host komatiite units, we favour a scheme that depends entirely on the readily observable relationship of sulfides to the host body.

In this scheme, proposed in slightly different forms by Lesher (1989) and Hill and Gole (1990), most deposits can be subdivided into two major types: Type I high-grade massive and net-textured (matrix) sulfides at or near the basal contacts of komatiite units, which may be lava flows (e.g. Alexo, Kambalda), deeply erosive flows (e.g. Raglan), or feeder sills (e.g. Thompson); and Type II low-grade disseminated sulfides in the cores of thick dunite bodies (e.g. Mt. Keith) (Hopf and Head, 1998; Rosengren *et al.*, 2005; Grguric *et al.*, 2006). Lesher and Keays (2002) refined the scheme and defined three additional types: Type III stratiform PGE-enriched layers in differentiated flows, lava lakes, or sills; Type IV mineralization hosted by sulfidic sediments in close proximity to komatiitic rocks; and Type V tectonically-modified mineralization displaced from its original location. Types I and II overwhelmingly dominate the metal budget in Archean deposits, but Types IV and V mineralisation together comprise a large proportion of the ore in the Proterozoic deposits in the Thompson belt.

Komatiite-associated Ni–Cu–PGE deposits share a number of common features with most other types of magmatic Ni–Cu–PGE deposits:

(1) Type I ores occur at or near the basal contacts of the host units and are typically localized in embayments that transgress footwall rocks (Lesher, 1989; Naldrett, 2004).
(2) Type II ores exhibit interstitial disseminated magmatic textures (Eckstrand, 1975; Groves and Keays, 1979; Grguric *et al.*, 2006).
(3) Ore compositions vary in a consistent and predictable manner with the compositions of the host magmas (Barnes and Naldrett, 1987).
(4) S isotopic compositions vary from deposit to deposit, but are commonly similar to the country rocks (Lesher and Groves, 1986; Ripley, 1986).
(5) The ore mineralogy is dominated by pyrrhotite, pentlandite and chalcopyrite with minor pyrite, magnetite, ferrochromite and platinum-group minerals (Naldrett, 2004).

However, komatiite-associated Ni–Cu–PGE deposits have several characteristics that distinguish them from other types of magmatic sulfide deposits. These features, illustrated in the examples to follow, relate to the unique physical and geochemical properties of komatiites, which exerted critical controls on ore genesis, ore emplacement and ore chemistry:

(1) Most deposits are extrusive (e.g. Kambalda, Alexo-Dundonald, Damba-Silwane) or deeply erosive (e.g. Raglan), compared to deposits associated with mafic magmas, virtually all of which are intrusive (e.g. Duluth, Norilsk, Voisey's Bay).
(2) The ores are almost exclusively associated with the thickest and most magnesian units in the host sequence, typically comprising thick olivine mesocumulate to adcumulate rocks. These units are interpreted to represent major magma conduits, lava conduits and channelized parts of sheet flows; the 'pathway facies' described in the previous chapter. The largest deposits tend to be associated with the most olivine-rich host units (Figure 10.1).
(3) The mineralized units in extrusive or invasive settings normally occur at or near the base of the stratigraphic sequence associated with sulfidic sediments and are overlain by thinner, less magnesian spinifex-textured compound flows with no interflow sediments, indicating that the initial eruptions were episodic but more voluminous and fed by major pathways, whereas the later eruptions were less voluminous but more continuously emplaced (Lesher, 1989).

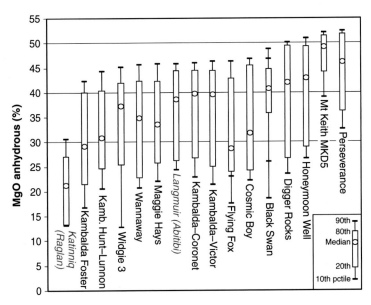

Fig. 10.1. MgO contents of host komatiite units (defined as a contiguous body of predominantly komatiitic rocks) to selected nickel sulfide deposits, shown as a box-and-whisker plot. Data sources summarized in Barnes (2006).

(4) The extrusive ores often occur as linear ribbons ('shoots' in Australian usage), reflecting their original control by linear lava conduits or pathways (Lesher *et al.*, 1984).

(5) Where massive sulfides are present, they normally occur at or near the base of the host unit and are often overlain by net-textured ('matrix') ores (Figure 10.2), the latter of which are unique to this deposit type.

(6) The sulfide component of the ores is typically very rich in Ni, commonly over 10% and as high as 21% in 100% sulfides (Lesher and Campbell, 1993), and even higher in Type II ores owing to post-magmatic processes (Grguric *et al.*, 2006).

(7) Sulfide orebodies in extrusive komatiite-associated deposits are typically only slightly fractionated into Cu–Pt–Pd-rich and -poor portions (Barnes and Naldrett, 1986), unlike many slowly-cooled intrusive deposits derived from more evolved magmas, in which the ore magmas have fractionated strongly (e.g. Noril'sk, parts of Sudbury: see Naldrett (2004)).

In the following sections, we cover the essential features of the major deposit types by reference to some well-preserved and well-studied examples, which illustrate critical features and provide evidence of ore-forming processes. The examples of Type I and Type II mineralization are associated with typical Archean high-MgO komatiite sequences, but we have included the Proterozoic Katinniq deposit, which is associated with komatiitic basalts. The deposits in the Proterozoic Thompson nickel belt are associated with low-Mg komatiites, but are dominated by Type V mineralization.

10.3 Type I deposits

Type I deposits are accumulations of immiscible magmatic sulfide liquid at or near the basal contacts of olivine-enriched komatiite flows or subvolcanic intrusions. They are relatively sulfide-rich, commonly containing massive ore with 75–100 vol % sulfides, 'matrix' (Australian usage) or 'net-textured' (North American usage) ore, or both. Matrix ore comprises 30–70% olivine grains (or serpentine pseudomorphs) within a continuous network of sulfides (Figure 10.2). Where both textural types occur in undeformed areas, massive ores are normally overlain by matrix ores with sharp contacts, and matrix ores are typically overlain and flanked by halos of low-grade disseminated sulfides with more gradational contacts. However, some deposits (e.g. Katinniq, parts of Alexo) contain more complex sequences with multiple horizons of massive, net-textured and/or disseminated ores. The location of Type I ores at or near the bases of the host units indicates that they segregated at a very early stage in the crystallization histories of the host units.

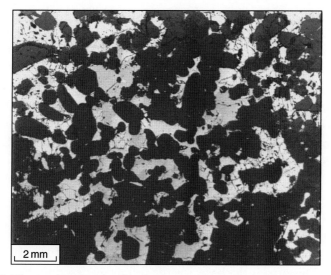

2 mm

Fig. 10.2. Photomicrograph of typical 'net-textured' or 'matrix' ore from the Katinniq deposit in Canada, showing serpentine pseudomorphs after olivine in a continuous matrix of pyrrhotite and pentlandite.

Example: Kambalda (Western Australia)

The Kambalda deposits in the southern part of the Norseman–Wiluna belt in Western Australia (Chapter 2, Figures 2.17, 2.18,) are the type examples of komatiite-associated Ni–Cu–PGE deposits (Lesher, 1989). The general aspects of the deposits have been described by Cowden (1988), Beresford *et al.* (2002), Gresham and Loftus-Hills (1981), Stone and Masterman (1998), and Stone and Archibald (2004) and there have been dozens of more detailed studies on various aspects of their geology, volcanology, geochemistry and petrogenesis, reviewed most recently by Barnes (2006).

The host komatiite sequence

The Kambalda Komatiite Formation was briefly introduced in Chapter 2. To summarize, it is up to 1 km thick, underlain by pillowed and massive tholeiitic basalts of the Lunnon Formation and overlain by pillowed and massive komatiitic basalts of the Devon Consuls and Paringa Formations (Cowden and Roberts, 1990; Gresham and Loftus-Hills, 1981). It is made up of a lower Silver Lake Member, composed of thick differentiated cumulate lava conduits and sheet flows (Lesher *et al.*, 1984; Lesher, 1989), and an upper Tripod Hill Member, composed of thin non-cumulate massive and differentiated flow lobes.

The Silver Lake Member contains two facies: a thicker (up to 100 m, possibly more), more magnesian (up to 45% MgO) olivine-rich 'channel-flow facies' and a thinner (10–30 m), less magnesian (up to 36% MgO) olivine-rich 'sheet-flow facies' (Lesher *et al.*, 1984; Lesher, 1989; Cowden and Roberts, 1990). The former represents a lava pathway and is characterized by the presence of basal Fe–Ni–Cu sulfide mineralization, an absence of underlying contact sediment and, in most cases, an absence of overlying inter-flow sediments (Figure 10.3). The lava pathways are very linear and are closely associated with the ribbon-like ore shoots and linear footwall troughs. The lava pathways and flanking sheet flows correspond to the 'ore' and 'non-ore' environment recognized early in the development of the camp (Gresham and Loftus-Hills, 1981).

The komatiites at Kambalda are pervasively metamorphosed to talc–carbonate–chlorite and serpentine–chlorite assemblages, which have commonly obliterated the primary textures. However, primary textures are preserved to varying degrees in most areas and igneous mineralogy is preserved in a few areas, most notably around the Victor Shoot (Lesher, 1983, 1989; Beresford *et al.*, 2002). This area is the basis for the following descriptions.

There is no correlation between flow thickness, the thickness of the A zones, the relative thicknesses of the A and B zones, or the degree of differentiation;

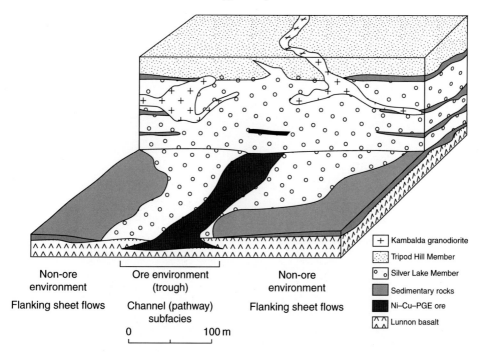

Fig. 10.3. Three-dimensional block diagram of a 'typical' Kambalda ore environment (modified from Gresham and Loftus-Hills (1981) and Stone and Masterman (1998)).

but thicker flows typically have thicker cumulate zones and flows with thicker A zones are typically more strongly differentiated (with aphyric flow tops, random and platy spinifex zones, and sometimes gabbroic rocks) than flows with thinner A zones (with only aphyric flow tops and random spinifex-textured zones) (Lesher *et al.*, 1984; Lesher, 1989; Beresford *et al.*, 2002). Channel-flow facies are dominated by medium-grained polyhedral mesocumulate olivine but sometimes contain layers of very coarse-grained branching dendritic (harrisitic) crescumulate olivine (Figure 10.4), whereas sheet-flow facies often contain abundant medium-grained more 'massive' dendritic (chevron) crescumulate olivine (Lesher, 1983, 1989; Beresford *et al.*, 2002). The flows are much thinner in the overlying Tripod Hill Member, although thicker flows often occur over the ore-bearing pathways (Gresham and Loftus-Hills, 1981), which have been interpreted as lava pathways that have been localized along the same volcanic pitch line (Lesher *et al.*, 1984). Internal differentiation within thicker flows has been described in Chapter 2.

Flow tops are typically coherent or cut by networks or curved fractures passing down into the random spinifex zone. Jigsaw-fit hyaloclastite breccias formed by quench fragmentation occur in a minority of flow tops.

Fig. 10.4. Photograph of a thin section of branching (harrisitic) crescumulate olivine in the central–upper part of the thick basal mesocumulate lava conduit facies at Victor Shoot, Kambalda. Diamond drill core KD6042A/555.0 m. Width of photo is 4 cm.

Vesicular layers (labelled VH on Figure 2.19) occur within many parts of the Silver Lake Member. Beresford *et al.* (2002) distinguished amygdales, segregation vesicles filled with spinifex-textured komatiite melt and/or sulfide liquid, and less common vesicle cylinders or chains formed by ascending bubble traces. Vesicle layers occur within both A and B zones, and multiple layers are found within the same flow unit, suggesting endogenous growth and prolonged flow through long-lived lava pathways. Segregation vesicles containing sulfides formed significant ore zones at Otter Shoot (Keele and Nickel, 1974) and Durkin Shoot (WMC Resources, unpublished data).

The relationship between the ore-bearing embayments, thick channelized basal flow, lateral flanking units and interflow sediments has been interpreted differently by different authors. Most workers agree that the thick basal host units are only partly confined to the ore-bearing embayments, and it has been recognized that although most basal host units appear to correlate with single flanking units (e.g. Lunnon, Victor), some correlate with multiple flanking units (e.g. Durkin). Ross and Hopkins (1975), Gresham and Loftus-Hills (1981) and Brown *et al.* (1999) suggested that the basal host units erupted from linear fissures, over which sulfides and olivine accumulated and sediments were inhibited from forming, but despite almost continuous exposure of the footwall rocks, no feeders have been identified. Naldrett and Campbell (1982) suggested that olivine was 'riffled' into the embayments, but this is not consistent with the mineralogical, textural and geochemical differences between the channel and sheet-flow facies (Lesher and Groves, 1986; Lesher, 1989; Lesher and Arndt, 1995; Lesher *et al.*, 2001). Cowden (1988) suggested that central lava pathways overflowed laterally to produce multiple flanking flows, but as noted by Lesher (1989) and Beresford *et al.* (2002), it is unlikely that a lava pathway could remain active long enough for the very fine-grained interflow sediments to accumulate between breakouts. We can reject the possibility that lava from the channel invaded laterally into the sediment, because except for a few rare occurrences of invasive flows described by Beresford and Cas (2001), there is no evidence (transgressive upper contacts, baking of sediments along upper contacts) suggesting that any of the flanking sheet flows are sills. This leaves at least two possible interpretations: (1) the basal unit may have fluidized the contact sediment rather than completely assimilating it (see discussion by Bavinton (1981) and Williams *et al.* (1998)), allowing it to accumulate more rapidly between breakouts, or (2) subsequent lava pathways localized along the same volcanic pitchline may have eroded the interflow sediment(s) forming a composite unit, thereby accounting for the presence of harrisitic crescumulate olivines and blebby (vesicular) sulfides in compound units (see Figure 5.40 in Lesher (1989)). The latter interpretation was also accepted by Beresford *et al.* (2002). Finally, Lesher *et al.* (1984) and Beresford *et al.* (2002) also noted that the lava pathways had inflated and might have had some positive topographic relief, like the flows on Gilmour Island described in Section 2.5.

The orebodies

The orebodies themselves, commonly referred to as 'shoots', typically have strongly linear, ribbon-like geometries, reaching up to 3 km in length, up to 300 m in width and usually less than 5 m in thickness, and range from less than 0.5 Mt to 10 Mt in mass (Cowden and Roberts, 1990; Gresham and Loftus-Hills, 1981; Stone and Masterman, 1998) (Figure 10.3). Most of the ores occur as 'contact ore' shoots within embayments ('troughs') in the upper surface of the underlying Lunnon basalts, but some occur as 'hanging-wall ore' shoots at the bases of overlying lava conduits. Importantly, the thin sulfidic sediment layers that occur along the basal and interflow contacts outside the embayments are absent within and adjacent to the ore-bearing embayments (Marston and Kay, 1980; Bavinton, 1981; Gresham and Loftus-Hills, 1981). The hanging-wall ores at Lunnon and Coronet display unequivocal evidence for melting and partial erosion of underlying spinifex-textured flow tops (Groves *et al.*, 1986; Beresford *et al.*, 2005).

Many of the least-deformed ore shoots contain massive sulfides overlain by matrix and disseminated sulfides, but some contain only massive and disseminated sulfides, and some contain only matrix and disseminated sulfides. Some ore zones (e.g. Lunnon) are overlain by a contiguous zone of low-grade disseminated sulfides, whereas others (e.g. Otter, Durkin) contain spatially-separate internal zones of disseminated spherical 'blebby' sulfides (see below), but many host units (e.g. Victor) are otherwise essentially barren.

Ni, Cu and PGE tenors (metal concentrations in 100% sulfide) vary considerably between (Ross and Keays, 1979; Cowden *et al.*, 1986; Lesher and Campbell, 1993) and to a lesser degree within (Keays *et al.*, 1981; Heath *et al.*, 2001; Stone *et al.*, 2004; Seat *et al.*, 2004) different ore shoots. The significance of this is discussed further below.

Origin of the 'troughs'

Almost all of the ore-localizing embayments or 'troughs' have been overprinted during polyphase deformation, giving rise to a long-standing controversy regarding their origin (see reviews by Lesher (1989), Lesher and Keays (2002) and Barnes (2006)). Cowden (1988) and Stone and Archibald (2004) suggested that the trough features are exclusively tectonic in origin, that the footwall embayments were produced entirely by late-stage thrust-faulting and related deformation, and that the original topography on which the Silver Lake Member komatiites was deposited was flat. However, this interpretation focusses entirely on the geometry of the komatiite–basalt contact which is assumed *ab initio* to have been planar. It ignores the plethora of petrographic, geochemical and isotopic evidence that the lava pathways were considerably thicker and more magnesian (Gresham and Loftus-Hills, 1981), that the B zones contained more olivine and more forsteritic olivine (Lesher and Groves, 1986), that the A zones are less contaminated than the flanking flows (Lesher and Arndt, 1995) and that the A zones are significantly less depleted in PGEs than those of the flanking flows (Lesher *et al.*, 2001). It also ignores the stratigraphic information presented by Lesher (1983, 1989) and Evans *et al.* (1989) showing that volcanic structures (pillow facings, flow-top breccia horizons) in the underlying Lunnon basalt preclude faulting and/or folding and require initial topographic relief in many areas. This debate goes to the heart of genetic models for the origin of the orebodies, and is discussed in more detail below.

Leaving aside the purely structural theories, there is a general consensus that the thick portions of the Kambalda basal flows represent some kind of flow channel, lava conduit or lava tube that was continuously fed by olivine-saturated komatiite over a long period of time. The mechanism by which this may have occurred is discussed in detail in Chapter 9. If this interpretation is correct, the Kambalda sequence represents a komatiite flow field that was fed through multiple linear feeder pathways during prolonged eruption and is therefore composed of multiple compound flows. The upward progression from thick, cumulate-rich flow units to the thin layered spinifex-rich flow lobes of the Tripod Hill Member, a common relationship found in many komatiite sequences, can be attributed to declining emplacement rates with time at a particular point in a flow field (Lesher, 1989; Lesher and Arndt, 1995).

Geochemistry of the Kambalda komatiites

The majority of the komatiitic rocks at Kambalda are Al-depleted (Munro-type) komatiites that define geochemical trends consistent with fractional crystallization from or accumulation of olivine (up to Fo_{94}) into a parental magma containing ~31% MgO (Lesher *et al.*, 1981; Lesher & Groves, 1986; Redman & Keays, 1985; Arndt & Jenner, 1986; Lesher, 1989; Lesher and Arndt, 1995). Most of the cumulate and non-cumulate rocks in the lava pathway facies of the Silver Lake Member and most of the cumulate rocks in the sheet-flow facies of the Silver Lake Member have normal Ni and PGE contents and are depleted in highly-incompatible lithophile elements (HILE: Th, U, Nb, Ta, LREE) relative to moderately compatible lithophile elements (MILE: MREE, Zr, Ti, HREE), consistent with derivation from a sulfide-undersaturated magma formed via a high degree of partial melting of (normal) depleted mantle (Lesher and Arndt, 1995; Lesher *et al.*, 2001). In contrast, many of the non-cumulate rocks in the sheet-flow facies of the Silver Lake Member are moderately to strongly depleted in Ni and PGE, and are variably enriched in HILE relative to MILE, consistent with contamination and sulfide segregation during emplacement (Lesher and Arndt, 1995; Lesher *et al.*, 2001). Komatiites in the Tripod Hill Member are uniformly slightly enriched in HILE relative to MILE, consistent with up to 5% crustal contamination during ascent (Lesher and Arndt, 1995), but have normal Ni and PGE contents, indicating that they were not saturated in sulfide (Keays, 1982; Lesher *et al.*, 2001). Komatiitic basalts in the Devon Consuls and Paringa Basalt Formations are moderately to strongly enriched in HILE relative to MILE, and have anomalously low Nb/Th values, consistent with up to 30% crustal contamination during ascent

(Arndt and Jenner, 1986; Lesher and Arndt, 1995), but they have normal Ni and PGE contents, indicating that they were not saturated in sulfide (Redman and Keays, 1985; Lesher *et al.*, 2001). The geochemical variations at Kambalda are therefore consistent with increasing degrees of contamination by rocks of the upper crust upward through the stratigraphic sequence (Silver Lake Member < Tripod Hill Member < Devon Consuls basalt < Paringa basalt), as the magmas gradually heated up their conduits, with only local contamination by interflow sulfidic sediments in the Silver Lake member (Lesher and Arndt, 1995).

As noted by Lesher *et al.* (2001), there is no systematic correlation between the degrees of contamination (e.g. LREE enrichment, Nb depletion) and depletion in chalcophile elements in the komatiites at Kambalda, Perseverance, Raglan and Thompson. For example, the Paringa basalt at Kambalda is interpreted to be contaminated by up to 30% upper crust, but is undepleted in PGE (Redman and Keays, 1985), indicating that contamination by itself is not sufficient to induce sulfide saturation in komatiitic magmas. Importantly, the non-cumulate portions of the sheet-flow facies at Kambalda are weakly contaminated (Lesher and Arndt, 1995), but significantly depleted in Ni and PGE, indicating that those rocks (and only those rocks) equilibrated with sulfides. The amounts of contamination and PGE depletion produced during the ore-forming process depend on several factors (Lesher *et al.*, 2001): (1) the stratigraphic architecture of the system (e.g. thickness and physical accessibility of the contaminant), (2) the fluid dynamics and thermodynamics of the lava or magma, (3) the physical, chemical and thermal characteristics of the contaminant, (4) the amount of contaminant melted and incorporated (e.g. the amount of silicate partial melt: see Lesher and Burnham (2001)), (5) the sulfur and metal content of the contaminant, (6) the initial sulfide saturation state of the magma, (7) the assimilation:crystallization ratio, (8) the amount of lava replenishment, and (9) the effective magma:sulfide ratio (R factor) of the system. Because these processes vary independently from deposit to deposit, from area to area within a deposit, and within a single area with time, there are many opportunities to decouple contamination from depletion in chalcophile elements.

Timing of ore formation and flow crystallization

Detailed mineralogical and geochemical data profiles through flows at Victor and Lunnon (see Chapter 2, Figure 2.20) provide constraints on the relative timing of ore formation, olivine accumulation and crust growth in these flows. The middle and lower parts of the spinifex-textured A zone crystallized after the majority of the cumulate-textured B zone and therefore well after any erosion (Lesher, 1989). The upper and lower parts of the lava pathways and the lower parts of adjacent sheet-flow facies are depleted in LREE, have normal Nd isotopic signatures and are undepleted in PGE, whereas the upper parts of sheet-flow facies are enriched in LREE, have non-radiogenic Nd isotopic signatures and are strongly depleted in PGE (Lesher and Arndt, 1995; Lesher et al., 2001). Sulfidic sediments occur beneath sheet-flow facies, but are absent beneath and adjacent to conduit facies. Sheet-flow facies adjacent to conduits sometimes contain sediment xenomelts (Frost and Groves, 1986). Together, these data suggest the following sequence of events:

(1) Emplacement of a broad sheet flow with focussing of flow along pre-existing embayments (see discussion in Lesher (1989)), disaggregation and melting of unconsolidated sulfidic sediments in and along the margins of the embayments, segregation of a Fe–Ni–Cu sulfide melt at the base of the channel facies and generation of contaminated/depleted geochemical signatures in the lava. No permanent crust was present at this stage, but the flow was likely capped by an ephemeral crust.
(2) Crystallization of a permanent upper crust on the sheet facies, preserving contaminated/depleted geochemical signatures and, rarely, trapping buoyant xenomelts that were not dissolved in the more rapidly-flowing channel facies.
(3) Flushing of the liquid core of the sheet facies and the entire conduit facies, replenishing those zones with uncontaminated lava (see discussion in Lesher and Arndt (1995)).
(4) Crystallization of a thin permanent upper crust on the conduit facies and accumulation of olivine in the B zones of both conduit and sheet facies. Flow velocities were greater in the conduit facies than in the sheet facies, accounting for the more magnesian cumulus olivines in the former (see discussion in Lesher (1989)).
(5) Ponding of the conduit facies, earlier in some areas (e.g. Lunnon) and later in other areas (e.g. Victor), and crystallization of the A_2 spinifex zone.

Example: Alexo (Ontario)

The Alexo Ni–Cu–PGE deposit occurs in Dundonald–Cleague Townships in northern Ontario (Naldrett, 1966; Barnes and Naldrett, 1986; Davis, 1999; Houlé et al., 2002) (Chapter 2, Figure 2.10). Although the deposits in this area are smaller, individually and as a group, than those at Kambalda, critical contact relationships are superbly exposed in a series of hydraulically-stripped and/or glacially-polished outcrops (Figure 10.5), and highlight many of the genetic relationships that are rarely or incompletely observed at other deposits.

Four sequences of komatiitic rocks are present in the Alexo area, each comprising multiple units ranging in thickness from several hundred metres to less than 2 m and each containing basal Ni–Cu–PGE mineralization,

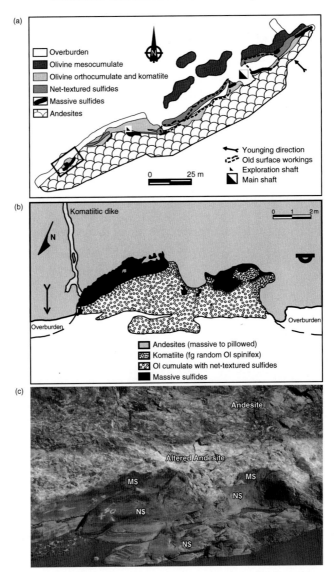

Fig. 10.5. (a) Simplified map of the Alexo mine area showing old surface workings and newly-stripped outcrops (from Houlé *et al.* (2002), based on mapping by K. Montgomery and M. G. Houlé). Note the highly irregular and transgressive nature of the contact between komatiitic rocks and footwall andesites, which define a broad first-order embayment and several smaller second-order embayments. (b) Map of second-order embayment at the SW end of outcrop in (a), looking SE (mapping by M. G. Houlé). Note the highly irregular nature of the contract between the ores and footwall andesites. (c) Photograph of the area shown in (b) (photograph by M. G. Houlé). Note the silicification of the andesites adjacent to contact and the komatiitic dyke extending downward into the andesites. (M$ – massive sulfides; N$ – net-textured sulfides.)

separated by sulfide-rich calc-alkaline felsic–mafic massive, pillowed and fragmental volcanic rocks and graphitic metasedimentary rocks (Davis, 1999). The komatiitic rocks range in composition from komatiitic basalt to dunite, and are typically (but not always) differentiated with flow-top breccias, upper spinifex-textured zones and lower olivine orthocumulate, mesocumulate and adcumulate zones (Arndt, 1986a; Barnes *et al.*, 1983; Puchtel *et al.*, 2004b). Thin interflow graphitic argillites occur between the thinner komatiitic spinifex-textured flows in the third komatiite sequence. The thicker more olivine-rich units are interpreted as lava conduits (Barnes *et al.*, 1983).

The Fe–Ni–Cu sulfide mineralization at Alexo is hosted by a ~600 m thick komatiitic peridotite–dunite unit at the base of the fourth komatiite sequence, which is overlain and flanked by thinner spinifex-textured flows and inter- preted as a lava conduit (Davis, 1999). The sulfides occur as massive to semimassive lenses that range in thickness from a few centimetres to greater than 12 m and grade upward into net-textured and disseminated sulfides (Naldrett, 1966; Davis, 1999).

Detailed mapping of the stripped lower contact by Houlé *et al.* (2002) (Figure 10.5) revealed that the orebody occurs within series of nested embay- ments on scales from 100 m to 20 cm, producing a very irregular contact that clearly transgresses underlying pillowed andesites. Narrow (<5 cm), irregular, aphyric komatiite dykes extend downward from the host unit into fractures within the andesite footwall and then spread laterally as very thin (<1 cm) sills, providing a 'snapshot' of what may have earlier been a thermomechanical erosion process but which later became primarily a mechanical erosion pro- cess. There is a continuous contact metamorphic areole that extends ~30 cm from the lower contact into the footwall andesites. Together, these features provide unambiguous physical evidence for thermomechanical erosion of the footwall during ore emplacement.

The Alexo deposit was the birthplace of Naldrett's (1973) 'billiard ball' model for the origin of the massive-matrix-disseminated ore profiles that are common in this deposit type. According to this model, the configuration is the result of static buoyancy relationships between dense sulfide liquid (~4000 kg m^{-3}), less dense komatiite melt (~2700 kg m^{-3}), and a column of close- packed olivine crystals of intermediate density (~3300 kg m^{-3}), analogous to 'mercury', 'water' and 'billiard balls' in the model. Archimedes' Principle dic- tates that the mercury (molten sulfide) should form a basal layer, in which the billiard balls (olivines) should float; however, the opposing tendencies of the billiard balls (olivines) to sink in the water and float in the mercury (molten sulfide) force some proportion of the billiard ball (olivine) column down into the mercury (molten sulfide) layer, producing a matrix or network of mercury (molten sulfide) interstitial to the billiard balls (olivines). This model has been modified considerably with the recognition of the dynamic nature of the host units and, more importantly, of multiple thin units of massive, net-textured and disseminated ores at several areas, including Alexo, which could not have formed this way, but it may apply in other areas.

Example: Black Swan (Western Australia)

The Black Swan succession about 100 km north of Kambalda (see Figure 2.17) is a particularly well-preserved example of bimodal komatiitic and calc-alkaline dacitic volcanism, contrasting with the basaltic and andesitic substrates in the previous examples. The complex komatiite stratigraphy hosts a number of small orebodies, including the Silver Swan massive ore shoot, and the larger disseminated Black Swan and Cygnet orebodies (Barnes *et al.*, 2004a; Dowling *et al.*, 2004; Hill *et al.*, 2004).

The Black Swan sequence is a complex assemblage of mainly cumulate-rich flows (Figure 10.6). A thick basal unit up to 200 m thick, capped by a persistent 50–100 cm thick spinifex layer and predominantly orthocumulate with pods of olivine mesocumulates and adcumulates, passes up into a complex intercalation of highly discontinuous and possibly invasive flows with thin spinifex tops, and dacite lavas or tuffs which were evidently erupted simultaneously. This sequence is flanked by a 500 m thick pile of coarse-grained and locally vesicular olivine orthocumulates containing spinifex veins in the upper part, which itself is flanked to the south by a series of typical thin differentiated pyroxene–spinifex-textured flow lobes with intervening dacitic and sulfidic tuffs. Disseminated Type IIa blebby ores, described below, are localized within the thick orthocumulate pile.

The Silver Swan orebody is a vertically-plunging ribbon-shaped orebody up to 20 m thick, 20–75 m wide, and at least 1 km long and open at depth. It lies at the base of the lower ultramafic unit, and is locally separated from the mesocumulate rocks above by a thin discontinuous layer of fine-grained quenched orthocumulates. The orebody itself has a convex-up morphology, and a primary upper contact with essentially barren olivine orthocumulate. The basal contact varies from being relatively planar overall with cuspate projections to being highly irregular, with downward projections of sulfide into irregular cracks in the footwall and upward projections of dacite showing transitions from angular blocks to partially molten 'plumes' connected to the floor.

The most remarkable feature is the presence of abundant xenolithic inclusions of dacite in the ores (Dowling *et al.*, 2004). These inclusions

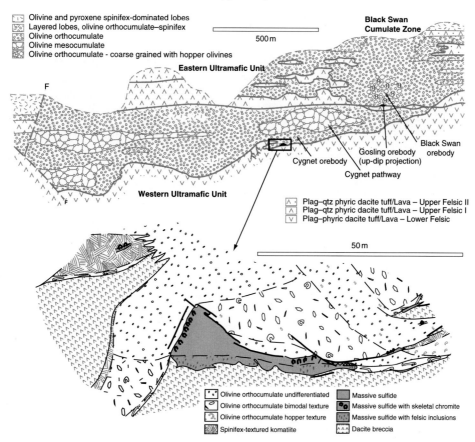

Fig. 10.6. Schematic stratigraphic reconstruction of the Black Swan komatiite complex, and simplified level plan of the Silver Swan orebody (after Hill *et al.* (2004) and Dowling *et al.* (2004)).

show clear evidence of partial melting at their margins and, in the interior of the orebody, are disaggregated into irregular sinuous blobs and plumes (Figure 10.7). The morphology, internal textures and compositions of the inclusions leave no doubt that they are partially molten fragments of the immediate dacite footwall. They extend throughout the lower 5 m of the massive ore shoot; where the total thickness of the shoot is less than 5 m they extend all the way through, and in places accumulate to form a layer of 'xenomelt' separating the top of the massive ore from overlying ultramafic rock.

These features leave little doubt that the Silver Swan shoot formed a sheet of molten sulfide magma overlying a floor which was undergoing thermomechanical erosion, partial melting and incorporation into the overlying pool of magma.

(a)   (b)   (c)

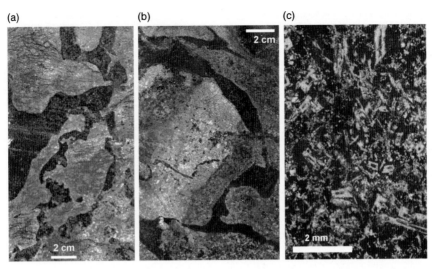

Fig. 10.7. Felsic inclusions ('plumes') in the Silver Swan massive ore shoot. (a), (b) Polished drill core slabs showing irregular cuspate morphologies of partially molten 'plumes' of dacite; (c) thin section of a plume margin showing dendritic plagioclase crystals formed by remelting and subsequent recrystallisation of original porphyritic dacite (modified from Dowling *et al.* (2004)).

Example: Raglan (New Quebec)

Ni–Cu–PGE mineralization in the Raglan area of the 1.9 Ga Cape Smith Belt in New Quebec has been described by Barnes *et al.* (1982), Giovenazzo *et al.* (1989), Barnes and Barnes (1990), St. Onge *et al.* (1997), St-Onge and Lucas (1993), Lesher (1999) and Lesher (2007). All of the currently economic mineralization occurs in the east-central part of the belt, at or near the contact between the Chukotat group, which comprises a series of (from base to top) olivine–phyric, pyroxene–phyric and plagioclase–phyric basalts, and the Povungnituk group, which comprises (in the upper parts) tholeiitic basalts, semipelitic sulfidic graphitic metasediments and peridotite–pyroxenite–gabbro sills. All of the rocks have been regionally metamorphosed to lower greenschist facies, but preservation of palimpsest textures and volcanic structures is excellent. Exposure is excellent because of the lack of vegetation and although the outcrops are variably frost-heaved, the field relationships are clear in many areas.

The Ni–Cu–PGE mineralization in the Raglan area occurs at or near the base of a series of thick, relatively massive and undifferentiated olivine mesocumulate bodies that transgress underlying slates and gabbros (Figure 10.8). The host units are gradational in terms of composition and texture with overlying olivine–phyric basalts, but are much thicker and exhibit

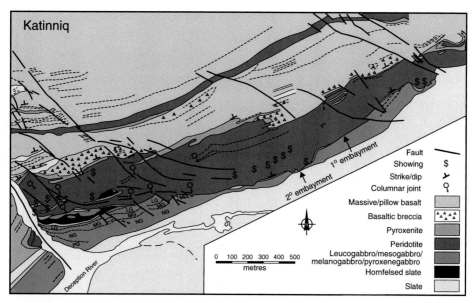

Fig. 10.8. Geological map of the Katinniq peridotite complex (Lesher, 2007) showing the transgressive nature of the lower contact, including first-order (1°) and second-order (2°) embayments, and the irregular nature of the upper contact. LG = leucogabbro, MG = melanogabbro, NG = mesogabbro. PG = pyroxene-rich gabbro (plagioclase-rich pyroxenite). Surface showings are indicated, but most of the ore zones are not exposed on surface.

columnar-jointed upper and lower parts, polyhedral-jointed upper parts, random platy olivine or acicular pyroxene spinifex-textured upper margins and/or flow-top breccias. Most units are composed primarily of mesocumulate peridotite, but contain discontinuous horizons of orthocumulate peridotite, pyroxene oikocrystic peridotite, spinifex-textured komatiite, komatiitic basalt, gabbro, or hornfelsed sediment, defining an internal volcanic stratigraphy contiguous with adjacent basalts. Such horizons are beheaded by, and correlate with, horizons of strata-bound disseminated nickel sulfide mineralization, which appear to define reactivated lava conduits. Footwall rocks (basalts, argillaceous sediments and gabbros) have been thermally eroded, forming first- and second-order embayments which localize stratiform Ni–Cu–PGE mineralization. Footwall rocks are strongly hornfelsed (recrystallized, bleached, spotted) beneath the peridotite complexes, but unmodified elsewhere. Overlying sediments and basalts are normally conformable and unhornfelsed (Figure 10.8), but the lateral parts are locally transgressive (Gillies, 1993; Lesher 2007).

The host units have been interpreted previously as sills (Barnes *et al.*, 1982) and as lava ponds (Barnes and Barnes, 1990), but the features above suggest that they are deeply erosive lava conduit complexes, analogous to the lunar, Martian, and Venusian rilles (Lesher, 1999, 2007).

10.4 Type II deposits

Type II ores are centrally-disposed accumulations of disseminated sulfides within olivine-rich komatiite units, typically lenticular dunite bodies. The major Type-II-dominated deposits occur within the 150 km long Agnew–Wiluna trend of the Norseman–Wiluna greenstone belt (Figure 2.17), which includes the type example, the Mt Keith MKD5 deposit (Grguric *et al.*, 2006; Rosengren *et al.*, 2007; Fiorentini *et al.*, 2007) (Figure 10.9). Mount Keith is currently one of the world's largest sulfide nickel mines, and along with the neighbouring Yakabindie deposits accounts for about 60% of known Ni sulfide reserves in komatiites worldwide. Smaller examples are known in the Abitibi belt (e.g. Dumont (Brugmann *et al.*, 1989; Duke, 1986a,b)), in the Thompson nickel belt (e.g. William Lake (Layton-Matthews *et al.*, 2007)), and in Zimbabwe (e.g. Hunter's Road (Prendergast, 2001)).

There are two main varieties of Type II mineralization (Lesher and Keays, 2002). Type IIa deposits are defined by disseminated spherical or subspherical sulfide aggregates referred to as 'blebs' or 'globules'. Type IIb deposits characteristically contain disseminated sulfides interstitial to and moulded around olivine grains (or their pseudomorphs) (Figure 10.10). Type IIb ores are typified by the giant Mt Keith and Yakabindie deposits.

Type IIb ores may occur in units that contain significant amounts of Type I mineralization (e.g. Perseverance), but more commonly occur in units that contain very little if any Type I mineralization (e.g. Mt Keith, Yakabindie, Dumont, Hunter's Road, some deposits in the Thompson nickel belt). The sulfides are also present in far greater abundances (1–5%) than are capable of being dissolved in the minute amount of interstitial liquid, indicating that the sulfides accumulated as cumulus sulfide liquid during olivine crystallization. The proportion of cumulus sulfide liquid in the rock is at least a factor of 3 greater than would be expected from purely cotectic crystallization and liquation, implying that a component of the sulfide was transported and mechanically deposited (Barnes, 2007).

Type IIb deposits typically contain very Ni-rich sulfide assemblages (pentlandite with or without heazelwoodite); a common grade of ~0.6% Ni; and centimetre-scale layering defined by variations in sulfide abundance and olivine grain size (Figure 10.10). The extremely Ni-rich nature of the sulfide assemblages, in which much of the ore is pure pentlandite, is the critical factor that makes these very low-grade deposits economically exploitable. It results

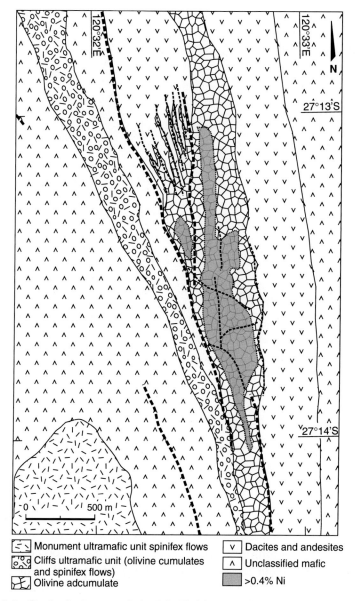

Fig. 10.9. Geological map of the Mt Keith area showing the location and extent of the MKD5 orebody (after Grguric *et al.* (2006)).

from addition of Ni originally in olivine during subsolidus equilibration (Binns and Groves, 1976; Stolz and Nesbitt, 1981) and from upgrade by oxidation and nickel addition during serpentinization (Eckstrand, 1975; Groves and Keays, 1979; Donaldson, 1981; Grguric *et al.*, 2006).

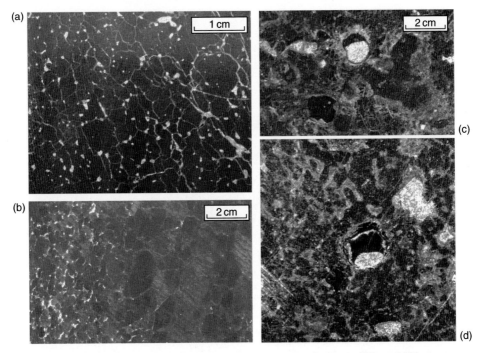

Fig. 10.10. Photographs of typical sulfide textures in Type IIa and IIb ores. Photographs of polished slabs of drill core: (a) typical Mt Keith ore showing sulfide blebs (light colour) interstitial to serpentinized olivine (black) in well-developed adcumulate texture; (b) alternating sulfide-rich and sulfide-poor layers in fresh olivine adcumulate, Perseverance. Olivine grain size is substantially coarser in the sulfide-poor layer; (c), (d) blebby ores from Black Swan. Sulfide blebs within amygdales, now partially occupied by serpentine and chlorite after original segregated komatiite liquid, within serpentinized olivine orthocumulate.

Type IIa 'blebby' ores often occur within units that also contain Type I mineralization (e.g. Black Swan, Dundonald and several Kambalda deposits). Type IIa blebs are commonly 1 cm or more in diameter, too coarse to remain suspended in the magma except at very high flow rates and, at up to 30 modal % sulfide, too abundant to have exsolved *in situ*. They must have formed during flow-through crystallization of the host unit (Lesher *et al.*, 1984; Lesher and Groves, 1986). The blebs at Black Swan and Kambalda share spherical domains with segregated komatiite melt, which itself shows rarely-preserved microspinifex textures, with a hemispherical meniscus developed between the original silicate and sulfide melts. These textures indicate the origin as partially or completely filled original segregation vesicles (Beresford *et al.*, 2000; Dowling *et al.*, 2004), and suggest that sulfides were drawn into gas bubbles after vesiculation of the komatiite lava.

Example: Mt Keith (Western Australia)

The Mt Keith deposit occurs within one of the lenticular dunite bodies of the Agnew–Wiluna trend (Figure 2.17), a cryptically layered body up to 700 m thick of extremely coarse-grained olivine meso- and adcumulates, ranging up to some of the purest olivine adcumulates found anywhere in komatiites. The body is overlain and underlain by dacite lavas and tuffs, and has apophyses of chilled margin material extending into the hanging wall, indicating a subvolcanic intrusive (Rosengren *et al.*, 2005) or invasive origin (Figure 10.9).

No fresh olivine is preserved, but whole-rock compositions of adcumulates imply original olivine compositions in the range Fo_{92-94}, with most of the ore hosted by $Fo_{93.5}$ adcumulates (Rosengren *et al.*, 2007; Fiorentini *et al.*, 2007). Much of the dunite body contains minor proportions of distinctive lobate or occasionally poikilitic chromite with distinctive Al-rich, ferric-iron-poor compositions (Barnes, 1998).

The ore zone is centrally disposed within the dunite lens, and has a broadly lenticular shape with a maximum thickness of nearly 200 m (Grguric *et al.*, 2006). A distinctive feature of the centrally-disposed mineralized interval is a heterogeneity and 'clumping' of disseminated sulfides on a scale of a few centimetres, and a related variation in host olivine grain size such that mineralized patches typically have slightly smaller olivines. The distribution of sulfide abundance is discussed further below, in the context of the genesis of this deposit type.

10.5 Type V deposits

Although most Type I deposits have been deformed to some degree and contain Type V ores with brecciated '*durchbewegung*' textures that have been mobilized to various degrees from their original locations and in some cases from their original host rocks, some deposits have been so pervasively deformed that they contain primarily Type V mineralization. The deposits in the Thompson nickel belt are particularly good examples, but are also of interest because they occur in an intrusive setting.

Example: Thompson (Manitoba)

The Ni–Cu–PGE deposits in the 1.9 Ga Thompson nickel belt of north-central Manitoba have been described by Bleeker (1990), Layton-Matthews *et al.* (2003), Burnham *et al.* (2003) and Layton-Matthews *et al.* (2007), and are reviewed by Naldrett (2004). The Thompson nickel belt lies within the Circum-Superior Boundary Zone and has been complexly deformed and metamorphosed to upper amphibolite facies, which has complicated the geological and stratigraphic relationships, dismembered and separated ores and host rocks, and modified the geochemical and isotopic compositions of the ores and host rocks.

The ores in the Thompson nickel belt are similar to many other komatiite-associated Ni–Cu–PGE deposits in many respects: (1) they are associated with the most magnesian cumulate ultramafic units in the sequence; (2) they are closely associated with S-rich country rocks; and (3) the ore mineralogy and chemistry are similar to most other deposits. However, they are also very different in many respects: (1) the ores are hosted by subvolcanic feeder sills rather than lava conduits; (2) many of the ores formed at low mass ratios of silicate to sulfide liquid; (3) massive ores have been extensively deformed and in some cases dislocated from the ultramafic host rocks; and (4) the ores are systematically depleted in Au, Cu and Pt relative to the other chalcophile elements.

The host rocks are thick, weakly-differentiated, coarse-grained dunite and peridotite bodies that have been intruded into metapelites of the Ospwagan group (Figure 10.11). Although ultramafic bodies occur within all but the uppermost parts of the Ospwagan group, only those associated with sulfide facies iron formations in the Pipe Formation are mineralized. The ultramafic bodies have been interpreted as sills on the basis of the following features (Bleeker, 1990; Layton-Matthews *et al.*, 2007):

(1) They occur at multiple stratigraphic levels in the Ospwagan group (Figure 10.11).
(2) None exhibits any positive evidence for an extrusive origin (e.g. quench textures, volcaniclastic textures, pillow structures or interflow sediments).
(3) They are frequently extremely thick, partially differentiated and contain minerals with coarse textures and uniform compositions (olivine $\pm$ chromite), consistent with slow cooling.
(4) Only thin chill zones occur at the top of the bodies and the upper contact is not associated with any particular sedimentary rock.

The Ni–Cu–PGE ores in the Thompson nickel belt include: Type V tectonic 'offset' mineralization, which has been dislocated from the ultramafic host rocks (e.g. Thompson 1D, parts of Pipe I and Birchtree); Type IV sediment-hosted mineralization, in which barren sedimentary sulfides have been upgraded via magmatic/metamorphic processes (e.g. Thompson 1 A, 1B, 1 C, and 1D, parts of Pipe II); Type II primary magmatic disseminated sulfides within the ultra-mafic host rocks (e.g. William Lake, Pipe II, Mystery Lake, Moak Lake); and Type I primary magmatic stratiform basal mineralization (e.g. Pipe II, Pipe Deep, parts of Birchtree, Thompson 1 A, and Thompson 1B).

The high forsterite (up to Fo_{93}) and Ni (up to 4500 ppm) contents of relict igneous olivines in the ultramafic host rocks indicate that the magmas from

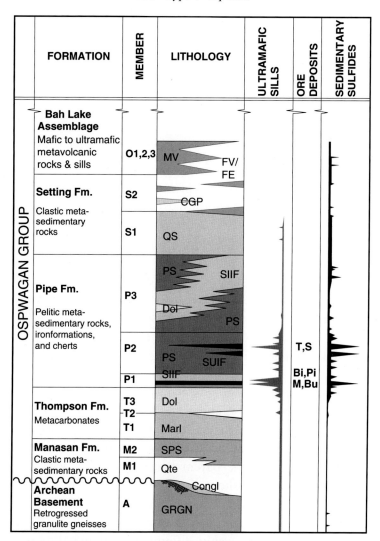

Fig. 10.11. Stratigraphic column for the Thompson Nickel Belt (modified after Bleeker (1990)), showing the preferential localization of ore deposits with ultramafic sills and sulfide facies iron formations. Lithologies: MV – mafic to ultramafic metavolcanic rocks; FV/FE – rare felsic metavolcanic and felsic epiclastic rocks; CGP – metaconglomerates, metagreywackes, and minor metapelites; OS – interlayered quartzites and schists (metaturbidites); PS – pelitic schists; SIIF – silicate facies iron formation; Dol – dolomitic marble; SUIF – sulfide facies iron formation; Marl – impure calcareous metasediments; SPS – semipelitic schists; Qte – quartzite; Congl – local basal metaconglomerate; GRGN – Archean granitic gneisses; locally retrogressed; Deposits: T – Thompson; S – Soab; Bi – Birchtree; Pi – Pipe; M – Manibridge; Bu – Bucko.

which the ores formed were low-Mg komatiites with up to 22–23% MgO (Burnham *et al.*, 2003). However, the ores in the Thompson nickel belt have relatively low Ni, Ir, Os and Ru contents and very low Cu, Pt, Pd and Rh contents, indicating that they formed at relatively low R factors (20–500) (Naldrett *et al.*, 1979).

10.6 Compositions of komatiite-hosted sulfide ores

The compositions of komatiite-hosted sulfide ores are prime pieces of evidence for their magmatic origin. The metal–sulfur ratios of sulfide liquids in equilibrium are constrained within narrow bounds by the composition of the host silicate magma (Naldrett, 1969). As a consequence, magmatic sulfide liquids have bulk metal–sulfur ratios close to unity, and this determines their mineralogy on solidification. Their metal contents are also consistent with a magmatic origin: komatiites and komatiite-associated ores are both enriched in Ni–Co–IPGE (IPGE: Ir, Ru, Os) relative to Cu–Au–PPGE (PPGE: Pt, Pd) (Barnes *et al.*, 1985; Lesher and Keays, 2002), whereas less magnesian magmas and ores derived from them have lower Ni–Co–IPGE and higher Cu–Au–PPGE contents (Barnes *et al.*, 1985; Naldrett, 2004).

Type I ores have high Ni tenors (weight per cent metal in 100% sulfides), typically between 10 and 20%, and low Cu tenors, normally less than 1%, compared with other classes of magmatic sulfide ores. This is a consequence of the higher Ni and lower Cu contents of the more primitive komatiite parent magmas compared with mafic parent magmas. Tenors vary considerably, sometimes between adjacent shoots in the same camp, reflecting the superimposition of several independent controls as discussed below.

The S isotopic compositions of komatiite-associated Ni–Cu–PGE ores vary from deposit to deposit, are normally different from magmatic sulfides and are often similar to the S isotopic compositions of the country rocks (Lesher and Groves, 1986).

The Re–Os isotopic compositions of komatiite-associated Ni–Cu–PGE ores vary from chondritic (mantle-like) to highly radiogenic (e.g. Raglan (Shirey and Barnes, 1995); Kambalda (Foster *et al.*, 1996); Thompson (Hulbert *et al.* 2005); Abitibi (Lahaye *et al.*, 2001)). Although some of these data were originally used to suggest that the S in the deposits could not have been derived by incorporation of S from associated sediments (Shirey and Barnes, 1995; Foster *et al.*, 1996; Lambert *et al.*, 1998), it has subsequently been shown that the observed Os isotopic compositions can be attributed to variations in the R factor, resulting from mixing of highly radiogenic crustally derived Os with less radiogenic mantle-derived Os (Lesher and Stone, 1996; Lesher and Burnham, 2001; Lesher *et al.*, 2001; Lahaye *et al.*, 2001).

Influence of magma composition

The compositions of mafic–ultramafic magmas exert a first-order control on the compositions of magmatic Ni–Cu–PGE deposits. As a consequence of having formed by higher degrees of partial melting and consuming greater amounts of olivine, komatiitic magmas have higher Ni, Co and IPGE contents and lower Cu, Au and PPGE (Pt and Pd) contents than komatiitic basalts, picrites or basalts (Barnes and Naldrett, 1986). Because these metals all partition strongly ($D^{Co} \sim 30$, $D^{Ni} \sim 100$–200) or very strongly ($D^{Cu} \sim 600$, $D^{PGE} \sim 10^5$) into sulfides, komatiite-associated ores also have higher Ni, Co and IPGE contents and lower Cu, Au and PPGE contents than ores associated with less magnesian magmas (see review by Naldrett (2004)).

Influence of the R factor

Because Ni, Cu, Co, Au and PGEs partition so strongly into the sulfide phase, magmas that equilibrate with large amounts of sulfides may become strongly depleted in these metals, thereby reducing the amounts of metals in both the magma and the sulfides compared to the amounts that would be present if the magmas had equilibrated with only small amounts of sulfides (Campbell and Naldrett, 1979). This effect is quantified in terms of the mass ratio of silicate magma to sulfide liquid (the *R* factor) and the partition coefficient *D* for any element between the sulfide and silicate liquids. The *R* factor becomes a dominant control when $R < 10D$. This process explains why different ore shoots at Kambalda (Lesher and Campbell, 1993), different ore zones in the Katinniq deposit (Barnes and Picard, 1993; Gillies, 1993), and different deposits in the Thompson nickel belt (Naldrett *et al.*, 1979; Layton-Matthews *et al.*, 2007) have very different ore tenors, despite having formed from broadly similar magma compositions.

Calculated *R* factors for Type I massive–net-textured ores range between 100 and 500 at Kambalda (Lesher and Campbell, 1993), 300 and 1100 for Raglan (Barnes and Picard, 1993; Gillies, 1993; Lesher *et al.*, 1999) and 10 and 500 for Thompson (Layton-Matthews *et al.*, 2007). These are similar to the 100–1200 range for Sudbury (Naldrett *et al.*, 1979, 1999), but much lower than the 500–100 000 range for Norilsk (Naldrett *et al.*, 1999).

Calculated *R* factors for Type II disseminated ores are ~ 140 for Mt Keith and ~ 370 for Dumont (Lesher and Keays, 2002), and 40–600 at Thompson

(Layton-Matthews *et al.*, 2007). Although Type II sulfides might be expected to have formed under higher R conditions than Type I massive sulfides, especially if transported (see above), the effective R factors were relatively low. This interpretation is supported by the depletion of Ni in the olivines at Dumont (Duke, 1986a,b) and Perseverance (Barnes *et al.*, 1988).

In theory, because komatiite magmas are only slightly denser than basaltic magmas (2800–$2700\,\mathrm{kg\,m^{-3}}$ vs 2700–$2600\,\mathrm{kg\,m^{-3}}$), but the viscosities are much lower than basaltic magmas ($\sim 0.1\,\mathrm{Pa\ s}$ vs $\sim 100\,\mathrm{Pa\ s}$), sulfides should settle much more rapidly and therefore have less opportunity to equilibrate with komatiitic magma, which would reduce the effective R factor. The geometry and dynamics of the ore-forming environment (e.g. lava conduit vs feeder sill vs magma conduit) are probably the more important controls, but this may explain why some deposits associated with picritic magmas (e.g. Norilsk) were able to form at very high R factors and to generate very high tenor ores, but most komatiite-associated deposits seem to have formed at intermediate to low R factors.

Monosulfide solid solution fractionation

The first phase to crystallize from a sulfide melt is a Ni-rich pyrrhotite phase, monosulfide solid solution (MSS) (Naldrett, 1969; Ebel and Naldrett, 1996). Because $D^{Cu}_{MSS/\text{sulfide liquid}}$ and $D^{PPGE}_{MSS/\text{sulfide liquid}}$ are normally much lower than $D^{Ni}_{MSS/\text{sulfide liquid}}$ and $D^{IPGE}_{MSS/\text{sulfide liquid}}$ (Barnes and Lightfoot, 2005), sulfide melts become progressively richer in Cu and PPGE and poorer in IPGE as they crystallize. Large sulfide deposits associated with mafic magmas in slowly-cooled environments, as at Sudbury and Noril'sk, contain spectacular zones of Cu–PPGE-rich mineralization, but most komatiite-associated deposits exhibit only limited amounts of fractionation.

There are two reasons for the lower degree of MSS fractionation in komatiite-associated ores: (1) the sulfide liquids were richer in Ni and poorer in Cu, and therefore solidified over a much narrower temperature range than the Cu-rich ores associated with mafic magmas and (2) extrusive and invasive deposits cool more quickly than subvolcanic and plutonic deposits.

Nevertheless, there is evidence for some degree of MSS fractionation in some komatiite-associated deposits: massive sulfides are commonly somewhat enriched in IPGEs and depleted in PPGEs relative to matrix and disseminated ores, which is most easily explained if the massive ores are preferentially enriched in cumulus MSS (Barnes and Naldrett, 1986). A good example of this is seen in the Silver Swan massive ore shoot at Black Swan, which has evidently undergone significant fractionation and accumulation of MSS (Barnes, 2004).

10.7 Origin of komatiite-associated ores

Sulfide saturation

The formation of magmatic sulfide deposits requires the host silicate magma to be saturated in sulfide, such that immiscible sulfide and silicate liquids are present and able to react with one another. Early models for the origin of komatiite-associated ores (Naldrett and Turner, 1977; Lesher *et al.*, 1981) assumed that mineralized komatiites erupted containing mantle-derived sulfides. However, komatiites are derived by very high (~50%) degree partial melting of the mantle (Arndt, 1986a; Herzberg and O'Hara, 1998). S is present in low abundances (125–250 ppm) in the mantle (McDonough and Sun, 1995) and is incompatible during melting, so that all of the S is consumed at relatively low degrees of partial melting (Naldrett and Barnes, 1986), and the abundance of S in the komatiite melt progressively declines with further melting. As a consequence, komatiites should have very low S contents (~250–500 ppm (Lesher and Stone, 1996)), which is well below the ≥1600 ppm S capacity of komatiitic magmas at sulfide saturation (Shima and Naldrett, 1975; Barnes, 2007).

Experimental studies by Helz (1982), Mavrogenes and O'Neill (1999) and Wendlandt (1982) have shown that there is a strong negative pressure dependence on sulfide solubility in basaltic melts, and that sulfide saturation isopleths are subparallel to anhydrous silicate liquidi. As discussed in earlier chapters, komatiites should ascend adiabatically along steep $P–T$ trajectories, and should therefore arrive at the surface both superheated and substantially undersaturated in sulfide (Lesher and Groves, 1986).

This is consistent with the high (normal) PGE contents of komatiites, which indicate that they have not undergone any fractional segregation of sulfide liquid (Keays, 1982, 1995; Sproule *et al.*, 2005). Lesher and Stone (1996) showed that most komatiites do not reach sulfide saturation until they fractionate considerable amounts of olivine. Even komatiites that have assimilated up to 30% continental crust and fractionated to basaltic compositions have not reached sulfide saturation (Lesher and Arndt, 1995). The presence of *local* contamination and *local* PGE depletion in the host units (Barnes *et al.*, 1995; Lesher and Arndt, 1995; Lesher *et al.*, 2001) indicates that the mineralization in many Type I deposits formed during eruption and emplacement, and involved incorporation of S from the country rocks.

The thermomechanical erosion model for Type I deposits

Following the suggestions in the early 1980s that some komatiite flows represented dynamic lava conduits (Lesher *et al.*, 1984; Barnes *et al.*, 1985) and therefore might have been hot enough to melt the rocks beneath them (Huppert *et al.* 1984; Huppert and Sparks, 1985b), it was quickly appreciated that a ground-melting process might account not only for the absence of sediments beneath the lava conduits at Kambalda but also for the S in the ores (Lesher *et al.*, 1984; Lesher and Groves, 1986). With the subsequent advent of high-precision PGE data for unmineralized komatiites, it also became evident that the ground-melting model could also explain the paradox of how such large amounts of sulfides could have formed from magmas that contained undepleted PGE contents and must have been erupted undersaturated in sulfides (Lesher *et al.*, 2001; Barnes *et al.*, 2004b).

Current models suggest that sulfide saturation was achieved on a local scale as a result of thermal or thermo-mechanical erosion of the sulfidic substrate at the base of dynamic lava conduits (Lesher, 1989; Lesher and Keays, 2002; Barnes, 2006). Some of the sulfide in the country rocks may have dissolved and subsequently exsolved from the magma during olivine crystallization, but because of the limited solubility of S in silicate magmas, most of it must have simply melted to form a basal layer of sulfide 'xenomelt' that, depending on the fluid dynamics, was transported by and equilibrated with the overlying magma (Lesher and Campbell, 1993). In this model, the S in the ore is derived from the sediment, whereas the Ni, PGE and Cu in the ores are derived from the komatiite lava as a result of their high partition coefficients into sulfide liquid.

Although this model has been largely accepted within the nickel exploration industry, it has been debated to the present day on geochemical and volcanological grounds (Foster *et al.*, 1996; Cas *et al.*, 1999; Cas and Beresford, 2001; Rice and Moore, 2001; Stone and Archibald, 2004). However, most of the geochemical issues have since been resolved (Lesher and Stone, 1996; Lesher and Burnham, 2001; Lesher *et al.*, 2001) and the geochemical objections of Lambert, Foster and coworkers on the basis of Re–Os systematics have since been retracted (Lahaye *et al.*, 2001). Objections to the model on volcanological grounds have largely been refuted as concepts and models derived from modern basaltic flow fields have become widely applied to komatiites.

Substrate erosion in modern basalts, and by extension in komatiitic flow conduits, involves several stages: (1) emplacement of a relatively thick sheet flow, fed by even thicker well-insulated lava tubes, by the processes of inflation and radial cooling discussed in Chapter 9; (2) prolonged lava flow through the tube, raising the temperature of the floor basalt to its liquidus; and (3) mechanical plucking and entrainment of blobs and fragments of partially-molten floor material into the eroding lava. The critical observation is that thermomechanical erosion is strictly limited to the major long-lived lava pathways, and none takes place at the base of newly propagating pahoehoe lobes.

Direct evidence for the operation of the thermomechanical erosion process in komatiites is rare but mounting (see reviews by Lesher *et al.* (2001), Lesher and Keays (2002) and Barnes (2006)). Critical observations include: (1) unequivocal transgressive relationships with footwall rocks at Kambalda, Raglan and Alexo; (2) the presence of xenoliths and xenomelts in ores and host rocks at Kambalda, Black Swan, Raglan, Alexo, Digger Rocks and Hunters Road; (3) the similarity of S isotopes between ores and adjacent sedimentary sulfides; and (4) geochemical evidence for local contamination of country rocks at Perseverance, Digger Rocks, Raglan, Thompson, Kambalda and Black Swan.

Origin of Type II ores

A feature of many deposits containing Type IIb mineralization is the consistency in ore grade at about 0.6% Ni. This led Duke (1986a) to propose an origin by cotectic precipitation of sulfide liquid and olivine from a cooling komatiite melt. According to this model, Type II ores are fundamentally distinct from Type I ores, in that sulfide liquid is physically transported to the depositional site in Type I ores, whereas sulfide liquation (i.e. the nucleation and growth of new droplets of sulfide liquid) and deposition occur simultaneously *in situ* in Type II ores. This is consistent with a close correlation on the centimetre scale between the grain size of olivine and the presence of sulfide blebs (Figure 10.11) and the presence of multiple ore horizons in some deposits indicating that sulfide saturation occurred during crystallization of the host rocks.

A cotectic crystallization mechanism implies that the sulfide abundance in these deposits should show a pronounced peak at a proportion corresponding to the average cotectic ratio of olivine to sulfide liquid; i.e. it should relate to the slope of the S saturation curve during olivine crystallization. Analysis of assay data from the Mt Keith and Yakabindie deposits (Barnes, 2006; Grguric *et al.*, 2006) indicates that this is true in some cases, but not others. Mount Keith shows a peaked sulfide abundance, whereas the Goliath deposit at Yakabindie and 'halo' Type II mineralization within dunites that also host Type I ores, such as Perseverance, show no such peak. The Six Mile deposit shows a sulfide peak corresponding to about 2.5% sulfide, or an olivine–sulfide ratio of 40:1. Barnes (2007) calculated the expected cotectic olivine–sulfide ratio for a fractionating komatiite magma, based on the empirically derived equation of Li and Ripley (2005) relating S solubility to magma composition, pressure and temperature, and obtained values around 100 for appropriate silicate melt compositions.

The implication is that at least some Type II mineralization is the result of an accumulation of transported sulfide droplets. Despite the high density of sulfide liquid and the low viscosity of the komatiite lava, sulfide droplets several millimetres in size were capable of being transported and entrained within major flow pathways. It also implies that there is a spectrum of variability between Type I and II ores. At least some Type II orebodies may well form downstream from Type I ores.

10.8 Exploration guidelines

Not all magma pathways are mineralized, but all komatiite-associated Ni–Cu–PGE deposits are hosted by cumulate rocks that represent pathways: lava conduits (e.g. Kambalda), channelized sheet flows (e.g. the Cross Lake deposit in the Raglan area), invasive lava conduits (e.g. the Katinniq deposit in the Raglan area), feeder sills (e.g. Thompson) and magma conduits (e.g. Shangani). The critical feature appears to be flow rate and access to S: magma conduits that have access to S from country rocks are normally mineralized, whereas lava/ magma conduits that do not have access to S and non-channelized facies (i.e. volcaniclastic flows, pillowed flows, lava lobes, most sheet flows: see Chapter 9) are barren. The non-channelized facies represent less voluminous eruptions or more distal portions of komatiite lava piles (Lesher, 1989; Hill *et al.*, 1995; Prendergast, 2001). Greenstone belts where the komatiite stratigraphy is dominated by thin flow lobe facies, such as Barberton, have low potential for Fe–Ni–Cu sulfide mineralization.

Effective exploration for komatiite-associated Ni–Cu–PGE deposits requires a sound understanding of their physical volcanology and lithogeochemistry. The following exploration guidelines have been modified from Lesher (1989), Lesher and Stone (1996), Lesher and Keays (2002) and Barnes (2006):

Province selection

(1) Any tectonic setting, although most prospective provinces exhibit evidence of rifting during the volcanic stage.
(2) Provinces where mafic–ultramafic magmas (komatiites, komatiitic basalts, picrites–ferropicrites, high-Mg basalts) were generated by high-degree partial melting of the mantle.
(3) Any type of MgO-rich magma (i.e. Al-depleted, Al-undepleted, Al-enriched, ferropicritic).

Area selection

(1) Sequences that are thicker than other sequences in the same province and may represent loci for lava/magma channelization.

(2) Sequences that contain a greater proportion of thicker, more olivine-rich, poorly differentiated units.

(3) Sequences where some, but not all, of the lavas and/or intrusions are enriched in HILE relative to MILE and depleted in Nb relative to Th, indicating local contamination by upper crustal rocks.

(4) Sequences where some, but not all, of the lavas and/or intrusions are depleted in PGE, indicating local segregation of sulfides.

Unit selection

(1) Units that have mesocumulate–adcumulate textures. Komatiitic dunites will be characterized by olivine ± chromite ± sulfide assemblages, komatiitic peridotites by olivine ± chromite ± pyroxene ± sulfide assemblages, and gabbros by pyroxene + plagioclase ± olivine ± sulfide assemblages.

(2) Units that exhibit poor across-plunge continuity, but good down-plunge continuity, although establishing this will depend on the physical volcanology (e.g. lava conduit vs channelized sheet flow vs feeder sill), orientation and deformation.

(3) Units that are rich in MgO. Serpentinized units will contain abundant magnetite and will exhibit high magnetic susceptibilities, whereas talc–carbonated units will not.

(4) Units that appear to transgress sulfidic sediments.

(5) Units that appear to have thermally-eroded footwall rocks and/or contain xenoliths or xenomelts.

(6) Units that are enriched/depleted in chalcophile elements and/or contaminated/uncontaminated, relative to associated units, depending on the physical volcanology.

11

The hydrous komatiite hypothesis

11.1 Introduction

Over the past decade Tim Grove and coworkers Maarten de Wit, Steve Parman and Jesse Dann have published a series of papers (de Wit and Stern, 1980; de Wit et al., 1983, 1987; Grove et al., 1994, 1996; Parman et al., 1996; Grove et al., 1997; Parman et al., 1997; Grove et al., 1999; Parman et al., 2001, 2003; Grove and Parman, 2004; Parman et al., 2004) in which they develop a model for the nature and origin of komatiite that is very different from the majority view. Most geologists and geochemists who have worked with komatiites appear to accept that these ultramafic magmas formed in unusually hot mantle plumes (Chapter 13) but Grove and coinvestigators have proposed instead that komatiites are hydrous magmas that formed at temperatures not much higher than those in ambient upper mantle. In their early papers they suggested that the volatiles in the source might have come from undegassed primordial mantle, but in later papers they advocated dehydration of oceanic crust in an Archean subduction zone. An important element of their hypothesis, at least as presented in the earlier papers, is the hypothesis that the komatiites of the Barberton belt are intrusive. According to their model, hydrous ultramafic magma did not erupt at the surface but intruded as dykes or sills into the upper parts of the basaltic volcanic pile. Under such conditions the presence of water, or its escape during degassing, could have had an important influence on the textures and mineralogy of komatiites. Magmatic water remains dissolved in the magma only under pressure, as in an intrusive context, and under these conditions its presence could have been crucial to the formation of spinifex texture. According to the hydrous komatiite model, the presence of water influenced both the sequence in which the magmatic minerals crystallized and the compositions of these minerals.

The issue of whether or not a komatiite is hydrous is important in its own right, particularly as an element into any investigation of the origin of mafic–ultramafic magmas and mechanisms of mantle melting. But it also has far broader implications. It is crucial to any discussion of temperatures in the terrestrial mantle, and, in particular, how these temperatures have varied through Earth history (Bickle, 1978, 1986; Richter, 1988; Nisbet *et al.*, 1993; Grove and Parman, 2004).

In this chapter I first review in detail the development of the hydrous komatiite model, from its first mention in papers by Maarten de Wit and colleagues, to its formulation in the review paper by Grove and Parman (2004). I also discuss a parallel hypothesis, developed particularly by Japanese experimental petrologists, in which the source of komatiite is thought to have contained a small amount of volatiles, but not necessarily enough to have had a major effect on the emplacement and crystallization of komatiite magma. Then I present my own point of view: I accept that some komatiites intruded as high-level sills, and that a few compositionally unusual types contained a small amount of water, but I maintain that most komatiites are essentially anhydrous and erupted as lava flows.

11.2 Development of the hypothesis

The Jamestown ophiolite complex and intrusive komatiites: early papers by Brooks and Hart and by de Wit and coauthors

Many of the more important elements of the hydrous komatiite model first appeared in a paper by Brooks and Hart (1974). These authors noted: (1) that komatiites are commonly associated with siliceous intermediate–felsic volcanic rocks, an association that 'strongly indicates an arc origin for komatiitic rocks'; (2) the high temperature required to generate anhydrous komatiite magmas 'can be lowered somewhat by involving significant quantities of water in the melt'; (3) 'On extrusion, rapid release of this water would ... lead to rapid crystallization by what might be termed "dehydration quenching"'. They proposed that the latter process could explain spinifex textures. Allègre (1982) developed these ideas and proposed that komatiites formed by melting of hydrated oceanic crust in a subduction zone.

In papers published by de Wit and coworkers in the 1980s (de Wit and Stern, 1980; de Wit *et al.*, 1987), the mafic and ultramafic rocks in the lower stratigraphic units of the Barberton greenstone belt were interpreted as parts of an ophiolite complex, a section of oceanic crust that formed at an Archean mid-ocean ridge. In this model, only some of the tholeiitic basalts in the stratigraphically lower formation of the belt are considered to be extrusive and many mafic pillowed units and all of the ultramafic rocks are interpreted as intrusive.

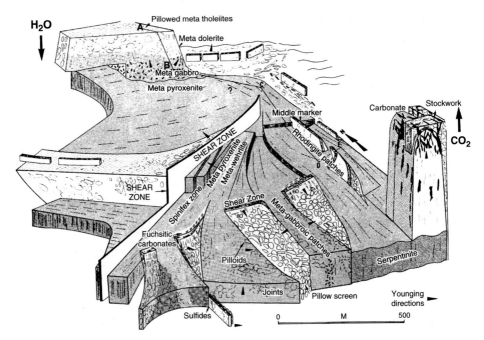

Fig. 11.1. Block diagram illustrating structures in part of the Barberton belt (from de Wit *et al.* (1987)). The ultramafic units are interpreted as a series of near-vertical intrusions that cut up through the volcanic and sedimentary units. Spinifex-textured units are shown as near-vertical dykes.

Certain pillows are described as 'pilloids' – structures that form in shallow-level intrusions. Cross-cutting relationships between spinifex-textured units and the surrounding rocks are cited as evidence of intrusion: the thin spinifex-textured units of the classic 'Spinifex Stream' locality (Viljoen and Viljoen, 1969a, b; Smith and Erlank, 1982; Viljoen *et al.*, 1983: see also Chapter 2) are interpreted by de Wit *et al.* (1987) as near-vertical dykes that cut up through a series of basaltic flows, cherts and intrusive ultramafic rocks (Figure 11.1). Some spinifex textures are assigned a secondary (metamorphic) origin. Others are recognized as magmatic, but are described as symmetric structures, coarse in the centre and becoming finer to towards both margins, which formed within steeply dipping dykes. These dykes intruded as thin sheets into olivine-rich rocks of cumulate or restitic origin. In this paper, de Wit *et al.* (1987) also advanced the degassing or 'dehydration quenching' model for the formation of spinifex according to which '... (the texture) can best be explained if the spinifex growth is related to rapid change in fluid pressure (specifically PH_2O and/or PCO_2) and associated temperature fluctuations, in a predominantly multiple intrusion environment ...'

Other themes that are closely associated with the hydrous komatiite model were developed in the same paper. For example, de Wit *et al.* (1987) used correlations between the MgO and H_2O contents of komatiites, and relations between olivine and bulk-rock compositions, to argue that the parental komatiite magma had a 'picritic' composition and contained no more than 20% MgO.

Finally, de Wit *et al.* (1987) argued that the ultramafic portions of some of the large layered units in the Barberton belt were not the lower ultramafic cumulate portions of differentiated intrusions but were tectonites – the olivine-rich residues left in the mantle after extraction of partial melt.

Formulation of the hydrous komatiite model: papers by Grove et al. *and Parman* et al. *published between 1996 and 1999*

Although the hydrous komatiite model was the subject of several talks given at international meetings in the early 1990s, its formal presentation was first made in a chapter by Grove *et al.* (1997) in de Wit and Ashwal's (1997) book on Archean greenstone belts, and in a paper by Parman *et al.* (1997). In Grove *et al.* (1997) we find further development of the intrusive komatiite hypothesis, although here the thin ultramafic units were described as dykes or sills that intruded more or less conformably into a series of basaltic lava flows, rather than as near-vertical intrusions within a peridotite complex. In Grove *et al.'s* (1997) sketch of a thin komatiite unit, the lower massive portion is identified as part of a single, differentiated ultramafic unit. In their detailed petrographic description, they described spinifex as a primary magmatic texture and highlighted the variations in the size and orientation of the olivine crystals: from fine, randomly oriented grains in the upper and lower parts of the zone to coarse crystals in the centre. They emphasized the presence of small spinifex-textured dykelets within the massive portions of some komatiite units (Figure 11.2) and used them as an argument for an intrusive origin of the komatiites in general. In their model, the massive peridotite represents either portions of thick dunitic intrusions or residual mantle peridotite.

Grove *et al.* (1997) defined what has come to be known as the 'spinifex paradox'. Donaldson (1976,1979,1982) had earlier pointed out that the morphology of olivine crystals in spinifex-textured komatiite was rather like those of crystals grown in petrological experiments at moderately high rates of cooling or high degrees of supercooling. These crystals do not have the delicate dendritic habits of olivines within liquids cooled extremely rapidly, and on this basis Donaldson argued that spinifex is not a quench texture. Grove *et al.*

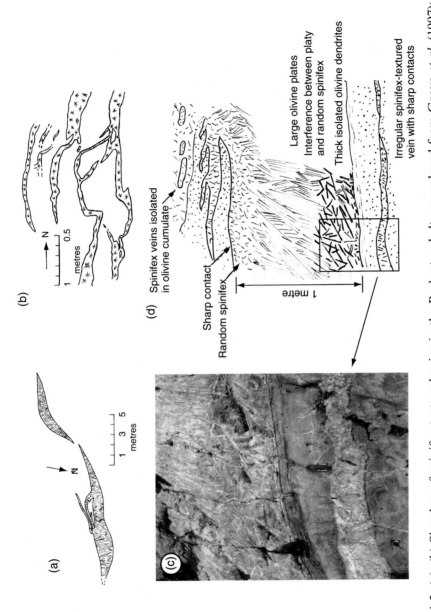

(a)

(b)

N

1 0.5
metres

(c)

(d) Spinifex veins isolated
in olivine cumulate

Sharp contact

Random spinifex

1 metre

Large olivine plates

Interference between platy
and random spinifex

Thick isolated olivine dendrites

Irregular spinifex-textured
vein with sharp contacts

N

1 3 5
metres

Fig. 11.2. (a), (b) Sketches of spinifex-textured veins in the Barberton belt, reproduced from *Grove et al.* (1997); (c), (d) photograph and sketch of spinifex veins in komatiitic units on 'Serpentine Mountain' in McArthur Township (47°9698 N, 53°4146 W) in the Abitibi belt of Canada (Houlé *et al.* 2008b). All three diagrams show spinifex veins that intrude massive olivine cumulates. De Wit *et al.* (1987) and Grove *et al.* (1997) interpret these veins as evidence that the komatiites are intrusive, but such veins are common in komatiite flows in many localities. The veins probably form through mobilization of intercumulus liquid and injection of this liquid into fractures within the cumulate portions of the flows.

(1996) conducted cooling-rate experiments on an anhydrous komatiite composition and found that olivine with the morphologies and size of crystals in spinifex-textured komatiite formed at cooling rates around $88\,°C\,h^{-1}$. They calculated that this rate is far higher than that expected in the interior portions of komatiitic units, where the largest spinifex blades are found, and concluded that 'the distribution and texture of olivine crystals within the olivine spinifex komatiites appear inconsistent with high temperature and extrusive origin'. They proposed that the development of the texture requires that 'some change in physical conditions induced supersaturation, promoted rapid crystal growth and inhibited nucleation rate'. In their view, the agent that promoted these changes was water, which could have acted in two ways. First, 'dehydration quenching' or the rapid degassing of wet magma as it ascends to the shallow crust could induce large undercooling and rapid crystal growth. Second, 'the presence of H_2O in the melt impedes the formation of crystal nuclei and increases melt diffusion rate and, therefore, crystal growth rate. The combination of these two effects is to lower nucleation rate and increase growth rate and small undercoolings'. The result, according to the hypothesis, is the formation of large skeletal crystals in slowly cooling central portions of the komatiite units. In a talk at the annual meeting of the American Geophysical Union, Grove *et al.* (1996) described experiments that produced spinifex textures during relatively slow cooling of hydrous komatiite.

Seemingly one of the strongest arguments in favour of hydrous komatiites was mentioned by both de Wit *et al.* (1987) and Grove *et al.* (1997), then developed fully by Parman *et al.* (1997). These authors undertook a detailed experimental study on a material of komatiite composition, under both anhydrous and hydrous conditions, at equilibrium and during rapid cooling. They paid particular attention to the compositions of the pyroxenes that crystallized under these different conditions and they produced the Mg# vs. Wo diagram shown in Figure 11.3. As has been discussed in Chapter 4, in this diagram we see that under anhydrous conditions, two or three pyroxenes crystallize: orthopyroxene, pigeonite and augite. Under anhydrous conditions, the augite has relatively low Wo contents. When the experimental charge is rapidly cooled, the augite has still lower Wo contents and a wide range from moderate to very low Mg/(Mg + Fe) ratios. Hydrous komatiite, in contrast, does not crystallize low-Ca pyroxene and the augite has distinctively high Wo contents, the consequence of crystallization at the lower temperatures in the water-rich silicate liquid. The latter characteristic is shared by the cores of augite grains in spinifex-textured and cumulate komatiites from the Barberton belt, which, together with an apparent absence of Ca-poor pyroxene, constituted, in the

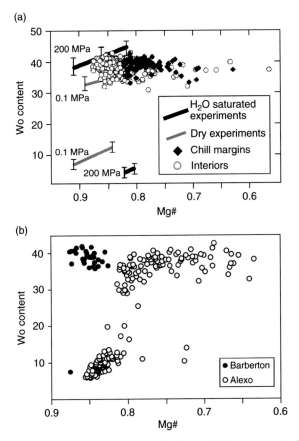

Fig. 11.3. Compositions of pyroxenes in komatiitic units: (a) from Grove *et al.* (1997); (b) compilation of analyses of pyroxenes in the Alexo komatiite (from Bouquain *et al.* (2008)).

opinion of Parman *et al.* (1997), a strong argument for a hydrous intrusive origin of the host komatiites.

Parman *et al.* (1997) reiterated the argument that the high MgO contents of many spinifex-textured komatiites resulted from Mg uptake during metamorphism. Using the composition of the 'least altered' komatiites, they proposed a parent liquid composition containing only 23% MgO. Using the pyroxene compositions and the experimental phase relations, they estimated that this magma contained at least 6% H_2O and proposed that it was emplaced at a pressure of 190 MPa at a depth of around 6 km within the volcanic pile. In Chapter 12, I discuss in detail how the MgO contents of komatiite liquids can be estimated and show that maximum MgO contents in the Barberton komatiites, like other Archean komatiites, approach, or in some cases exceed, 30%.

As for the source of water, Parman *et al.* (1997) preferred a subduction origin but did not rule out primordial volatiles inherited from accretion. They emphasized that this amount of water would decrease the temperature at the source by about 250 °C, bringing it down to levels close to those in the convecting upper mantle.

Grove *et al.* (1999) developed further many of the important arguments of the previous papers. They related differences in pyroxene compositions to variable cooling rates in different parts of the komatiite units. At the margins of these units, a high cooling rate results in the crystallization of augite with only moderate Wo, relatively low $Mg/(Mg + Fe)$ and high Al_2O_3; lower cooling rates in the interior result in high Wo and $Mg/(Mg + Fe)$ and low Al_2O_3. They also presented the results of experiments designed to reproduce conditions during intrusion and crystallization of hydrous 23% MgO komatiite at 6 km depth within the volcanic pile. They found that this magma is saturated with both olivine and orthopyroxene and concluded that the melt formed through high degrees of partial melting of a hydrous mantle source, and left a residue of harzburgitic composition when it escaped towards the surface. This information then formed the basis of a model in which the magma is produced in a subduction zone by 'flux melting', a term used to describe a process in which a 'H_2O-bearing melt separates from its residue and ascends through the mantle by porous flow. The melt continuously reacts and requilibrates with shallower mantle as it ascends . . .'

The Boston Creek komatiitic basalt

Stone *et al.* (1987, 1993, 1997) describe an unusual unit of Fe-rich komatiite or ferropicrite, interpreted either as a flow or a sill, in the southeastern part of the Abitibi belt in Ontario, Canada. This unit is about 100 m thick and contains an unusually thick upper layer of clinopyroxene spinifex. Close inspection of the upper contact of this unit shows what appears to be an intrusive relationship: layers in the finely banded chert that overlie the unit are cut by the chilled upper border and veins of chilled material penetrate up into the chert (Houlé *et al.*, 2001). Spinifex in the upper zone is texturally very similar to that in thick, layered komatiitic flows and, although the mineralogy and chemical composition is unusual, it seems reasonable to apply the term komatiitic (basalt) to the unit (see discussion in Chapter 2).

Chemically the intrusion is distinguished by unusually high Fe and Ti, relatively low Al–Ti ratios (due only in part to high TiO_2) and high concentrations of incompatible trace elements. The MgO contents are only modest, reaching a maximum of 17% in the upper part of the spinifex zone. The unit is described in this chapter because of the presence of hydroxyamphibole, which constitutes up to 25% of some samples from the intrusion. Stone *et al.* (1997) found amphibole inclusions in crystals of olivine, the liquidus mineral, and argued on this basis that amphibole crystallized early. Using experimental data on the amount of water required to stabilize amphibole, they estimated that the parental magma contained 1.2% water. This important paper demonstrates clearly that some komatiites contain moderate amounts of water. Nonetheless, this conclusion cannot be extended to all komatiites because the Boston Creek komatiitic basalt has a very unusual composition and formed from a mantle source of unusual composition, or under unusual conditions, or both (see Chapter 13).

Experimental studies advocating a hydrous deep mantle source of komatiite

A group of mainly Japanese scientists (Inoue, 1994; Hirose and Kawamoto, 1995; Kawamoto *et al.*, 1996; Hirose, 1997; Asahara *et al.*, 1998; Inoue *et al.*, 2000; Litasov and Ohtani, 2002) have conducted high-pressure melting experiments on hydrous peridotites, experiments that led them to conclude that komatiites could have formed through melting of a deep hydrous mantle source. In most of their papers, it is said explicitly or implicitly that the source is a mantle plume; the model is therefore very different from the subduction model of Grove, Parman and de Wit in which melting is said to take place at shallower levels in the mantle wedge above a subduction zone. Inoue and Sawamoto (1992), Inoue (1994), Kawamoto and Holloway (1997), Ohtani *et al.* (1997) and Asahara *et al.* (1998) used two main arguments to support their model. They conducted experiments on natural mantle peridotite or synthetic analogues under high pressures (8–23 GPa) and under both water-saturated and undersaturated conditions, and they showed that under these conditions, the melt had a composition very like that of Al-depleted or Al-undepleted komatiite. Small differences between the compositions of experimentally produced melts and natural komatiites had been observed during experiments conducted on anhydrous peridotite (Walter, 1998), but, as discussed in Chapters 7 and 13, it is difficult to establish whether these differences pose a real objection to anhydrous melting or whether they arise from difficulties in estimating the composition of parental komatiite magmas on the basis of the compositions of variably altered komatiite samples. Melting under both hydrous and anhydrous conditions seems equally capable of producing melts with compositions close to those of natural komatiites. In this situation, the only real argument in favour of hydrous melting is that the temperatures involved are lower, closer to those of ambient mantle. Whether this constitutes an advantage is debatable. In any convecting system, some regions will be far hotter than others, and even the group of scientists that questions whether plumes exist (e.g. Anderson (1994)) accepts that the komatiite source was hotter than that of normal upper mantle. Furthermore, the source of Gorgona komatiites was moderately to extremely depleted in incompatible elements (Arndt *et al.*, 1997b), and since water behaves like an incompatible element in deep mantle processes, the source of at least these komatiites should have been very dry (see Chapters 2 and 5). Herzberg and O'Hara (2002) estimate that the source of Gorgona picrites was about 400 °C hotter than ambient upper mantle.

The second argument advanced by the Japanese experimentalists was based on the realization that high-pressure olivine polymorphs such as wadsleyite and ringwoodite can contain significantly higher water contents than minerals stable at lower pressures. This led to a model in which komatiite forms through dehydration melting: as the high-pressure minerals break down at about 400 km depth during the ascent of a hot hydrous plume, the release of water causes partial melting, and because the melting takes place at great depth, the liquids that form have komatiitic compositions. This model seems quite reasonable, but whether this process is responsible for the formation of many komatiites is open to question. As discussed in the following section, there is no convincing evidence that common komatiites such as those from the Barberton or Abitibi belts were hydrous, and there are strong theoretical arguments that lead to the conclusion that hydrous ultramafic magma would not erupt as mobile fluid lava, as seems to be the norm for komatiites.

Responses to the hydrous komatiite model: papers by Nisbet et al., McDonough and Danushevsky, and Arndt et al.

The aim of the paper published by Nisbet *et al.* (1993a) was to place some constraints on the eruption temperatures of komatiite magmas. They did this by considering the MgO contents of the upper parts of komatiite flows or by using the compositions of preserved olivines to estimate the MgO contents of parental komatiite liquids, and then, by using Nisbet's (1982) equation that links MgO to temperature, they estimated the eruption temperature of komatiites from various parts of the world. This procedure is discussed in detail in Chapter 12. On this basis Nisbet *et al.* (1993a) concluded that Al-undepleted Munro-type komatiites erupted as liquids containing up to 30% MgO at temperatures approaching 1600 °C. Because the Al-depleted character of Barberton-type komatiite requires that they come from a deeper source, it was proposed that these magmas could have been even hotter. In these calculations it was assumed that the magmas were essentially anhydrous.

In a short paper published in *Geology*, Arndt *et al.* (1998a) first summarized the principal arguments that had been advanced up to that time in support of the hydrous komatiite model (essentially those given above), then presented some arguments against this hypothesis. It was noted that the field and petrological characteristics of Barberton komatiites are identical in all important respects to those of komatiites from other regions, regions in which good exposure and textural preservation leave no doubt that the units are extrusive lava flows. Arndt *et al.* then explained the problems associated with the eruption of hydrous komatiite. On approach to the surface, hydrous magma would exsolve water leaving the magma at a temperature well below its new anhydrous liquidus temperature. The probable consequence would be crystallization of abundant olivine and if this magma erupted, it would do so as viscous, olivine-pheocryst-charged picritic lava, not the mobile, phenocryst-poor ultramafic liquid documented in numerous studies of komatiite, including those from the Barberton region (see Chapters 2 and 3). Arndt *et al.* also argued that komatiite, a high-degree mantle melt, would most likely have formed by fractional melting, and that during fractional melting any water initially present in the source would be extracted in the first departing melt fractions. This paper was written before Grove *et al.* (1999) published their flux-melting model, which to some extent obviates the problem that fractional melting poses for the hydrous model. Arndt *et al.*'s conclusion at the time was that although some rare komatiites, such as those from Boston Creek, contained a small amount of water, most komatiites were essentially anhydrous.

Direct evidence on the water contents is obtainable, at least in theory, by measuring the compositions of melt inclusions in phenocrysts in komatiite magmas. McDonough and Ireland (1993) and McDonough and Danyushevsky (1995) analysed inclusions in olivine in komatiites from Zvishavane and Alexo and calculated that these olivines had crystallized in magmas that contained no more than about 0.2% water. In addition, they noted that some of this water could have come from hydrous rock that had been assimilated by the komatiite during its passage through the crust. This approach has been pursued by Mary Gee, Leonid Danushevsky and co-workers, who have obtained similar results (private communication).

Shimizu *et al.* (2001) measured higher water contents in inclusions in chromite grains in komatiites from Zvishavane, and they estimated that the parental magma contained 0.4–0.8% water. They argued that the mineral chromite preserves the original composition of trapped melts better than olivine, a conclusion that seems at odds with the major-element compositions that they report for the glass within the inclusions. In these compositions, elements that are relatively immobile during hydrous alteration (Al, Ti) plot on olivine control lines, but Ca shows a marked depletion in the more magnesian samples, a feature characteristic of altered komatiite samples (Chapter 5). It appears that the inclusions in chromites are altered, and if this is true, then high water contents probably do not represent those of the original magma.

More recently, Sylvester *et al.* (2000) and Kamenetsky *et al.* (2003) measured moderate water contents in melt inclusions in olivines from Gorgona komatiites. As mentioned above, these rocks have large to extreme depletion of incompatible trace elements, a feature best explained if the magmas formed through advanced fractional melting (Arndt *et al.*, 1997b). Any volatiles should have been removed during the segregation of early-formed melts leaving a dry and refractory source. Kamanetsky *et al.* (2003) concluded that the discrepancy between a high water content and low levels of incompatible trace elements points to the addition of water during assimilation, at crustal depths, of hydrated wall rocks.

11.3 Extrusive Barberton komatiites: papers by J. Dann published in 2000 and 2001

Two papers published by Jesse Dann (2000, 2001) destroyed the very foundation of the hydrous komatiite model as originally formulated. In these papers, Dann presented the results of several years of detailed mapping of the Komati Formation of the Barberton belt, the type area of komatiite and the region where the hydrous model was developed. He described in detail vesicular komatiites and elaborated on the role of inflation in the eruption of komatiites (see Chapters 3 and 10). More importantly, at least in the context of this chapter, he stated unequivocally that 'all the komatiites in the Komati Formation, including those with the 2-kbar augite compositions, were submarine lava flows.' Using field observations and detailed maps, he refuted many of the earlier arguments for an intrusive origin such as cross-cutting relations and growth of spinifex across magmatic boundaries (de Wit *et al.*, 1987). He recognized that some of the massive komatiite units may be intrusive and he developed a series of criteria that can be used to distinguish them from their extrusive counterparts. But there is no question, in his opinion, about the nature of the units that had formed the basis of the hydrous komatiite model: he stated unequivocally that many or most of the Barberton komatiites, like their counterparts in other Archean greenstone belts, erupted as lava flows.

He appeared reluctant to elaborate on the consequences of his findings and his sole substantive comment came within one of the last paragraphs in his 2000 paper.

'The high emplacement pressures determined by Parman et al. (1997) appear irreconcilable with the emplacement of the komatiites as submarine lava flows at lower pressures (<1 kbar or <9 km water depth). However, MORB magmas erupt on the ocean floor too rapidly for dissolved volatiles to exsolve and equilibrate with the magma (Dixon and Stolper, 1995). Boninites, hydrous arc magmas with noted similarities to komatiites, have interstitial glass with high water contents. Similarly, the komatiites may have erupted in deeper water more rapidly, and rapid solidification may have preserved the disequilibrium between liquid and volatiles (Parman, pers comm.)'.

In other words, the unusual compositions of augite in Barberton komatiites cannot have resulted from crystallization at moderate pressures in komatiite intrusions but instead may have resulted from crystallization in lavas that erupted and cooled rapidly and metastably preserved their high water contents.

11.4 Elaboration of the subduction zone model: papers by Parman *et al.* and Grove and Parman published between 2001 and 2004

In their papers of 2001 and 2003 Parman *et al.* dramatically changed their line of argument. There was little further mention of the conditions of emplacement and emphasis was placed on geochemistry and tectonic setting. In the first paper they used major- and trace-element geochemistry to address the question of whether komatiites were derived from a mantle plume or in a subduction zone. In one of their figures, reproduced as Figure 11.4, they showed that the SiO_2 and TiO_2 contents of many (though not all) Barberton komatiitic basalts have compositions like those of modern boninites. In another diagram, reproduced as Figure 11.5, they showed that selected samples of Barberton komatiitic basalt have trace-element patterns rather like those of selected boninites. They concluded on this basis that Barberton komatiitic basalts are the Archean equivalents of modern boninites and that these magmas formed through hydrous melting in a subduction zone. They recalled the arguments of Parman *et al.* (1997) to defend the interpretation that the ultramafic equivalents, Barberton komatiites, formed in the same setting, then emphasized how melting in the subduction zone would require mantle temperatures of only 1500–1600 °C, temperatures that are far lower than those needed for anhydrous melting in a mantle plume.

Parman *et al.* (2003) reported ion microprobe analyses of the trace-element contents in clinopyroxene in komatiites. The idea is sound – clinopyroxene is

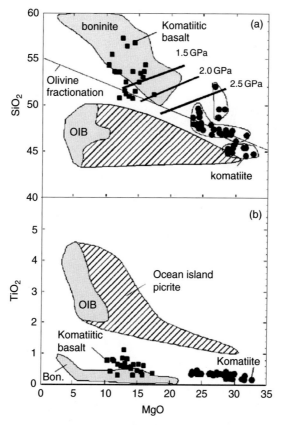

Fig. 11.4. Diagrams comparing the SiO_2 and TiO_2 contents of komatiites and komatiitic basalts from Barberton and the Abitibi belt with those of modern boninites and ocean island basalts and picrites (from Parman *et al.* (2003)).

the sole magmatic mineral preserved in komatiites that contains high-enough contents of REE and HFSE to be measured accurately using *in-situ* methods. By using these data, it should be possible to reconstruct the compositions of the komatiitic liquids and to establish which elements were mobile during metamorphism and which were immobile. The underlying idea was to identify elements that could be used as tectonic tracers to establish whether komatiites formed in a plume or in a subduction environment.

Parman *et al.* (2003) compared the trace-element patterns measured in clinopyroxene from Barberton komatiites with those calculated for a clinopyroxene that formed in equilibrium with liquid with the bulk composition of the komatiite sample. The results are summarized in Figure 11.6. They were not able to measure REE and HFSE simultaneously and therefore they were not able to use ratios such as Nb/La or Ti/Gd, which best discriminate magmas

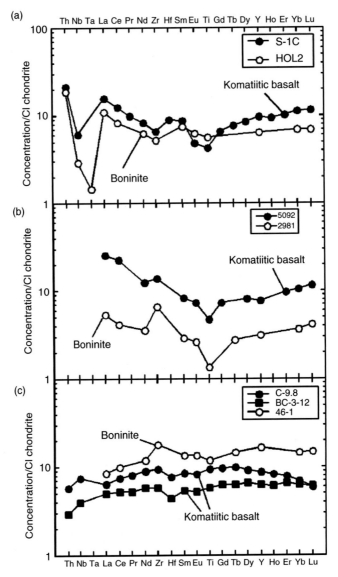

Fig. 11.5. Comparison of mantle-normalized trace-element patterns of selected komatiitic basalts from the Barberton belt with those of selected modern boninites (from Parman *et al.* (2003)). The labels C9.8, S-1C, HOL2 etc. are the names of the analysed samples.

from subduction settings. The ratios of REE in measured and calculated patterns agree well, but ratios involving the HFSE, such as Ti/Zr, diverge. The authors called on the variable partitioning behaviour of Ti in augite, or the removal of Ti in fractionating chromite, to explain the differences between measured and calculated trace-element patterns. In the last part of the paper,

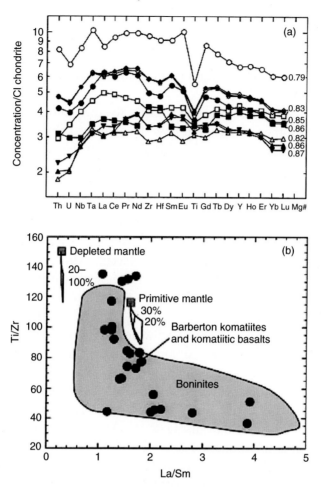

Fig. 11.6. (a) Trace-element contents, normalized to Cl chondrites, of augites in Barberton Komatiites (from Parman *et al.* (2003)). (b) Comparison of the trace-element ratios in augites from Barberton komatiites with the field of augites from boninites (from Parman *et al.* (2003)).

they used the La/Sm vs Ti/Zr diagram (Figure 11.6(b)) to infer the tectonic setting of komatiites, and they concluded that these magmas formed in a subduction setting.

Parman *et al.* (2004) discussed relationships between komatiite and the orthopyroxene-rich harzburgite of the lithospheric mantle underlying the Kaapvaal Craton in South Africa. They presented very useful new analytical data (but only for major elements) for the chill margins of Barberton komatiites. After some discussion of possible Fe diffusion during metamorphism, they concluded that the olivine cores retain their original compositions

(Fo$_{93-94}$) and they used this composition to estimate that the parental komatiite liquid contained between 25 and 30% MgO, about the same range as in some of the chill margins (23–32%), and far higher than the 23% MgO proposed as the maximum MgO content in Barberton komatiitic liquids in their earlier papers.

In one of the more recent papers published by the group, Grove and Parman (2004) developed their ideas about how the hydrous komatiite model changes our view of the thermal evolution of the earth over the past four billion years.

11.5 Other papers discussing the hydrous komatiite model

Arndt (2003) showed that the trace-element contents of most komatiites and komatiitic basalts are in fact very different from those of boninites. Although it is possible to select individual komatiitic basalts whose compositions superficially resemble those of individual boninites, when large data sets are used and when appropriate trace-element ratios are employed, the two types of rock are seen to be quite distinct. Figures 11.7–11.9 are updated versions of diagrams from Arndt (2003), diagrams in which the concentrations and ratios of elements such as Nb, Zr, Ti, Sr and the REE indicate clearly that most komatiites lack the 'subduction signature' that is so conspicuous in boninites. The data come mainly from the GEOROC database, augmented by my unpublished analyses. To quantify the presence or absence of this signature, I used ratios of elements that display contrasting behaviours during the formation of magmas in subduction zones. In subduction-related magmas, LILE, such as Rb, Ba and Sr, are enriched, and the HFSE are depleted, relative to the REE and Th. Ratios such as Nb/La or Sr/Sm, which measure the magnitude of positive or negative anomalies in mantle-normalized trace-element patterns, are effective discriminants of magmas from subduction settings. Problems arising from the well-known mobility of the LILE during metamorphism can be avoided by using the concentrations of trace elements measured in melt inclusions, which appear to be less susceptible to alteration than whole-rock samples. In Figures 11.7–11.9, it is clearly seen that the komatiites, including those from the Barberton region, plot in fields distinct from those of boninites.

Another aspect of the composition of Barberton komatiites discussed in Arndt (2003) is the geochemical evidence that points to their having formed under conditions in which garnet was retained in the residue of melting. As discussed in Chapter 7, high CaO/Al_2O_3 or low Al_2O_3/TiO_2 and high Gd/Yb are features characteristic of Barberton komatiites, features known from the very start of komatiite research: the diagnostic major-element ratios were first

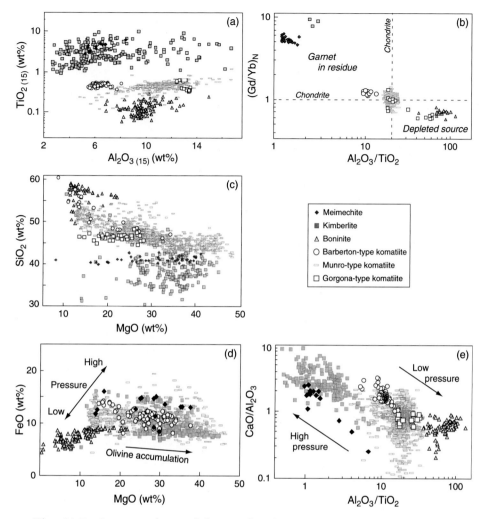

Fig. 11.7. A comparison of key major-element concentrations in rocks crystallized from ultramafic magmas. Figure 11.7(c), MgO vs SiO$_2$, shows the wide range of SiO$_2$ in these rocks, from high values in boninites and some komatiites, to low values in kimberlites. In Figure 11.7(a), the data were normalized to 15% MgO to eliminate the effects of the fractionation or accumulation of olivine ($\pm$ orthopyroxene). For the komatiites, meimechites and kimberlites, olivine of appropriate composition was added or subtracted; for the boninites, a mixture of olivine and orthopyroxene was used. This diagram shows the large differences in TiO$_2$ of the different series, and the large variations in Al$_2$O$_3$ contents of the three types of komatiite. Other differences between komatiites and other types of ultramafic rock are illustrated in (b), (d) and (e). Data are from numerous sources including: http://georoc.mpch-mainz.gwdg.de/ (boninites); Mitchell (1995) and a compilation of: C. B. Smith (kimberlites); Arndt *et al.* (1995, 1998b) (meimechites); Aitken and Echeverría (1984), Sun (1984), Gruau *et al.* (1990a,b), Xie *et al.* (1993), Lahaye *et al.* (1995), Kerr *et al.* (1996), Arndt *et al.* (1997a) and Sproule *et al.* (2002) (komatiites). (Diagram from Arndt (2003).)

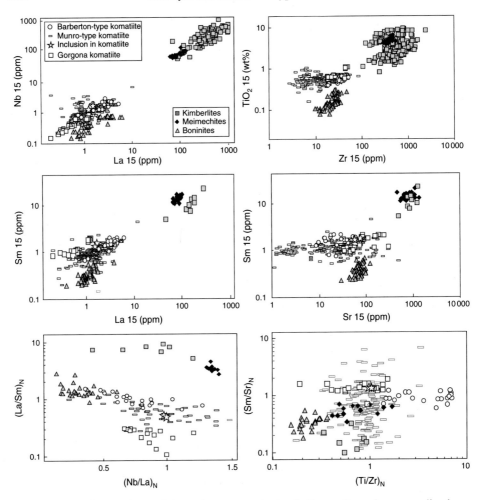

Fig. 11.8. Comparison of trace-element contents of ultramafic rocks, normalized to 15% MgO. Note the large differences in absolute concentrations and trace-element ratios of the various types of ultramafic rocks. Boninites are distinguished by low TiO_2, and low Nb/La, La/Sm and Sr/Sm. The data identified as 'inclusion in komatiites' represent analyses of melt inclusions in olivine from Zimbabwe komatiites (McDonough and Ireland, 1993). These analyses are included because they provide a reliable indication of the Sr and Ba contents of komatiites, data that are not available in all whole-rock compositions because of the mobility of these elements during metamorphism. Data sources as for Figure 11.7; diagram from Arndt (2003).

recognized by Viljoen and Viljoen (1969b,c), and the trace-element ratios were identified by Sun and Nesbitt (1978) and Green *et al.* (1974). Although there is some question about the interpretation of the high CaO–Al_2O_3 ratios because of the mobility of Ca, the relation between the other ratios and residual garnet

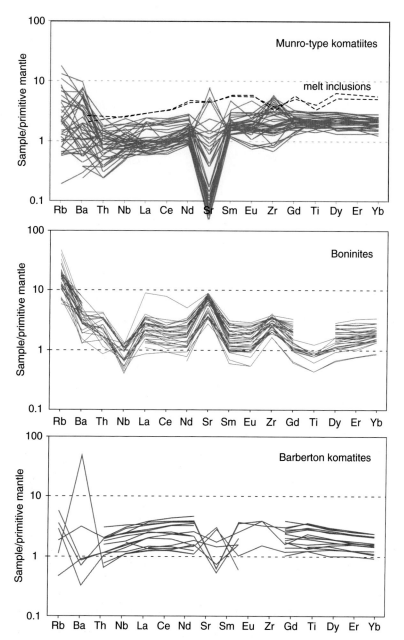

Fig. 11.9. Trace-element concentrations, normalized to the primitive mantle of Hofmann (1988). Note the large differences in trace-element concentrations and ratios of komatiites and boninites, and the distinctive negative Nb and Ti anomalies, and the positive Sr anomalies, in the boninites. The Sr anomaly in komatiites is highly variable, from positive to negative. The levels of trace elements in the melt inclusions in Zimbabwe olivines are higher than in the whole rocks because the trapped liquid crystallized olivine during cooling. This process does not change the ratios of incompatible elements and the absence of Sr and Ba anomalies in the melt inclusions suggests that the anomalies in the whole rocks probably result from mobility of these elements during alteration. Diagram from Arndt (2003).

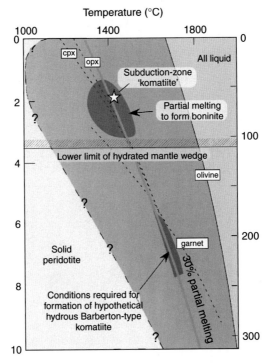

Fig. 11.10. Phase diagrams of hydrous mantle peridotite drawn using the data of Asahara *et al.* (1998). Boninites form by melting in the hydrated mantle wedge overlying subduction zone, at depths less than about 100 km. To produce liquid with the composition of Barberton-type komatiite, containing 30% MgO and with the geochemical signature of residual garnet, the depth of melting must have been greater than about 250 km. This depth is far greater than that of the hydrated mantle wedge, effectively eliminating the notion that Barberton komatiite forms in a subduction environment.

is undisputed. For garnet to remain in the residue of melting that produces magma of ultramafic composition, the pressure must be high. Estimates by Herzberg and Ohtani (1988) and Herzberg (1995) place these pressures around 6–8 GPa, or at depths of 180–240 km. These depths are far greater than those in the mantle wedge above a subduction zone (Figure 11.10).

The origin of spinifex textures was treated by Faure *et al.* (2006), who described experiments that resolved the 'spinifex paradox'. They reproduced spinifex textures by cooling liquids with compositions analogous to anhydrous komatiite within a thermal gradient such as existed in the upper margin of a komatiite lava flow. At moderate cooling rates of 2–10 °C h^{-1}, rates like those calculated for within the crust of a komatiite flow, they grew large skeletal crystals of olivine oriented perpendicular to the cooling surface, just as in platy spinifex. They therefore showed that water is not required to form spinifex texture.

11.6 A critical evaluation of the hydrous komatiite model

In the above sections I have summarized the arguments for and against the hydrous komatiite model. In the decade that has passed since the hydrous komatiite model was first formulated, almost every line of evidence advanced in its support has been called into question. The most telling arguments against the hydrous komatiite hyphothesis are:

- Barberton komatiites did not intrude as dykes or sills but erupted as submarine lava flows.
- Spinifex textures does not require the presence of water, but forms in the thermal gradient at the margins of komatiite flows.
- The geochemical characteristics of komatiites worldwide do not point to melting in a shallow subduction setting but rather to melting in a deep, hot essentially anhydrous source.

In my opinion, the situation is cut and dried: the evidence cited for the hydrous komatiite model simply does not hold water. In this section I first re-evaluate the current status of the hydrous model, then present some arguments for its alternative. A fuller discussion of the origin of komatiite follows in Chapter 13.

The hydrous komatiite model should have lost steam with the publication of Jesse Dann's (2000, 2001) papers. As originally formulated, the model was anchored by the notion that wet komatiite intruded deep within the lava pile and crystallized during or before losing its water. Crystallization at moderate pressure in an intrusive setting was considered essential if water was to remain in the magma and thereby influence both the growth of spinifex texture and the composition of augite; the drop in pressure accompanying eruption at the surface should have led to the loss of any water in the magma because the solubility of water in silicate liquids is minimal at low pressure. In papers published after 2000, the advocates of the hydrous model have responded by referring to boninitic glass that contains up to 7% water. According to Parman *et al.* (2001), '... while it is clear that our understanding of devolatilization is incomplete, the submarine eruption of komatiites could allow them to preserve high H_2O contents without fractionation or vesiculation.' But would the water be retained in the magma during lengthy flowage and protracted crystallization? As explained in Chapters 2 and 3, in regions of good outcrop, komatiite flows can be traced for hundreds of metres and komatiite formations for tens–hundreds of kilometres (e.g. Perring *et al.* (1995)). Hill *et al.* (1995) and Lesher and Keays (2002), and Barnes and Lesher in Chapter 9, reconstruct a volcanic setting in which komatiite erupted as vast, rapidly flowing rivers of lava that flowed out over basaltic plains for many tens of kilometres, thermo-mechanically eroding its substrate as it did so. In the inflation model for komatiite emplacement, lava invades beneath the crust of early flow units. Vesicles are indeed present in many komatiites, but they are rare: even in the Barberton region, where Dann (2001) highlighted their presence, vesicular komatiite makes up only about 1% of the komatiite sequence. The presence of vesicles in Gorgona komatiites has been re-emphasized by Kerr (2005), but these structures are restricted to a thin layer at the base of the spinifex layer. The 10–20 vol% of vesicles represent less than 1 wt% of water, and, in view of the high Cl and B contents measured by Kamenetsky *et al.* (2003), the trapped fluid was seawater. The fluid most probably was introduced into the komatiite through the assimilation of hydrothermally altered oceanic crust.

Fragmental komatiite is known in several areas (Chapter 3), but in each of these regions these rocks formed largely or completely as hyaloclastites. In those regions where hydrovolcanic processes are suspected (e.g. Schaeffer and Morton (1991)), the water is considered to have come from hydrated wall rocks assimilated into the komatiite magma.

On balance, the field characteristics and textures of komatiites, even those in the Barberton type area, conform to the model in which volatile-poor lava erupted at the surface. None of the arguments for the existence of hydrous

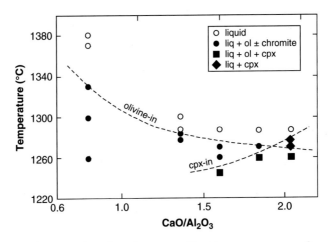

Fig. 11.11. Diagram showing the crystallization sequences of a series of experimental charges with compositions like those of komatiitic basalts but with with variable CaO/Al_2O_3. The temperature of the first crystallization of augite increases with increasing CaO/Al_2O_3. Data from Arndt (unpublished) and Bouquain *et al.*, (2008).

komatiite is watertight, except perhaps for the high-Ca augites and the apparent lack of low-Ca pyroxene documented by Parman *et al.* (1997). How can these be explained?

In Figure 11.3, we see that the Barberton augites do indeed plot in a field distinct from those of most other komatiitic pyroxenes, but rather than being distinguished by abnormally high Wo contents, as was emphasized by Parman *et al.* (1997), their outstanding feature is their high Mg/(Mg + Fe) ratios. Compare, for example, the compositions of augites from Barberton komatiites with those from the Alexo flow: the Wo contents of pyroxenes from the two regions broadly overlap. There can be no doubt that the Alexo unit is extrusive because this unit is capped by hyaloclastite whose composition matches that of the underlying chill zone and random spinifex lava. In the compositions of the pyroxene from this flow we have clear evidence that high Wo is not indicative of crystallization from hydrous magma in an intrusive context.

What remains to be found is an explanation of the high Mg# of augites from the Barberton komatiites. Figures 11.11 and 7.3(a) present summaries of an experimental study conducted in the Geophysical Laboratory, Carnegie Institution of Washington, some 30 years ago (Arndt, 1976a). Figure 11.11 shows the crystallization sequences in a series of experimental charges with the composition of a komatiitic basalt from Munro Township, doped to give a range of different $CaO–Al_2O_3$ ratios. Figure 7.3(a) shows the composition of liquids produced at various temperatures from another series of experimental charges whose compositions also represent komatiitic basalts with a wide range of $CaO–Al_2O_3$ ratios. Details of the experiments are given in Bouquain *et al.* (2008). What is significant in the figures is the relationship between the liquidus temperature of augite and the $CaO–Al_2O_3$ ratio: the higher the ratio, the earlier the crystallization of augite. It is difficult to compare the measured compositions of experimentally grown augites with those of natural pyroxenes because Fe loss to the Pt wires on which the experimental charges were suspended has increased, to variable extents, the Mg–Fe ratios of the samples. Nonetheless, the data plotted in the figures show clearly that the pyroxenes in those experimental charges with high $CaO–Al_2O_3$ ratios crystallized at higher temperatures and tend to have higher Mg# than those that crystallized from liquids with lower $CaO–Al_2O_3$. More experimental work must be done, but on this basis it can be proposed that the combination of high Wo content and high Mg# of the augites analysed by Parman *et al.* (1997, 2001) may simply be related to the high $CaO–Al_2O_3$ ratio of Barberton komatiites and is probably not an indication of crystallization of the pyroxene from hydrous magma.

12

Compositions and eruption temperatures
of komatiitic liquids

12.1 Introduction

For various reasons it is essential to know the compositions of the silicate liquids that crystallized into komatiite lavas. Of particular importance is the maximum MgO content, because it is this parameter that best distinguishes komatiites from other types of magma. In this chapter I start by discussing the methods that have been used to estimate the liquid composition and then show that some komatiites formed from liquids containing at least 30% MgO, and perhaps as much as 34% MgO. This MgO content is far greater than that of other magma types (Table 12.1), and because most komatiites are essentially anhydrous, as we saw in Chapter 11, they must have formed in exceptionally hot parts of the mantle.

12.2 Methods used to estimate liquid compositions

If we need to establish the composition of modern basaltic lava, we can ask an adventurous colleague to go to the nearest active volcano to sample directly a still-active lava flow. With komatiites this is impossible and indirect methods must be adopted to estimate the liquid composition. The situation is further complicated by two factors: (a) alteration, which affects all komatiites and is particularly prevalent in the fine-grained or glassy samples that most closely represent quenched liquids; and (b) the facility with which olivine crystals are redistributed in cooling komatiite flows, which results in a wide spectrum of liquid compositions even in single flows and produces an even greater range of compositions of crystal-liquid mixtures. To circumvent these problems, a variety of methods have been used, most of which are mentioned in the remarkable paper published more than 30 years ago by Green et al. (1974) and developed further in Bickle (1982). In this section I describe the most commonly used methods.

Table 12.1. *Compositions of selected komatiites and other high-MgO liquids*

	K4–1BA, Barberton	M666, Alexo	Zvishavane, Zimbabwe	Commondale, South Africa	Weltreden, South Africa	Murphy Well	Meimechite	Boninite
SiO_2	46.80	45.90	49.4	49.9	47.2	45.1	40.3	50.8
TiO_2	0.38	0.35	0.41	0.11	0.14	0.24	3.16	0.16
Al_2O_3	4.43	7.10	8.66	7.6	4.61	5.2	3.26	7.22
FeO(tot)	11.45	10.92	11.6	5.4	8.3	10.1	15.4	9.3
MnO	0.18	0.17	0.1	0.12	0.14	0.20	0.24	0.15
MgO	30.28	28.21	20.0	30.8	34.8	34.90	26.1	23.50
CaO	5.78	6.57	8.6	5.4	3.73	4.58	9.6	8.15
Na_2O	0.07	0.29	0.92	0.34	0.41	0.05	0.2	0.54
K_2O	0.01	0.10	0.05	0.01	0.02		1.08	0.09
Volatile content	<0.5	<0.5	<0.5	<0.5	<0.5	1–2(?)	4.9	
Max Fo	93.9	94.1	91.3	96.3	95.6	93.1		94
Anhydrous liquidus[a]	1606	1564	1400	1616	1696	1698	1522	1470
Probable eruption temperature[b]	~1600	~1550	1430	~1600	~1640	≤1650[c]		1382[d]

Notes:

[a] Anhydrous liquidus temperatures are calculated using the expression $T(°C) = MgO*20 + 1000$ from Nisbet (1982).

[b] Eruption temperatures are assumed to be similar to the anhydrous liquidus temperatures unless there is evidence of high volatile contents.

[c] This komatiite may have been hydrous and the anhydrous liquidus is a maximum temperature.

[d] Temperatures estimated by Danyushevsky *et al.* (2002a).

Source of analyses:

K4–1BA, Barberton: Parman *et al.* (2004); M666, Alexo: Arndt (1986b); Murphy Well: Lewis and Williams (1973); Commondale, South Africa: Wilson (2003); Weltreden, South Africa: Kareem and Byerly (2003); Zvishavane, Zimbabwe: Danyushevsky *et al.* (2002a); Meimechite: Arndt *et al.* (1998); Boninite: Danyushevsky *et al.* (2002b).

Analysis of spinifex-textured lavas

Viljoen and Viljoen (1969b,c), in their original papers on komatiites, used the term 'crystalline quench texture' to describe the texture we now know as spinifex. They had noted the similarity between the natural texture and the textures that form when highly magnesian silicate liquid is quenched in the laboratory, and they suggested that spinifex might have the composition of the original komatiite liquid. This line of reasoning has since been brought into question by the recognition that spinifex-textured rocks form by growth of olivine or pyroxene crystals into the interior of the flow or sill and often contain an excess proportion of these minerals (Barnes, 1983). In other words, they are a type of 'coagulation cumulate' or 'crescumulate' and as such contain a higher proportion of olivine or pyroxene than was present, in solution, in the liquid from which the mineral crystallized. The effect is most pronounced in platy olivine spinifex or 'string-beef' pyroxene spinifex in which the dominant minerals are aligned perpendicular to the flow or sill margin. In the random spinifex lavas that form closer to the margin, the amount of excess mineral seems to be less, and these lavas probably have compositions closer to those of the liquid from which they formed. Walker *et al.* (1988), for example, have used the compositions of random spinifex lavas as proxies for liquid compositions of komatiites from Munro Township, and in general random spinifex-textured lavas have compositions close to those of liquid compositions established using other methods.

Chilled margins

In layered intrusions and other bodies of cumulate rocks, the best way of estimating liquid compositions is by sampling and analysing the chilled border zones of the intrusion. The same approach can be taken with komatiites, but, as mentioned above, these fine-grained or glassy samples are particularly susceptible to alteration. In addition, many samples of both the upper and lower chilled margins of komatiite flows contain olivine phenocrysts. If it could be shown that these phenocrysts grew during eruption of the lavas and had not been concentrated mechanically, then the sample could be taken to represent a liquid; but these conditions cannot normally be demonstrated. One exception, the lower chilled margin of a komatiite flow from the Zvishavane locality in the Belingwe belt in Zimbabwe, is shown in Figure 3.9. This rock contains abundant, small, equant to skeletal grains of olivine whose distribution, size and morphology point to their having nucleated and grown in the rock in which they are found. This sample is relatively well preserved, like most of the Zvishavane komatiites, and its composition can be taken as that of a liquid. The Zvishavane komatiites are not particularly rich in MgO, as indicated by the composition of the chill sample which contains only 25% MgO.

A more magnesian composition is recorded in the chilled margin of the Alexo komatiite flow, illustrated in Figure 2.14(b). As shown in this figure, the sample contains a small fraction, less than 1%, of small, euhedral and solid olivine phenocrysts in a matrix containing abundant, much smaller, stubby but skeletal olivine grains (Figure 2.14(c)). The morphology and size of the latter crystals again point to nucleation and growth within the chilled margin, and because the proportion of olivine phenocrysts is small, the rock composition represents that of the liquid from which it formed. Further evidence comes from the compositions of olivine, in both the phenocrysts and the skeletal matrix grains, as discussed below. This sample contains 28% MgO (Table 12.1), which probably is close to the MgO content of the parental magmas of most Archean komatiites.

Bulk compositions of komatiite flows

Another way of estimating the composition of a komatiite liquid is to calculate the bulk composition of a komatiite flow by averaging the composition of each component (flow top, spinifex, olivine cumulate) in the proportions they are found in the flow, as discussed in detail in Chapter 10. The procedure gives results in the same range as those from other methods, but is not reliable for several reasons. In some cases the bulk composition is more magnesian than that of the upper chilled margin or random spinifex layer, probably because olivine sedimented out from or grew at the top of the cumulate pile beneath magma that flowed through the interior of the flow. Either process results in the accumulation of olivine and produces a bulk composition that is richer in olivine than the parental liquid. In other cases the bulk composition is less magnesian than the chilled margin, probably because the latter contains an excess of olivine phenocrysts. And once again alteration plays a spoiling role because elements such as CaO are particularly mobile in the olivine cumulates whose inclusion in the average biases the calculated bulk composition of the flow.

Combined use of olivine compositions and whole-rock analyses

The most reliable means of estimating liquid compositions employs olivine compositions in combination with trends of whole-rock compositions. Bickle (1982) first outlined how this could be done and the approach was used again by Nisbet *et al.* (1993a). The basis of the procedure is the well-established relation between the Mg–Fe ratios of olivine and silicate liquid, as expressed by the distribution coefficient $K_D = (Mg/Fe)_{liq}/(Mg/Fe)_{ol} = 0.30$ (Roeder and Emslie, 1970).

Using a selected olivine composition, the MgO/FeO value of the equilibrium liquid can be calculated, and if the FeO content of the liquid can then be estimated, the MgO content follows. Complications stem from the following sources.

(1) *Uncertainty surrounding the value of the partition coefficient.* It has been argued that K_D for ultramafic liquids is higher than that determined initially by Roeder and Emslie (1970), who worked with basaltic liquids. Values up to 0.38 are quoted by some authors (see Chapter 8). However, more recent compilations for komatiite have reproduced the earlier value of 0.30 (e.g. Parman *et al.* (2004)) and I will use this value in my calculations.

(2) *Variable ratios of ferrous to ferric iron.* The FeO in the relation of Roeder and Emslie is ferrous iron (Fe^{2+}) but in natural lavas a certain proportion of the total iron is present as Fe^{3+}. In variably altered komatiites this proportion cannot be measured and must be estimated. The procedure usually adopted is to constrain the calculation using estimates of the oxygen fugacity at which komatiite crystallized. Using Canil's (1997) experimental results, we can assume that f_{O_2} was close to the magnetite–quartz–fayalite buffer, and under these conditions, about 10% of total iron would be Fe^{3+}. S. J. Barnes has pointed out that in olivine cumulates, most of the iron will be present in olivine as Fe^{2+} and therefore the assumption that a fixed proportion of iron is Fe^{3+} is incorrect. He developed a procedure Barnes *et al.* (2008) in which the proportion of Fe^{2+} decreases linearly with increased MgO content. This procedure is adopted here.

(3) *Variable total iron contents in komatiites.* The greatest source of error in estimated MgO contents (though not MgO–FeO ratios) comes from the variation in FeO contents in komatiites. As seen in Figure 12.1, in every suite, with the possible exception of the drill-core samples from Zvishavane, the FeO content at a given MgO varies widely. This uncertainty cannot be avoided and must be acknowledged when quoting estimated MgO contents.

(4) *Choice of an appropriate olivine composition.* Because olivine is the sole silicate mineral to crystallize over at least half the solidification path of a magnesian komatiite, the composition of olivine varies over a wide range. Most olivine grains are zoned from Fo-rich cores to Fo-poor margins (e.g. Figures 4.4 and 4.5). Bickle (1982) selected the most magnesian olivine compositions from the cores of grains, arguing that this value represented that of the first-crystallized olivine and was thus in equilibrium with the most primitive liquid. For many years de Wit, Grove, Parman and coauthors (e.g. de Wit and Stern (1980), de Wit *et al.* (1987), Parman *et al.* (1997), Parman *et al.* (2001)) questioned this procedure. They emphasized that in almost all komatiites, olivine has at least

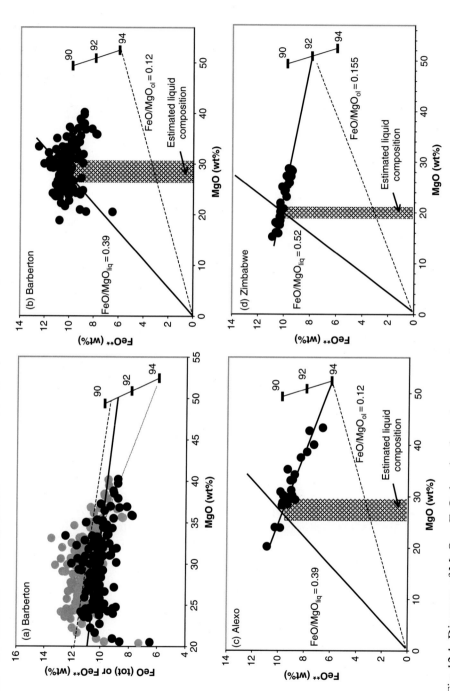

Fig. 12.1. Diagram of MgO vs FeO showing how the parental magma composition of various komatiites can be estimated. In (a), the grey circles represent uncorrected FeO(tot) contents for komatiites from the Barberton belt. The black circles in this and in (b)–(d) represent FeO**, which is the FeO content estimated using the procedure of S. J. Barnes. In (b)–(d), the lower dashed lines represent FeO–MgO ratios of the most Fo-rich olivine reported in the komatiite suite and the upper dashed lines represent the FeO–MgO ratio of silicate liquid in equilibrium with this olivine. The intersection between this line and the whole-rock komatiite data gives the MgO content of the parental liquids. Principal source of data are Smith and Erlank (1982), Parman et al. (2003), Arndt (1986b), Lahaye and Arndt (1996) and Bickle et al. (1993).

partially altered to serpentine or chlorite, particularly around grain margins. They argued that diffusion of Mg and/or Fe between the serpentinized margin and the fresh core would have changed the composition of the relict olivine, perhaps boosting its Fo content. In a more recent publication, however, they have retracted their objections and now appear to accept that the cores of olivine grains have preserved their magmatic compositions. Parman *et al.* (2004) estimated that the magma parental to the Barberton komatiites contained between 25% and 28% MgO, values similar to those estimated by other authors (Smith *et al.*, 1980; Smith and Erlank, 1982).

Five examples

Komatiites of the Barberton greenstone belt

Figure 12.1 illustrates many of the uncertainities that surround the calculation of the composition of the liquid parental to Barberton komatiites. In Figure 12.1(a), the MgO contents of lava flows from the type section along the 'Spinifex Stream' section of the belt are plotted against FeO(tot), which is total iron recalculated as FeO, and FeO(tot) **, which is the FeO content estimated using the procedure developed by Barnes *et al.* (2008). Also plotted are the compositions of stoichiometric olivine. A regression line through the FeO(tot) ** data intercepts the olivine trend at Fo_{91}, i.e., at about the same place as the intercepts in MgO vs SiO_2 and MgO vs Al_2O_3 diagrams (Figure 2.5). This intercept corresponds to the average composition of olivine in these komatiites.

The most magnesian olivine composition reported in these flows is $Fo_{93.9}$ (Parman *et al.*, 2004), which corresponds to an FeO–MgO weight per cent ratio of 0.116, shown as the dashed line in Figure 12.1(b). Liquids in equilibrium with this composition, calculated using a partition coefficient of 0.30, have an FeO–MgO weight per cent ratio of 0.386, represented by the upper line in the figure. This line intercepts the compositions of the Barberton komatiites between 27 and 31% MgO, values that constitute our estimate of the MgO content of the liquid parental to these komatiites. The values coincide with the MgO contents of the more magnesian chilled margins reported by Parman *et al.* (2004), such as the sample listed in Table 12.1. Some of these chilled margins have slightly higher MgO contents, due most probably to the presence of a small fraction of olivine phenocrysts (the addition of only 5% of olivine phenocrysts with the composition Fo_{93} increases the MgO content from 30 to 31%).

Alexo komatiite

Three separate lines of argument can be used to estimate the composition of the liquid that erupted to form the Alexo komatiite flow. The photomicrographs in Figure 2.14(b) and (c), show parts of the chilled margin of the flow, which formed at the top of the spinifex layer, just below the hyaloclastite breccia that caps the flow. In Figure 2.14(b), we see several olivine phenocrysts in a matrix containing abundant finer skeletal olivine grains. The olivine phenocrysts are virtually unaffected by serpentinization, which is limited to fine veins that cut across the grains. The compositions of olivine in the euhedral phenocryst in the centre of the photomicrograph can be measured from one margin to the other. The resultant profile, plotted in Figure 4.4, has a broad central plateau with the composition Fo_{94} surrounded on both sides by thin, more Fe-rich margins. This central plateau is thought to represent the part of the grain that grew immediately before or during eruption and the margins are thought to have grown during rapid cooling once the grains were trapped in the chilled margin. There is no change in Fo content of olivine near the serpentine veins. The cores of the small skeletal grains in the matrix (Figure 2.14(c)) have the same composition as the core of the olivine phenocryst (Arndt, 1986b). The habit, size and distribution of the latter grains suggest that they formed during quenching and it can be assumed that the composition of the matrix is that of the original liquid. Because the proportion of olivine phenocrysts is very low, less than 1%, the composition of the whole rock should also be essentially the same as that of the liquid (the presence of 1 vol% phenocrysts increases the MgO of the bulk rock by only 0.4 wt%). The bulk composition, listed in Table 12.1, contains 28.2% MgO, and the average of five broad beam analyses of the matrix contains 28.5% MgO (in both cases the analyses were normalized to 100% on an anhydrous basis).

We can test the hypothesis that the composition of the chilled margin is that of the parental liquid by calculating the FeO/MgO of a liquid in equilibrium with the most magnesian olivine, that found in the cores of the phenocrysts. Using a partition coefficient of 0.30, we calculate that this liquid should have a FeO/MgO of 0.37, which compares with the ratio of 0.35 measured in the chill sample. The hyaloclastite that overlies the chilled margin has slightly higher MgO content of 30.4%; as explained in Arndt (1986b), this originally glassy sample is strongly altered and it appears to have gained some Mg during the alteration.

A similar result is obtained using the intercept method, as presented in Figure 12.1(c). Whole-rock data from the flow intercept the olivine trend at the composition Fo_{94}, a value identical to that of the phenocrysts in the chilled margin. The MgO content of the initial liquid, estimated taking into account the overall variation of FeO contents of the rock suite, is $28 \pm 2\%$, identical within error to that of the chill sample.

In conclusion, the liquid parental to the Alexo komatiite flow contained about 28% MgO as shown by: (1) the bulk composition of the chilled margin;

(2) the composition of the matrix; (3) the correspondence between these compositions and that of a liquid in equilibrium with the most magnesian olivine; (4) the similar composition of the hyloclastite breccia.

Zimbabwe komatiites

As Figure 12.1(d) shows, the data from the fresh komatiites from the Zvishavane region plot on a well-defined trend that intersects the olivine trend at Fo_{92}. This value is slightly higher than forsterite contents of the majority of phenocrysts and cumulate grains in these flows, which vary between $Fo_{91.2}$ and $Fo_{91.6}$ (Renner *et al.*, 1994). It is also slightly higher than that obtained through the same method for Barberton komatiites (Figure 12.1(a), (b)), but the FeO contents of Zvishavane komatiites are lower than those of Barberton komatiites, and the estimated MgO content of the parental liquid is lower. Because of the small scatter of FeO contents, the MgO content of the parental liquid can be determined with precision at $20 \pm 1\%$. The value is similar to that estimated using different methods by Renner *et al.* (1994), Silva *et al.* (1997) and Bolhar *et al.* (2003). A small fraction of cumulus grains have higher forsterite contents, up to $Fo_{93.6}$. When this composition is used together with the whole-rock FeO content to calculate the MgO content, a higher value of around 25% is obtained. Renner *et al.* (1994) consider that the more magnesian grains are xenocrysts derived from a more magnesian precursor komatiite magma.

Three unusual high-Mg komatiites: Commondale, Weltevreden and Murphy Well

Wilson (2003) described the very unusual siliceous komatiites from the 3.3 Ga Commondale greenstone belt, which is located some 150 km south of the Barberton belt. These komatiite flows are distinguished by spinifex textures in which the dominant mineral is either olivine or orthopyroxene, the latter being the mineralogical manifestation of the high MgO and SiO_2 contents of these usual komatiites. Chilled margins and the upper random spinifex lavas contain between 29 and 31% MgO. The maximum forsterite contents of the olivine are very high, up to 96.5, but because the FeO contents of the rocks are low, between 3.5 and 5%, the calculated MgO contents of equilibrium liquids are not extreme. Using the same procedure as that illustrated in Figure 12.1, MgO contents are calculated to be between 29 and 31%, the same range as the MgO contents of the chilled margins and very similar to values estimated by Wilson (2003).

Another locality with similar komatiites is the 3.3 Ga Weltevreden formation in the central part of the Barberton belt. Here again spinifex textures containing both olivine and orthopyroxene are reported (Kareem and Byerly, 2003) and the olivines are also remarkably rich in forsterite, up to $Fo_{95.6}$. But these rocks have higher FeO contents than the Commondale komatiites, and the calculated MgO contents in parental liquids are commensurately high. The calculated MgO content of the parental liquid is about 32%.

The third locality where the existence of highly magnesian liquids can be directly inferred is Murphy Well in the east part of the Yilgarn province of Western Australia. Over 30 years ago Lewis and Williams (1973) provided a detailed description of this komatiite unit, probably a flow, whose thickness is estimated at about 162 m. The lower 140 m is composed of olivine cumulate like that of other thick differentiated flows, but the upper 20 m consists almost entirely of unusual skeletal olivine in a sparse, fine-grained groundmass (Figure 3.15(d)). Through the upper 20 m of the flow the texture remains similar – only the size of the skeletal olivine grains gradually increases from top to base – and throughout this zone the whole-rock composition varies very little. The MgO content remains essentially constant and remarkably high, at about 33–35%. In Lewis and Williams's (1973) interpretation, the flow was emplaced as a single pulse of 34% MgO liquid, and to explain the persistent unusual texture throughout, they proposed that the unit crystallized rapidly from a superheated liquid.

The maximum forsterite content reported by Lewis and Williams (1973) is Fo_{93} and if the whole-rock compositions are used in the procedure illustrated in Figure 12.1, a relatively low MgO content of around 29% MgO is calculated. The difference between the bulk rock MgO and the MgO calculated for the liquid might be explained if this komatiite had been hydrous. This idea is developed in a paper in preparation by Seigel, Henriot and Arndt.

Thick dunitic units

Thick dunitic flows and sills such as those at Mt Keith, Perseverance and the Walter Williams Formation (see Chapters 3, 9 and 10) also crystallized from highly magnesian liquids. Here the evidence is indirect because these units are differentiated and chilled marginal zones, where they occur, cannot confidently be identified as having formed from the parental magmas. The evidence comes from olivine in the cumulate zones, whose composition remains constant at around Fo_{94} for depth intervals of tens of metres (Barnes *et al.*, 1988; Rosengren *et al.*, 2007). No matter how this olivine accumulated – by crystal settling or *in-situ* growth – it represents the liquidus mineral and it must have been in equilibrium with the liquid from which it formed. If this liquid had an FeO content like the majority of komatiites of the Yilgarn block, between 10 and 12% FeO(tot), the MgO content calculated using the procedure described above varies from 29 to 34%.

Eruption temperatures

Eruption temperatures for komatiites and other highly magnesian lavas are listed in Table 12.1. These temperatures have been calculated assuming, in most cases, that the komatiites were essentially anhydrous and using Nisbet's (1982) equation that relates temperature to MgO content (Chapter 9). Calculated values range from 1400 °C for the MgO-poor komatiites of Zvishavane in Zimbabwe to around 1640 °C for the Weltevreden komatiites.

The volatile-bearing komatiites and basaltic rocks in the table erupted at lower temperatures. The anhydrous liquidus of the Boston Creek flow has been estimated at about 1350 °C; if the magma contained 1.2% water, as advocated by Stone *et al.* (1997), this temperature would have been depressed by about 100 °C. The meimechite parental magma was more magnesian, containing about 25% MgO according to Sobolev *et al.* (1991). Any CO_2 originally present in the magma would probably have escaped as the solubility of this component decreased in the lower part of the crust; water would also have exolved as the magma approached the surface. The loss of these phases would have caused olivine to crystallize, releasing latent heat that raised the magma temperature (Albarède, 1983). This magma would have erupted as a phenocryst-charged liquid at a temperature somewhere below about 1500 °C, the volatile-free liquidus. The phenocryst-rich character of the meimechites is consistent with this interpretation. Finally the boninite, whose anhydrous liquidus temperature is 1470 °C, may have erupted as hydrous lavas that metastably maintained high water contents. The temperature of eruption of such a lava is difficult to estimate.

13

Petrogenesis of komatiite

13.1 Nature of the komatiite source

If we accept the arguments developed in Chapters 11 and 12, we can conclude that high temperature is the defining feature of the komatiite source. This immediately rules out a number of possible settings for the formation of komatiites such as subcontinental lithospheric mantle, the mantle wedge above a subduction zone, and a mid-ocean ridge, where temperatures are most unlikely to have been high enough to generate the highly magnesian anhydrous parental magmas of komatiites. What is required is a source whose temperature is significantly higher than that of ambient mantle. Unless the effect of high temperature was counterbalanced by its composition – an elevated content of Fe or garnet, for example – the source would be less dense than surrounding mantle and it would have ascended at a rate that depended on its form and size and on the rheology of the surrounding mantle.

We have no real idea of the exact nature of a komatiite source but a mantle plume probably fits the bill. The source could have been a mantle plume of the type thought (by most authors) to be feeding Hawaii and other chains of oceanic islands – a cylinder of hot, buoyant material rising from deeper in the mantle – or it could have been a 'starting mantle plume' – a huge, roughly mushroom-shaped structure that probably rose from the core–mantle boundary. Given the uncertainty that surrounds the exact form and dynamics of modern plumes, we cannot hope to be more precise when we discuss the source of Archean komatiites.

The composition of the source of komatiite magmas is also uncertain. We can presume that the source was largely peridotitic, because only under very unusual conditions would a less magnesian or more mafic source be able to generate magmas with ultramafic compositions and very high Ni and Cr contents. Several lines of argument show, however, that this source may not have had the normal pyrolite composition commonly used as a starting point in the discussions of the origin of basaltic magmas. For example, Hanson and Langmuir (1978) and Francis (1995) argued on the basis of the major-element compositions of komatiites and picrites that the source of most komatiites had a higher Mg–Fe ratio, and probably a higher FeO content, than the source of most modern basalts. Don Francis, in his model, argued for secular variation in the composition of the source of komatiites and picrites, the Archean source being richer in FeO than the modern source. The high FeO content of Karasjok-type komatiites seems clearly to require an Fe-rich source. Another argument comes from Sobolev *et al.* (2005, 2007), who used the compositions of olivine in mafic and ultramafic rocks to estimate the proportions of peridotite and eclogite in the sources of these rocks. Sobolev *et al.* proposed that the sources of some komatiites contained a significant proportion of eclogite, up to 40% in the case of Proterozoic komatiitic basalts from Gilmour Island. For most Archean komatiites, as well as for the Cretaceous Gorgona komatiites, the proportion of the eclogite component was far less, averaging around 20%. Herzberg and O'Hara (2002) reached similar conclusions using experimental data.

The trace-element composition of the komatiite source is poorly con-strained. The depletion of incompatible trace elements that characterizes Munro-type komatiites (Chapter 5) indicates that the material from which they formed was similarly depleted in these elements. However, if, as is likely, these magmas formed through fractional melting, it is possible that the pre-melting source was less depleted or even enriched in incompatible elements and that the depletion recorded by the trace elements resulted from the separation of earlier melts. On the other hand, the ubiquitous high initial ^{143}Nd/^{144}Nd and low ^{87}Sr/^{86}Sr of Munro-type komatiites (Chapter 6) suggests that the source had undergone long-term depletion of incompatible elements. The same applies to other types of komatiite: as mentioned in Chapter 6, the garnet-subtraction signature of Barberton-type komatiite was almost cer-tainly generated during the melting process and is not an original feature of the source. In this case as well, the specific character of the source was generated during the melting process and was not a long-term feature. But once again the Karasjok-type Fe–Ti-rich komatiites are an exception: these magmas prob-ably inherited their unusual compositions directly from their source because it is difficult to imagine a syn-melting process that would boost the FeO content of the liquid.

With this preamble we can define the komatiite source as broadly peridoti-tic, perhaps slightly richer in Fe than the source of modern oceanic basalts and picrites, and in most cases moderately depleted in the more incompatible trace elements. In all of these respects, it is rather like that of modern plume-derived magmas such as ocean island basalts. The source was most probably hetero-geneous, but, because the degree of melting required to produce komatiite is generally high, and particularly if the source fractionally melted, the removal of early-formed melts would have preferentially extracted the more enriched components which generally melt at relatively low temperatures. The immedi-ate source of most komatiite therefore was the relatively refractory material left after extraction of early-formed melts.

Some of the basalts associated with komatiites formed through fractional crystallization of parental komatiite magma and others resulted from partial melting of cooler parts of the komatiite source. In some cases a direct relation-ship is evident, as in the case of the gabbroic interiors of differentiated mafic–ultramafic flows or sills, or spinifex-textured komatiitic basalts. Although these rocks were clearly derived through fractional crystallization of komatiite mag-mas, differences in trace-element and isotopic ratios often indicate that assim-ilation of crustal rocks accompanied the crystallization (Sun and Nesbitt, 1978; Kinzler and Grove, 1985; Arndt and Jenner, 1986; Sun et al., 1989). Other examples of komatiitic basalts and picrites have geochemical character-istics that are too dissimilar for a direct crystallization relationship and they probably formed from other parts of the same source or from sources distinct from that of the komatiite. Finally, the tholeiitic basalts that dominate Archean greenstone belts most probably were generated independently from the koma-tiites, from cooler parts of the mantle source.

13.2 What does it take to make an ultramafic magma?

Given a source of broadly peridotitic composition, what is required to produce a magma of ultramafic composition? The recipe is simple: either a large proportion of the source must melt, or it must melt at high pressure. The reasoning is explained in Figures 13.1 and 13.2. The first diagram (Figure 13.1) shows how the composition of a partial melt of mantle peridotite changes as the degree of partial melting increases. At low degrees (percentages) of partial melting at low pressure, from about 1 to 15%, the liquid is more or less basaltic because all four major mantle minerals (olivine, orthopyroxene, clinopyroxene and an aluminous phase) melt together to produce a liquid whose composition is far removed from that of its ultramafic source. Once the degree of melting exceeds 15–20%, the aluminous phase (plagioclase, spinel or garnet) and the calcic phase (clinopyr-oxene) are exhausted and from then on, as Mg-rich olivine and orthopyroxene enter the melt, the liquid becomes progressively more ultramafic. Munro-type Al-undepleted komatiites probably form somewhere near the point of exhaus-tion of orthopyroxene when the degree of melting approaches 50%.

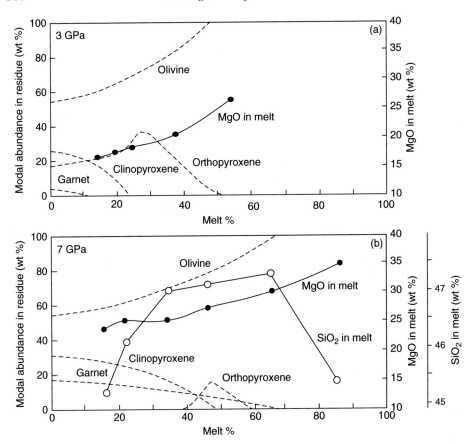

Fig. 13.1. Diagrams showing how, during partial melting of peridotite, the solid mineral phases are progressively exhausted and the composition of the melt changes as a function of the degree of melting and pressure. At moderate pressure (3 GPa, (a)) the composition of the melt changes from picritic at low degrees of melting to komatiitic at high degrees of melting; at high pressure (7 GPa, (b)), the melt composition is komatiitic even at low degrees of melting. (Drawn using data of Walter (1998).)

Figures 13.1 and 13.2 shows how a liquid produced at a given degree of melting becomes increasingly magnesian as pressure, or depth in the mantle, increases. As described in Chapter 7, experimental studies (O'Hara, 1968; Green, 1974; Takahashi and Scarfe, 1985; Herzberg and Ohtani, 1988; Ohtani *et al.*, 1988; Walter, 1998; Gudfinnsson and Presnall, 2000; Herzberg and O'Hara, 2002) have shown that at low pressures, a liquid produced by low-degree melting at temperatures near the solidus has a basaltic composition and contains 10–15% MgO; then, as the pressure increases, the near-solidus liquid changes first to picrite, then to komatiite. Broadly speaking, this

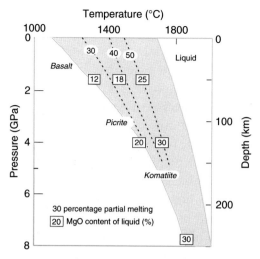

Fig. 13.2. Simplified phase diagram of mantle peridotite (from Herzberg and Zhang, (1996)). The MgO contents of melts in equilibrium with peridotite under various pressure–temperature conditions are shown. One series corresponds to 'near-solidus' melts and another corresponds to liquids produced at about 50% partial melting. In each series the MgO content increases with increasing pressure. Note that magmas with 30% MgO can be produced either by low-degree melting at great depth, or by high-degree melting at shallow depth.

comes about because high pressure stabilizes minerals with more compact tetrahedral coordination, like garnet, relative to olivine and orthopyroxene. At high pressure these ferromagnesian minerals more readily enter the melt than at low degrees of melting, and the magma consequently becomes richer in MgO and poorer in Al_2O_3. During melting at high pressure, garnet, which is Al-rich and Mg-poor, is more stable and is retained in the solid residue even at relatively high degrees of partial melting. Melt produced by moderate degrees of melting at very high pressures has a composition that reflects the retention of garnet in the residue, being particularly rich in MgO and depleted in Al_2O_3. Melting under these conditions accounts for the formation of Barberton-type Al-depleted komatiites (Green, 1974; Sun and Nesbitt, 1978; Green, 1981; Herzberg and Ohtani, 1988). As will be described below, the density of melt produced at very high pressure is similar to or higher than that of common mantle minerals. This phenomenon also plays a role in the formation of these komatiites. When the melting takes place at lower pressures and if the degree of melting is higher, the magma remains rich in MgO but is not depleted in Al_2O_3. These conditions give rise to Munro-type Al-undepleted komatiites.

13.3 Conditions of melting

In practice the formation of komatiite is far more complicated than described above. If we accept that komatiite magma forms in mantle plumes, we need to consider melting in a source that rises from a site deep in the mantle. The source follows a path that is steeper than the solidus of mantle peridotite and it melts as its temperature becomes progressively higher than the solidus. As explained above, it is possible to produce komatiite either by melting near the liquidus at high pressure, or by high-degree melting at similar or lower pressures. To establish which process is the more important, or, more to the point, which process produces which type of komatiite, we need to estimate the degree of melting. This information is not accessible using the major elements because of the competing influences of pressure and the degree of melting, but it is provided by the concentrations of certain trace elements. If we knew accurately the composition of the komatiite source, we could calculate directly the degree of melting from the concentration in the komatiite of a highly incompatible element, whose concentration varies inversely with the melt fraction. In practice this information is difficult to obtain, for two reasons: (a) most of the highly incompatible elements are also highly mobile during alteration, a process that strongly perturbs their concentrations in komatiite lavas; (b) the concentrations of these elements change dramatically during partial melting and other processes that have affected all mantle rocks. Even though the concentrations of highly incompatible elements in the source of mid-ocean ridge basalt in the present-day depleted upper mantle are reasonably well known (Hofmann, 1988; McDonough, 1990) (they are 5–10 times lower than in estimates of the primitive mantle composition), we do not know the concentrations in the komatiite source. Some indication of the possible variability is provided by studies of modern oceanic island basalts, whose sources are demonstrably heterogeneous.

We can circumvent the problem, at least partially, by using the concentrations of moderately incompatible elements. Hofmann (1988) has shown that the concentrations of Zr and the middle REE in mantle rocks are relatively insensitive to melt extraction, metasomatism and other such processes, and, in addition, these elements are relatively immobile during alteration. If we make some reasonable assumptions about proportions and types of minerals, their partition coefficients and the melting mechanism, we can place some bounds on the degree of melting that gave rise to the common types of komatiite. The results are shown in Figure 13.3 (details of the calculation are given in the figure caption). It is seen that about 30% batch melting produces Barberton-type komatiites, and about 50% batch melting yields Munro-type komatiites.

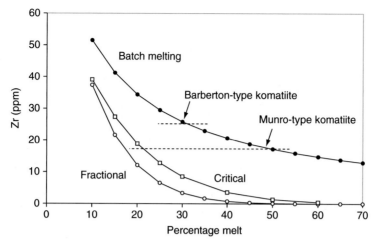

Fig. 13.3. Variations in Zr contents of liquids produced by batch, critical and fractional melting of a source of primitive mantle composition (9.4 ppm, Hofmann (1988)). Barberton-type komatiite, which contains about 25 ppm Zr (Figure 2.9), could have formed by 30% batch melting. Munro-type komatiite contains about 18 ppm Zr, requiring about 50% batch melting or 20% fractional or critical melting. The following values, presented in the order olivine, orthopyroxene, clinopyroxene and garnet, were used for the calculation: partition coefficients –0.0011, 0.001, 0.3, 0.6; percentages in the source 50, 20, 15, 15; percentages in the liquid –55, 20, 15, 10. The equations used are given in Gurenko and Chaussidon (1995).

Gorgona komatiites (not shown) form through lower degrees of melting and oddities like the Karasjok-type Fe–Ti-rich komatiites form from still lower degrees of melting from a source of unusual composition, as will be explained later in the chapter.

The assumption that all these komatiites formed by batch melting is unrealistic: under most circumstances melt separates from its source long before the degree of melting reaches 30–50%. If the melt is assumed to leave the source immediately it forms (fractional melting), the calculated degree of melting is much smaller. As shown in Figure 13.3, the concentrations of Zr in Munro-type komatiites correspond to that of a liquid extracted after about 20% of fractional melting. The temperatures and pressures required for 20% fractional melting are not very different from those appropriate for 50% batch melting because, as fractional melting proceeds, the source becomes increasingly refractory and the temperature required for a given degree of melting increases accordingly.

These results place immediate constraints on the depth of melting. Figure 13.4 is a simplified phase diagram of mantle peridotite showing the paths followed by komatiite melts as they ascend from their source in the

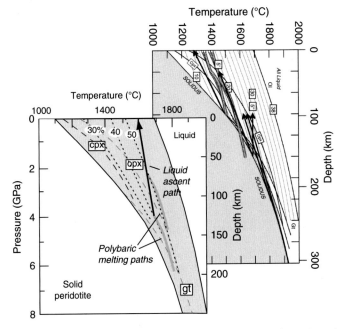

Fig. 13.4. Simplified phase diagrams of mantle peridotite, from Herzberg and O'Hara (2002). Two ascent paths are shown. The dashed lines labelled 'polybaric melting path' are those of peridotite that continually partially melts as it rises; the black arrow is an approximate adiabatic ascent path of komatiite liquid. The inset, copied directly from Herzberg and O'Hara (2002), shows a range of possible paths and illustrates the uncertainties in defining these paths.

mantle. It is assumed that the melt did not interact at all with the solid rocks of the mantle and crust. Although the exact path followed by the liquid is difficult to predict, as shown by the large range of possibilities in Herzberg and O'Hara's (2002) diagram (reproduced in the inset in Figure 13.4), the acceptance of 1600°C as the eruption temperature of a 30% MgO anhydrous komatiite sets some rigorous constraints. The path shown is steeper than Herzberg's preferred liquid ascent path (labelled '5' in the inset), and it intersects the 50% melting curve (that which corresponds to the formation of Munro-type komatiite) at about 70 km. The same path intersects the 30% melting curve (Barberton-type komatiite) at about 120 km. A shallower path more like that preferred by Herzberg would intersect the melting curves at still greater depths or not at all, as long as its upper end remains pinned by the 1600 °C eruption temperature. These inferences depend, of course, on the positions of the peridotite liquidus and solidus, but, as discussed by

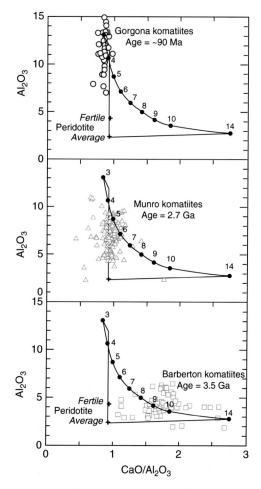

Fig. 13.5. Diagrams of CaO/Al_2O_3 vs Al_2O_3, from Herzberg (1995). The curves show how the composition of liquids produced by partial melting of mantle peridotite depend on the pressure of melting, which is indicated by the numbers to the right of the curve (in GPa). The positions of data from the three main types of komatiite record increasing pressures of melting.

Herzberg and O'Hara (2002), there seems to be good agreement about the reliability of the phase diagram.

Various geobarometers can be used to test the validity of these estimates. Herzberg (1995) has used the results of his experimental studies of the partial melting of peridotite to produce a series of equations that relate the MgO and Al_2O_3 contents or the $CaO–Al_2O_3$ ratio to the depth at which a mantle melt last equilibrated with its mantle source. This information is summarized in Figure 13.2, which shows how the MgO content of liquids produced by melting near the solidus varies as pressure or depth increases. Figure 13.5 shows how two additional parameters, Al_2O_3 and CaO/Al_2O_3, vary with pressure. Although the MgO content increases with increasing pressure, it is not particularly diagnostic of the depth of the site of melting for two reasons. First, the compositions of komatiite lavas depend so strongly on olivine fractionation or accumulation that the MgO content of the primary liquid is rather difficult to establish (see Chapter 12). Second, as Herzberg (1995) has shown, the MgO content of the melt that forms in a rising melting mantle source remains nearly constant because the effect of falling pressure, which decreases the MgO content, is counterbalanced by an increasing degree of partial melting, which increases the MgO content.

The CaO/Al_2O_3 vs Al_2O_3 diagram (Figure 13.5) provides more useful information. In this diagram, the compositions of near-solidus melts plot along the curved line, and liquids formed by higher degrees of melting plot between this line and the composition of the source (in this case the mantle peridotite KLB-1). Compiled compositions of the three main types of komatiite plot at different positions in the diagram. The compositions of Gorgona komatiites produce a near-vertical trend that extends from the solidus curve at positions between 3 and 4 GPa to about half of the distance to the source composition. This variation is interpreted by Herzberg (1995) to indicate that some of the rocks formed as near-solidus melts and others at higher degrees of melting, all at relatively shallow depths of around 100 km. Results for Munro-type komatiites scatter more widely, between positions that correspond to 5–6 GPa on the solidus curve to positions very close to the source. Part of the spread results from olivine fractionation or accumulation, or from CaO mobility during alteration, and another part, the rest of the spread, results from differing degrees of partial melting which took place at greater depths, of up to 150 km. Most Barberton komatiites plot around the 9–10 GPa positions, but again they scatter, most probably for the same reasons as the Munro komatiites. The Barberton compositions correspond to melting at greater depths, at up to 300 km. In a broad sense, these depths correspond to those inferred by extrapolation downward from the anhydrous eruption temperatures, as shown in Figure 13.4.

The depths discussed above correspond to those at which melt last equilibrated with its source; these depths are shallower than the range over which melting takes place, which starts much deeper in the mantle. In the case of modern basalts, melting starts at relatively shallow depths and continues until the source reaches the base of the lithosphere. Melting of the source of mid-ocean ridge basalts starts at 30–50 km and finishes close to the surface, just below newly formed oceanic crust (Langmuir *et al.*, 1992). Melting of the sources of oceanic island basalts starts and finishes at greater depths: it starts when the hot mantle plume intersects the mantle solidus and finishes at the base of the oceanic lithosphere at a depth of 50–100 km. In the case of komatiites, the depths of melting are far greater, well below all reasonable estimates of the thickness of the lithosphere. Under these circumstances, a lithosphere control on melting cannot be assumed, as discussed below.

The slope of the path followed by the mantle source as it partially melts is shallower than the liquid ascent path because heat is absorbed as solid is converted to liquid. The exact paths are difficult to calculate because they depend on the nature of the source and on the details of the melting mechanism, and our knowledge of the thermodynamic parameters needed to model melting even in the relatively simple mid-ocean ridge setting is inadequate (e.g. Asimov *et al.* (1995), Hirschmann *et al.* (1998), Asimov *et al.* (2001)). Nonetheless, at the broad scale of komatiite formation, the melting curves cannot be far from the paths shown in the lower part of the diagrams in Figure 13.4. When drawn at the scale of the diagram, both the liquidus and the solidus of mantle peridotite are curved because of the compressibility of silicate liquid. At depths greater than about 200 km, the melting curve is nearly parallel to both the solidus and the curves that represent a constant degree of partial melting. This means that the proportion of melt increases only slowly as the source ascends: as predicted by Miller *et al.* (1991) in their important paper on the equation of state of molten komatiite, the komatiite source must have started to melt at extreme depths. Most probably it was already partially molten when it traversed the transition zone between the upper and lower mantle.

13.4 The formation of various types of komatiite

The paths followed by ascending mantle plumes with different temperatures are shown diagrammatically in Fig. 13.6, and Table 13.1 summarizes the conditions of melting. The three main types of komatiite form at very different positions in the mantle and, as explained below, they result from different partial melting processes. Here I refer principally to the difference between equilibrium melting, the process during which the melt remains within the source as the degree of melting increases, and fractional melting, when the melt escapes from the source as soon as it forms (McKenzie, 1984; McKenzie and Bickle, 1988). Normal mantle melting is somewhere between these two extremes: once the degree of melting exceeds a low threshold, probably about 1–2%, the melt escapes the source as fast as it forms, but a small fraction of melt always remains in the source buffering the effect of the extraction of earlier-formed melts. The latter process, which has been called critical melting,

Table 13.1. *Conditions of formation of komatiites and related rocks*

Type	Conditions
Barberton-type komatiite	Batch melting of fertile peridotite at very high pressure (>9 GPa)
Munro-type komatiite	Critical melting of depleted peridotite at high pressure (~5–7 GPa)
Gorgona-type komatiite	Critical melting of depleted peridotite at moderate pressure (~3–4 GPa)
Gorgona picrite	Advanced critical melting of depleted peridotite at moderate pressure (~3–4 GPa)
Karasjok-type komatiite	Critical melting of depleted peridotite
Karasjok picrite	Partial melting of enriched peridotite
Commondale komatiite	Extreme critical melting of depleted peridotite
Komatiitic basalts	Fractional crystallization of parental komatiite, or crystallization plus crustal contamination, or independent melting of the mantle source

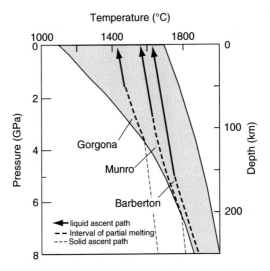

Fig. 13.6. Pressure–temperature diagram summarizing the paths taken by the mantle sources that melted to form the three main types of komatiite.

probably occurs during the formation of the majority of common mantle melts, which have basaltic or picritic compositions. Once the melt has escaped the source, it commonly mixes with other melts or pools in shallow magma chambers or magma conduits to produce hybrid magma whose composition is the average of all melts extracted from the source. This is the case for basalts at mid-ocean ridges and in many other modern tectonic settings (Langmuir *et al.*,

1992) but most probably does not apply to the formation of most komatiites, as discussed in the following sections.

Barberton-type komatiite

In Chapter 7 I described the experimental studies that demonstrate that the density of komatiite liquid may exceed that of common mantle minerals such as olivine and pyroxene at pressures corresponding to those in the lower part of the upper mantle. In a narrow pressure window between about 9 and 13 GPa (Ohtani, 1984; Ohtani *et al.*, 1998), a window that corresponds to mantle depths of 270–400 km, the density of komatiite is similar to or slightly greater than that of solid mantle peridotite. At shallower depths, melt escapes from its source because its density is lower than that of the residual crystals, but in the 9–13 GPa window, the melt has neutral buoyancy and may remain in the source.

If the mantle source is to melt in this depth range, its temperature must be far higher than that of the ambient mantle. This excess temperature causes the overall density of the region of partial melting – a mixture of partial melt and residual solid – to be less than that of the surrounding peridotite. It is this density difference that causes the source (the plume) to move upwards. Eventually, when it passes up through the 9 GPa threshold, the melt density falls below that of the solid residue and from this stage onwards the melt can escape upward. During the initial period of partial melting, the liquid remains in contact with the residual minerals but when the threshold is broached, the melt probably escapes the source in a single pulse: in other words, the formation of Barberton komatiite may constitute a rare example of equilibrium or batch melting of the mantle.

Evidence that Barberton komatiites are the products of melting deep in the mantle comes from their major-element compositions, as shown in Figure 13.5, and from their depleted HREE, illustrated in Chapters 2 and 5. These characteristics provide strong evidence that garnet (or majorite, a garnet–pyroxene solid solution stable at high pressure) separated from the source of the komatiite, or from the komatiite itself, at some stage during its formation. Almost all petrologists recognize that garnet is stable in the presence of melts of ultramafic composition only at very high pressure (Green *et al.*, 1974; Sun and Nesbitt, 1978; Ohtani *et al.*, 1988; Herzberg, 1992). The phase diagrams discussed in Chapter 7 show, for example, that garnet becomes a stable phase in partially molten peridotite at pressures greater than 3 GPa. At about 9 GPa, it replaces olivine as the liquidus phase. Any melt generated by low–moderate degrees of melting at pressures greater than about 5 GPa has a komatiite composition (MgO > 20%) and forms in equilibrium with residue containing garnet (Green, 1974; Ohtani, 1984; Herzberg and Ohtani, 1988). Such melts have the chemical compositions of Barberton-type komatiites.

In papers published in the 1970s and 1980s, it was suggested that the garnet-depletion signature of these komatiites might have been inherited from the source (Sun and Nesbitt, 1978; Smith *et al.*, 1985; Smith and Ludden, 1989). Cawthorn and Strong (1974), for example, proposed that the source may have originated as the upper, majorite-depleted layer of a solidified, differentiated magma ocean. More recently, however, this type of model has fallen into disfavour, for several reasons. First, komatiites do not show the large fractionation of element ratios such as Zr/Sm or Nb/La that would result from the fractionation of perovskite in the lower part of a magma ocean (Kato *et al.*, 1988). Second, the isotope compositions of komatiites do not have the extreme values predicted for a fractionated source that formed soon after the accretion of the planet. More recently Blichert-Toft and Arndt (1999) published Lu–Hf isotope data that indicated that the source of Barberton komatiites was characterized by long-term enrichment, not depletion, of garnet, and Boyet and Carlson (2005) used ^{142}Nd data to show that any stratification in the fossil magma ocean was obliterated within at most a few hundreds of million years of its formation.

The magmas that crystallized as Barberton-type komatiite may have formed as follows. We do not have enough information to say what happened to the source in the lower mantle where it must have originated in order to have a sufficiently high temperature to melt in the lower part of the upper mantle. Our story starts with a mixture of liquid and solid olivine, pyroxene and garnet in a plume rising through the interval from about 20 GPa (~600 km) to about 9 GPa (~270 km). Already at this stage (stage 1 in Figure 13.7), the melt had an ultramafic composition and, since it formed in equilibrium with residual garnet, it had already acquired the distinctive signature of Barberton-type Al-depleted komatiite. The density of the melt was close to that of olivine and pyroxene but less than that of garnet. If the proportion of melt was large enough, greater than about 20–30%, it is possible that the denser garnet lagged behind, leaving olivine and neutrally buoyant liquid to move ahead into the upper part of the melting region. This physical separation of garnet may have contributed to the development of the garnet-depletion signature.

In stage 2, the density of the melt had become slightly less than that of olivine. The melt started to separate from most of the solid residue, but it may have entrained some olivine. These crystals melted as the liquid pulse continued to rise, increasing the MgO content of the liquid. In stage 3, the level of

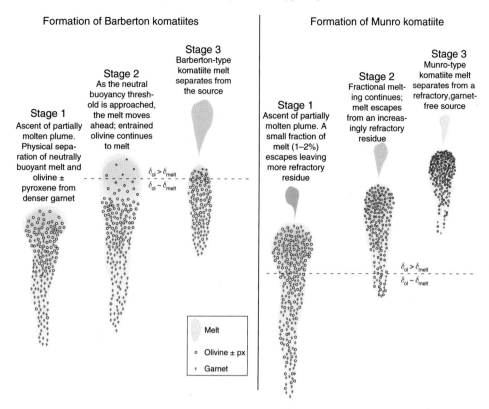

Fig. 13.7. Diagram showing the melting processes during the formation of Barberton- and Munro-type komatiites. (The formation of Gorgona-type komatiites was similar to that of Munro-type komatiites but took place at lower pressure.) The grey tadpole shapes represent the regions in a rising mantle plume where melting takes place and the dashed line shows the threshold below which the melt density is about the same as that of the solid residue.

neutral buoyancy had been broached and the melt separated completely. Its density was less than that of surrounding mantle, its viscosity was very low, and it moved rapidly to the surface to erupt as the flows we now see in the Barberton greenstone belt.

Munro-type komatiite

Other types of komatiite form at lower temperatures. Because of the marked curvature of the solidus and the lines of constant partial melting in the middle–lower part of the upper mantle, the less extreme temperatures imply melting at considerably shallower depths, with the important consequence that melting began above the neutral-buoyancy window. As a result, the melt escaped from its source almost as soon as it formed. Munro- and Gorgona-type komatiites probably formed by fractional (or, more precisely, critical) melting, as discussed by Arndt (1977c), Arth *et al.* (1977) and Sun and Nesbitt (1978) and developed by Arndt (1994; 2003), Herzberg and Ohtani (1988) and Herzberg and O'Hara (2002). We see in Figure 13.6 that the temperature of the mantle source for Munro-type komatiites is less than that of Barberton komatiites, and the amount of melt present as the plume passed through the neutral-buoyancy zone was relatively low; so low that garnet could not physically separate from the melt and other solid phases. Only when the plume rose past the neutral-buoyancy threshold did melt start to escape (Figure 13.7). Because the fraction of initial melt was low, it formed in equilibrium with four mantle phases. The composition of this melt could have been similar to that of a meimechite, a rare magma type that has been interpreted as the product of low-degree melting at great depth (Arndt *et al.*, 1998b; Elkins-Tanton *et al.*, 2006). Meimechites have high MgO, low Al_2O_3 and very high contents of TiO_2 and incompatible trace elements. The segregation of this melt left the residue depleted in incompatible elements and with high Al_2O_3/TiO_2. From this stage onwards, melt separated continuously from the source, driving the residue and the resultant magmas to more and more refractory compositions.

Walter (1998) has questioned the hypothesis that fractional melting produces Munro-type komatiites. He based his arguments on the observation that Munro-type komatiites have Al_2O_3/TiO_2 around 20 (lower than the chondritic ratios), CaO/Al_2O_3 around 1 (higher than the chondritic ratio) and chondrite-normalized Gd/Yb close to 1. He emphasized the contradictory aspects of these ratios: the non-chondritic Al_2O_3/TiO_2 and CaO/Al_2O_3 imply residual garnet but the near-chondritic Gd/Yb implies an absence of garnet.

As shown in Figure 13.8 and Table 13.2, a model of critical melting readily explains these characteristics. During critical melting of the komatiite source, the initial melt forms at extreme depths and at a very low degree of melting. This melt had an unusual composition in which a high MgO content and low Al_2O_3 content are combined with strong enrichment of incompatible trace elements (the meimechite signature). When a small fraction of such a melt is extracted, it leaves a residue with a relatively high Al_2O_3 content and a depleted trace-element pattern. In liquids produced by further melting of this residue, the effect of equilibration with the high-pressure mineral assemblage competes with the effect of the extraction of earlier melts. Add to this the fact that the komatiite source has undergone an earlier melt extraction event, as demonstrated by Nd and Hf isotope data (Chapter 6) and it becomes possible to reproduce the major- and trace-element composition of Munro-type

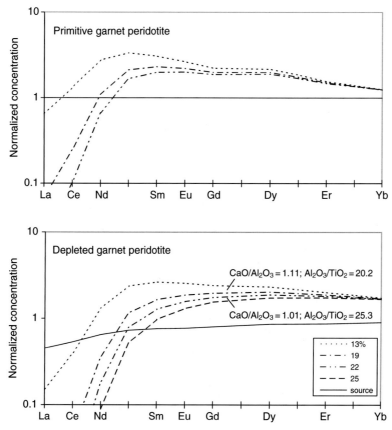

Fig. 13.8. Modelling of REE contents of Munro komatiites. These magmas are assumed to be the products of critical melting of peridotite and their compositions are calculated using the information given in Table 13.2. If the source is assumed to have been slightly depleted in light REE, as in the lower diagram, the heavy REE patterns in liquids produced by ~20% melting are flat. In addition, Al_2O_3–TiO_2 ratios are close to the chondritic value of 20. These values match those recorded in Munro-type komatiites.

komatiites. When these effects are modelled using standard procedures as outlined in Figure 13.8 and Table 13.2, it is seen that the melt extracted at about 20% critical melting of a slightly depleted source has Al_2O_3/TiO_2, CaO/Al_2O_3 and Gd/Yb very similar to those of Munro-type komatiites.

The relatively constant composition of the parental magmas of Archean komatiites, which, with few exceptions, seem to have MgO contents between 25 and 30%, may be related to the point at which orthopyroxene is lost from the residue of melting. The exact point when this happens depends on the pressure and the melting mechanism, but Figure 13.1 shows that during batch melting, as reproduced in experiments, it occurs at 50–60% melting. During

Table 13.2. *Parameters used to model the REE contents of Munro komatiites*

	La	Ce	Nd	Sm	Eu	Gd	Dy	Er	Yb
Partition coefficients[a]									
olivine	0.00003	0.0001	0.00042	0.0011	0.00075	0.00075	0.0014	0.013	0.03
orthopyroxene	0.00004	0.0001	0.00052	0.0016	0.00064	0.00064	0.0084	0.017	0.033
clinopyroxene	0.0536	0.0858	0.187	0.291	0.35	0.4	0.442	0.387	0.43
garnet	0.001	0.004	0.0057	0.625	1	1.5	2	3	4
Primitive mantle[b] (ppm)	0.6139	1.6011	1.1892	0.3865	0.1456	0.5128	0.6378	0.4167	0.4144
Depleted mantle (ppm)	0.2770	0.8569	0.7729	0.2940	0.1121	0.4102	0.5485	0.3620	0.3730

Proportions in the source: olivine = 50%; orthopyroxene = 20%; clinopyroxene = 20%, garnet = 10%
Proportions in the melt: olivine = 40%; orthopyroxene = 20%; clinopyroxene = 20%, garnet = 20%

[a] From the GERM data bank <http://earthref.org/>.
[b] From Hofmann (1988).

critical melting orthopyroxene will be exhausted earlier and the melt will contain 25–30% MgO.

Gorgona komatiites and picrites

As discussed in Chapters 2 and 5, the komatiites from Gorgona Island are characterized by moderate MgO contents, relatively high Al_2O_3/TiO_2 and strong depletion of incompatible trace elements. The picrites from Gorgona formed from liquids with still higher MgO contents, lower FeO and extreme depletion of incompatible elements. An important feature of both is their relatively uniform, moderately high initial $^{143}Nd–^{144}Nd$ ratios. Révillon *et al.* (2000) have shown that the large variations in the ratios of trace elements such as La/Sm or Sm/Nd are not matched by variations in initial $^{143}Nd/^{144}Nd$, a pattern that they interpreted to indicate that both types of magma came from the same source and that fractional melting produced the range in trace element ratios.

Herzberg and O'Hara (2002) used a combination of data from high-pressure experimental studies of the phase relations of fertile and depleted mantle peridotite, combined with the MgO and FeO contents of Gorgona lavas and the forsterite contents of olivine, to place constraints on the conditions of formation of Gorgona komatiites and picrites. They took into account different types of melting in an approach that is rigorous but complicated, as will be immediately evident to readers of the paper cited above. The more important of their results are presented in Figure 13.9. Gorgona komatiites are interpreted as accumulated fractional melts that formed at an average pressure of 3–4 GPa and, if the source were a fertile mantle peridotite with a composition like that of KLB-1, the average degree of melting would have been about 40%. The Nd and Sr isotopic compositions of these komatiites indicate, however, that the source was depleted peridotite, and if this type of source is adopted, the calculated degree of melting is around 30%. Melting takes place at relatively low pressures under conditions where garnet is absent from the residue and this accounts for the relatively high Al_2O_3/TiO_2. The advanced stage of fractional melting accounts for the strong depletion of incompatible trace elements.

The yet more depleted and more magnesian Gorgona picrites formed through a higher degree of fractional melting. Separation of earlier-formed liquids increased MgO, decreased FeO and extracted incompatible trace elements to produce the severe depletion that characterizes these rocks (Arndt *et al.*, 1997b; Herzberg and O'Hara, 2002). The average depth of melt extraction was about 120 km (4 Gpa) and the initial melting took place at much greater, but poorly constrained, depths.

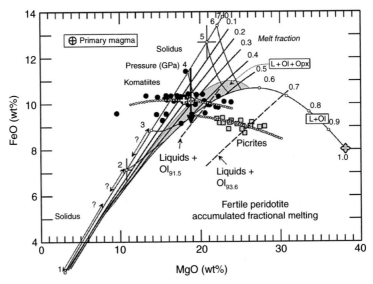

Fig. 13.9. Herzberg and O'Hara's (2002) modelling of the formation of Gorgona komatiites (black circles) and picrites (grey squares). The grid represents the compositions of liquids produced by accumulated fractional melting of fertile peridotite; melt fractions and the pressures of melting are indicated by the numbers adjacent to the curves. Primary liquid compositions can be inferred from FeO/MgO ratios in olivines, shown as dashed lines. The Gorgona komatiite could have formed as accumulated fractional melt at 4 GPa and 40% melting; the picrites form at lower pressure and higher degrees of fractional melting.

To account for variable degrees of melting in a single source region requires that the temperatures in this region varied: perhaps the komatiites came from parts with moderate temperatures and the picrites from the hottest interior of the core. In general, the formation of Gorgona lavas is well explained by melting in various parts of the mantle plume that fed the Caribbean flood basalts, as summarized by Kerr (2005).

Commondale and Weltevreden komatiites

These unusual komatiites come from two different greenstone belts in South Africa (Kareem and Byerly, 2003; Wilson, 2003; Wilson *et al.*, 2003; see also descriptions in Chapters 4, 5). They are rare, but are included here because they place interesting constraints on the more extreme conditions of melting, or on the more unusual mantle sources, from which komatiites are derived. Both the Commondale and Weltevreden komatiites are distinguished by the presence of olivine–orthopyroxene spinifex texture, the mineralogical manifestation of relatively high SiO_2 content in the liquid from which they crystallized. They contain olivines with dramatically high forsterite contents, up to Fo_{97}, and, as discussed in Chapter 12, they probably formed from some of the hottest and most magnesian komatiites ever recorded. I know of no trace-element data for the Weltevreden komatiites, but Wilson *et al.*'s (2003) data show that the Commondale rocks have extreme depletion of incompatible trace elements, even greater than in the Gorgona rocks.

Wilson *et al.* (2003) have proposed that the Commondale komatiites formed in two stages of partial melting. According to them, refractory depleted peridotite, perhaps the residue of melting that produced Barberton komatiites, became incorporated in subcontinental lithospheric mantle, then remelted when fluxed by the ingress of water from a nearby subduction zone. The model is rather like the one advocated by Parman *et al.* (2001, 2004) for the Barberton komatiites themselves (Chapter 12). Wilson *et al.*'s firmest argument is probably the high SiO_2 contents of the Commondale komatiites, which reach 50% in lavas containing 30% MgO. High SiO_2 contents are a characteristic feature of lavas from subduction settings, and modern boninites, like the Commondale komatiites, have the same combination of high SiO_2 and high MgO contents. The trace-element patterns in the Commondale komatiites, on the other hand, are very different from those of boninites (as are those of Barberton komatiites).

Two arguments militate against the subduction hypothesis for the Commondale and Weltevreden komatiites. The first is the very low water content inferred by Kareem and Byerly (2003) on the basis of their microprobe analyses of melting inclusions in olivines from the Weltevreden komatiites, which indicate that at least these did not form via hydrous melting. The second is more technical. Refractory, depleted harzburgite of the type invoked by Wilson *et al.* (2003) for the source of Commondale komatiites, melts only at very high temperatures, even in the presence of water. In any source of variable composition, including normal subcontinental lithospheric mantle or a variably metasomatized mantle wedge, the more fertile components would melt preferentially. The majority of magmas derived from partial melting of subcontinental lithospheric mantle have normal basaltic–picritic compositions, and those coming from the mantle wedge would be dominated by basalts or boninites with a distinctive 'subduction zone' trace-element signature (low Nb/La, high Sr/Sm and so on). The Commondale komatiites do not have this signature. The only plausible way of producing magma from an extremely refractory source without having it mix into and be diluted by more abundant basaltic magma is to have this source much hotter than the surrounding mantle. This is virtually impossible in a lithosphere or mantle wedge setting but is entirely normal for melting in a thermally heterogeneous mantle plume. In such a plume, the hottest parts start to melt first, at the greatest depths, and if melting can continue to shallow levels, the hottest parts arrive at these levels stripped of their low-melting fractions. The hottest, now highly refractory, parts of the plume undergo further melting to produce highly depleted liquids with compositions like those of Commondale komatiites.

But how do we explain the high SiO_2 contents? The moderately high degrees of fractional melting needed to form the Commondale komatiites (and to explain the extreme depletion of incompatible trace elements) coincide, at pressures of 3–4 GPa, with the point of exhaustion of orthopyroxene. This means that orthopyroxene, the mantle phase with the highest SiO_2 content, has completely entered the melt. Walter (1998) has shown that even the highly magnesian melts that are produced under such conditions are highly siliceous (Figure 13.1), with SiO_2 contents up to 49% (for 37% melting at 3 GPa). This SiO_2 content is similar to that of the majority of Commondale komatiites. It is not as high, however, as in some orthopyroxene spinifex-textured lava and aphric flow tops, which contain up to 52% SiO_2. Several possible origins can be suggested for the higher values: (1) fractional melting might produce liquids with higher SiO_2 content; (2) as with spinifex lavas from the Alexo region (Barnes, 1983; Arndt, 1986b) the spinifex lavas could contain a fraction of accumulated pyroxene, which would have boosted the SiO_2 content; or (3) the SiO_2 contents, particularly in the flow-top samples, could have increased during alteration.

Fe-rich Karasjok-type komatiites from Boston Creek and Lapland

The Karasjok-type Fe-Ti-rich komatiites are characterized by a combination of moderate MgO and high FeO contents combined with low Al_2O_3, high CaO/Al_2O_3 and strongly depleted HREE. As for Barberton komatiites, the latter features can be attributed to the removal of garnet, either during or before the magma-forming event. Herzberg and O'Hara (2002) used the Al_2O_3 vs CaO/Al_2O_3 diagram to infer a high pressure of melting for Boston Creek komatiites, between 10 and 14 GPa, which is similar to that for Barberton komatiites but unusual for late-Archean lavas. Another distinctive feature of Boston Creek komatiites, their high FeO contents, casts doubt on this interpretation. The majority of Boston Creek komatiites have FeO contents between 15 and 20%, and some samples contain up to 25%. When these values are plotted on the MgO vs FeO diagram, they fall well above the field of liquids issued from melting of normal peridotite. Rather than deep melting of a normal peridotite source, it seems more reasonable to consider that these

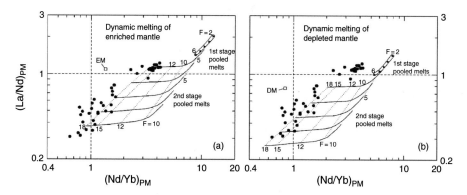

Fig. 13.10. Hanski *et al.*'s (2001) modelling of the formation of Karasjok-type komatiites and picrites. The curved lines show the trace-element ratios in liquids produced by partial melting of (a) enriched mantle (EM) and (b) depleted mantle (DM). The fraction of melting (*F*) is shown adjacent to the curves. The black circles represent the compositions of komatiites and picrites. The differences in trace-element ratios are too large for the different types to have formed through partial melting of a homogeneous source. The source of the picrites must have been more enriched in incompatible elements than the source of the komatiites. See Hanski *et al.* (2001) for details of the calculations.

komatiites may have come from a source of unusual composition. Various authors have come to a similar conclusion and they call on 'mantle metasomatism', an infinitely flexible and wonderfully useful process, to produce a source of appropriate composition. As is so often the case, metasomatism is said to have taken place in the lithosphere, the argument being that we have no real idea of the composition of the lithosphere so it can have whatever composition we want it to have.

Hanski *et al.* (2001) modelled in detail the melting processes responsible for the generation of the komatiites and picrites of Finnish Lapland (Figure 13.10). They concluded that the komatiites, in which the incompatible trace elements are depleted as in Munro-type komatiites, formed through fractional melting of a slightly depleted mantle source. However, to explain the high Fe and Ti of these magmas they proposed that the source was relatively enriched in these elements. The picrites associated with the komatiites are strongly enriched in Fe, Ti and in incompatible trace elements as well. Hanski *et al.* (2001) concluded that these magmas could not have come from a source with the same composition as the one that yielded the komatiites. In their opinion, all the magmas of the komatiite–picrite association formed in a mantle plume that was heterogeneous, perhaps because it contained a variable content of recycled eclogite. This type of explanation seems much more reasonable than simply invoking metasomatism of the lithosphere.

Francis and coworkers (Francis, 1995, Francis *et al.*, 1999) also concluded that the source of Archean and Proterozoic komatiites and alkali picrites was different from that of normal mantle. They found that the source of the Precambrian magmas was richer in Fe than that of modern plume-derived magmas and they proposed that the composition of their deep mantle source had changed through geological time.

Finally, the presence of early-crystallizing amphibole and the moderate water contents in melt inclusions in the Boston Creek komatiites (Stone *et al.*, 1997, 2003, 2005) indicate that the source was slightly hydrous. Whatever component contributed high levels of incompatible elements to the source evidently supplied a small amount of water as well. It is notable that the trace-element patterns of these komatiites show no trace of the 'subduction signature': in these rocks, critical ratios, such as Nb/La, are like those of enriched intraplate sources, not those of magmas from subduction zones. As explained above and in Chapter 11, the Fe–Ti-rich komatiites and picrites represent an example of melting of a source that contained elevated concentrations of Fe, Ti and incompatible trace elements and it is to be expected that the source was slightly hydrous as well.

13.5 The formation of komatiitic and tholeiitic basalts

The origin of komatiitic basalts, and particularly the question of whether they are the products of fractional crystallization of komatiite or formed independently as primary mantle melts, was a subject of considerable debate in the 1970s and 1980s (Jahn and Shih, 1974; Nesbitt and Sun, 1976; Arndt *et al.*, 1977; Nisbet *et al.*, 1977; Sun and Nesbitt, 1978; Nesbitt *et al.*, 1979; Jahn *et al.*, 1982; Smith and Erlank, 1982; Viljoen *et al.*, 1983; Arndt and Nesbitt, 1984; Sun, 1984). The close spatial association of komatiitic basalt and komatiite, their geochemical similarities, their spinifex textures and the presence of rocks with the compositions of komatiitic basalts in differentiated ultramafic–mafic flows and sills all argued for the first explanation (Francis and Hynes, 1979; Arndt and Nesbitt, 1982). However, when the geochemical data were subjected to close scrutiny, differences were found that ruled out a direct link through fractional crystallization between komatiite and komatiitic basalt. Particularly telling were the analyses of 'siliceous high-magnesian basalts', whose high levels of incompatible trace elements and low ^{143}Nd–^{144}Nd ratios are distinctly different from those of associated komatiites (Chauvel *et al.*, 1985; Arndt and Jenner, 1986; Barley, 1986; Cattell, 1987; Sun *et al.*, 1989). As discussed in Chapter 5, these basalts appear to have assimilated a component of old continental material and most authors interpret them as komatiites that assimilated granitoid crustal rocks on their way to the surface. In other areas some of the magnesian basalts (or picrites) spatially associated with komatiites display differences in Fe contents or in the ratios of immobile trace elements that also rule out derivation by fractional crystallization of komatiite while others are the products of a low–moderate degree of partial melting in different parts of the mantle source that produced the komatiites (Nesbitt and Sun, 1976; Arndt and Nesbitt, 1984). Another example of rock suites in which the relationship between the ultramafic and mafic lavas is ambiguous is the komatiite–picrite association in Finnish Lapland where the two rock suites have strongly contrasting major- and trace-element signatures. As discussed above, Hanski *et al.* (2001) have shown through detailed geochemical modelling that the komatiitic basalts not only cannot represent fractionally crystallized komatiites but must also have come from a different mantle source. The overall conclusion is that in any komatiitic province the komatiitic basalts have diverse origins: some certainly are the products of fractional crystallization of komatiite, some result from a combination of crystallization and contamination, and others are the products of a low–moderate degree of partial melting in different parts of the mantle source that produced the komatiites.

What do we know about the origin of the tholeiitic basalts that are the dominant component of most Archean greenstone belts? These rocks have relatively evolved compositions with low MgO contents between about 5 and 8%, flat mantle-normalized trace-element patterns, and isotopic compositions that indicate derivation from a moderately depleted mantle source (Nesbitt and Sun, 1976; Sun and Nesbitt, 1978; Shirey and Hanson, 1986; Condie, 1990; Laflèche *et al.*, 1992). In many respects their petrological and geochemical characteristics resemble those of tholeiitic basalts in oceanic plateaus and, if the effects of crustal contamination are ignored, continental flood basaltic provinces. Only slightly elevated Fe and Ni contents and lower levels of moderately incompatible trace elements, such as the HREE, distinguish the Archean tholeiites (Arndt, 1991; Condie, 1994). These differences may indicate secular variation in the composition of the mantle source, as advocated by Condie (1984, 1994) and Francis *et al.* (1999), or alternatively changes in the conditions of partial melting. Arndt *et al.* (1997a) suggested that the combination of high Fe–Ni and low HREE indicated that the melting that produced Archean tholeiites took place at greater depths and from a hotter source than the melting that produced basalts in modern volcanic plateaus.

The exact relationship between tholeiites and komatiites remains unclear. The two types of magma occur together – komatiite is always associated with tholeiite, even though tholeiites need not be accompanied by komatiite – and the two magmas alternate in many Archean greenstone belts indicating that they formed simultaneously, presumably by melting in different parts of the mantle. This pattern is seen on Gorgona Island, and in several continental flood basalt provinces. In the Siberian volcanic province, the tholeiites have a remarkably uniform geochemical and isotopic composition, in marked contrast with the wide variation in the compositions of the picrites and meimechites of this region (Lightfoot *et al.*, 1990; Wooden *et al.*, 1993). Likewise, the compositions of the tholeiites in oceanic plateaus such as Ontong Java (Fitton *et al.*, 2004) and the Caribbean (Kerr *et al.*, 1997) are equally monotonous. Cox (1980), O'Hara (1980), Arndt *et al.* (1997b) and Révillon *et al.* (2000) have proposed that flood basalts acquire their uniform compositions through mixing and crystallization in large open-system crustal magma chambers. For reasons that remain poorly understood, though are probably related to crustal architecture and density contrasts between magmas and crustal rocks, these chambers trap a large proportion of the picritic parent magmas and transmit to the surface only their fractionally crystallized and homogenized products. The komatiites and picrites, on the other hand, bypass the magma chambers and reach the surface in an essentially unmodified state.

13.6 Summary

(1) Ultramafic magmas form by high degrees of melting of a peridotite source and/or by melting at great depths in the mantle.
(2) The source was probably a mantle plume with a temperature significantly higher than that of ambient mantle.
(3) Barberton-type komatiites are the products of batch melting at pressures above about 8 GPa or depths greater than 240 km. Under these conditions the partial melt had about the same density as solid mantle minerals and it initially remained within the source. It escaped only when the plume ascended above the neutral-buoyancy threshold. Majorite garnet is a residual phase and its presence in the residue produced magma with the Al-depleted character of Barberton komatiites.
(4) Munro- and Gorgona-type komatiites formed through fractional melting at shallower levels. The departure of early-formed melts left a refractory residue that underwent further melting to produce these trace-element-depleted komatiites.
(5) Karasjok-type komatiites formed from a source that was richer in FeO and incompatible trace elements than that of other komatiites.

(6) Some komatiitic basalts are the products of fractional crystallization of komatiite. Others result from combined crystallization and crustal contamination, while others are the products of independent melting of the mantle source.

(7) The tholeiitic basalts associated with komatiites in Archean greenstone belts or with picrites in Phanerozoic volcanic plateaus have uniform, relatively evolved compositions. These basalts probably formed through melting in separate parts of the mantle source, and they acquired their uniform compositions through reworking in large open-system crustal magma chambers.

14

Geodynamic setting

14.1 Introduction

In what type of tectonic environment did komatiite erupt? For all komatiites, but particularly those of Archean age, the setting can best be estimated from the nature of associated rocks, and the broader tectonic setting must be inferred using general information about the structure of the region and the overall geological context. Given our poor knowledge of the geodynamic regime(s) that existed in the Archean, any attempt to infer the exact setting is subject to considerable uncertainty. In several regions, komatiite appears to have erupted onto continental crust that was covered by shallow seas, a setting that is rarely encountered in the modern tectonic environment. Any investigation of the eruptive setting must take into account the peculiarities of the Archean environment, including the hotter mantle (Bickle, 1978; Richter, 1988), thicker oceanic crust (Sleep and Windley, 1982; Vlaar, 1986) and hotter continental crust (because of enhanced radioactive heating) (Mareschal and Jaupart, 2006).

Many papers that treat the tectonic setting of Archean terrains dwell on the post-magmatic history and particularly the structural evolution of granite-greenstone belts (e.g. Sylvester *et al.* (1997), Wyman *et al.* (2002), Van Kranendonk *et al.* (2004), Benn *et al.* (2006)). Papers such as these discuss the accretion of fragments of oceanic crust and island arcs, and go in detail into the post-volcanic history of deformation, metamorphism and granite intrusion. While important in any general discussion of Archean tectonics, this history follows, and is largely independent from, the eruption of komatiite. In many cases, these ultramafic lavas erupted as part of a volcanic assemblage dominated by tholeiitic basalts, most probably in an oceanic setting as part of the oceanic crust or an oceanic plateau (Storey *et al.*, 1991; Bickle *et al.*, 1994; Arndt *et al.*, 1997a). In other cases, komatiite together with tholeiitic basalt appears to have erupted on submerged continental crust (Groves *et al.*, 1984; Compston *et al.*, 1986) or island arcs (Dostal and Mueller, 1997; Hollings *et al.*, 1999; Wyman, 1999). The cohabitation of plumes and subduction zones implied by the latter association raises intriguing questions about the geodynamic regime(s) that operated in the Archean.

An intriguing question is whether komatiite ever erupted subaerially. In most regions the presence of pillows in the basalts that dominate the volcanic sequences and the nature of intercalated sediments show clearly that the setting was subaqueous, and most probably submarine. Nonetheless, some komatiites could well have erupted onto dry land, because large areas of continental crust already existed at the end of the Archean, and komatiite, like modern continental flood basalt, probably erupted onto this crust. The problem is that we have very little idea of how a subaerial komatiite would have looked and no unambiguous criteria that might be used to identify komatiites that erupted onto land and not under water.

A survey of the various environments of komatiite produces the diverse list that is summarized in Table 14.1. With the exceptions of subaerial settings and mid-ocean ridges, every other geodynamic setting seemed to have contained komatiites. The latter two settings are excluded from the list not because they did not contain komatiite (they almost certainly did) but because we know of no convincingly documented examples of these settings in the preserved Archean record. Before attempting to explain the diversity of eruptive environments, these settings will be reviewed one by one.

'Mafic plains'

This term has been used by many authors (Dimroth *et al.*, 1985; Card, 1990; Condie, 1994) to describe the normal habitat of Archean komatiites. Most of the examples discussed in previous chapters come from this setting, one that is characterized by thick sequences of large, laterally persistent, originally flat-lying mafic and ultramafic flows. Very commonly pillow basalts alternate with massive flows, as illustrated in Figure 14.1. These volcanic units are interlayered with immature cherts (Bavinton, 1981; de Wit *et al.*, 1987) and terrigenous sedimentary rocks (Schau, 1977; Donaldson and de Kemp, 1998; Bleeker *et al.*, 2000), or in some places with thin bands of felsic volcanic rock (Barnes *et al.*, 1988; Gibson and Gamble, 2000) (Figure 14.2). Notable examples include: the lower part of the Onverwacht Group in the Barberton greenstone belt of South Africa (de Wit *et al.*, 1987; Dann, 2000); the Reliance Formation of the Belingwe belt in Zimbabwe (Bickle and Nisbet, 1993; Bickle *et al.*, 1994); the Kalgoorlie Terrain of the Norseman–Wiluna belt in Western

Table 14.1. *Tectonic settings of komatiites*

Locality	Age	Setting	Evidence	Reference
Gorgona Island	90 Ma	Upfaulted part of an oceanic plateau	Petrological and geochemical characteristics of associated basalts and deep-water sedimentary rocks; spatial association and similarity of age with Caribbean plateau	Gansser *et al.* (1979), Echeverría (1982)
Gilmour Island and Cape Smith Belt	1.9 Ga	Rifted continental margin of a small oceanic basin	Petrological and geochemical characteristics of associated basalts and deep-water sedimentary rocks; tectonic reconstruction of evolution of Circum-Superior belt	Baragar and Scoates (1981)
Lapland Fe-rich komatiites		Continental rift?	Characteristics of basalts and sedimentary rocks; tectonic reconstruction	Hanski *et al.* (2001)
Munro Township, Canada; Belingwe Belt, Zimbabwe	2.7 Ga	'Mafic plain' – probably part of an oceanic plateau	Petrological and geochemical characteristics of associated basalts and sedimentary rocks; paucity of associated intermediate to felsic volcanic rocks	
Hunter Mine Group, Canada	2.7 Ga	Eruption onto an island arc	Associated intermediate–felsic volcanic rocks	Dostal and Mueller (1997)
Kambalda, Australia	2.7 Ga	Submarine eruption onto a submerged continental platform	Interbedded pillow lavas and cherts provide evidence of submarine eruption; geochemical data and zircon xenocrysts indicate that the komatiites assimilated continental crustal rocks	Claoué-Long *et al.* (1988), Groves *et al.* (1984)
Prince Albert Group, Canada	2.7 Ga	Eruption into a fluviatile or shallow marine continental setting	Assciated quartz-rich arenites and other shallow-water sedimentary rocks	Schau (1977)
Komati Formation, Barberton belt, South Africa	3.5 Ga	'Mafic plain' – probably part of an oceanic plateau	Petrological and geochemical characteristics of associated basalts and sedimentary rocks; paucity of associated intermediate to felsic volcanic rocks	de Wit *et al.* (1992)

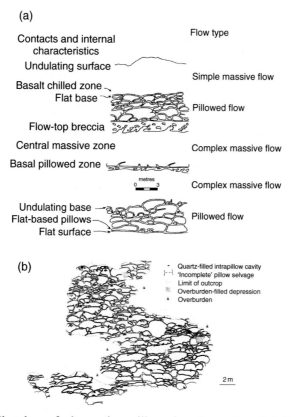

Fig. 14.1. Sketches of alternating pillowed and massive tholeiitic basaltic flows from the Utik Lake region of Manitoba, part of the Superior Province of Canada (from Hargreaves and Ayres (1979)). These sequences are typical of the 'mafic plain' environment.

Australia (Gresham and Loftus-Hills, 1981; Cowden and Roberts, 1990); and parts of the Staughton–Roquemaure Group of the Abitibi belt in Canada (Card, 1990; Mueller, 2005).

Although deformation complicates the interpretation of the original form and disposition of the lavas, in regions of continuous outcrop both mafic and ultramafic types are seen to form large, laterally continuous units. In some regions, even thin units of basalt of distinctive lithology can be traced for many kilometres along strike: one notable example is the V8 flow, a conspicuous variolitic plagioclase phyric flow that can be traced for about 20 km, from one side of the Timmins mining camp to the other (Pyke, 1982). Another example is the 150 km long and 30 km wide Walter Williams Formation in the central Yilgarn Craton (Hill *et al.*, 1995), described in Chapters 3 and 9. Units of this type are generally interpreted to indicate that large volumes of low-viscosity mafic–ultramafic lava erupted rapidly onto a relatively flat surface to form an

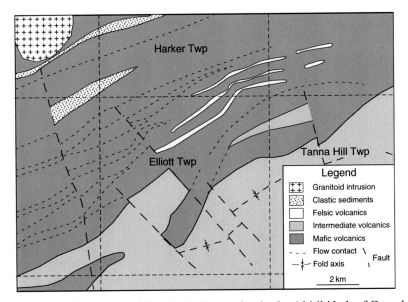

Fig. 14.2. Tuff bands of the Kenojevis Formation in the Abitibi belt of Canada.
These bands of felsic volcanic rock indicate that the mafic plains sequence
contains a bimodal volcanic assemblage, not a homogeneous mafic sequence as
in a Cretaceous oceanic plateau (from Satterly (1951)).

extensive submarine volcanic plain; hence the name used by Dimroth *et al.*
(1985).

The basaltic flows very commonly are pillowed, and they alternate with
hyaloclastic tuffs, cherts and, in some places, clastic sediments. These features
provide firm evidence that the eruptions were subaqueous, in all probability
submarine. Thick massive sheet flows form a high proportion of most basaltic
sequences, providing further evidence for high eruption rates.

The along-strike persistence of the units indicates that the setting was unlike
that of a modern mid-ocean ridge, where relatively short flows are largely
confined to the median valley or extend only short distances along the ridge
flanks. However, we have very little idea of the morphology of a mid-ocean
ridge composed of thick oceanic crust, as would have been produced by partial
melting of a hotter Archean mantle (Sleep and Windley, 1982) and we cannot
rule out the possibility that some parts of Archean greenstone belts formed in
such an environment. Nevertheless, the modern tectonic setting that most
closely resembles that of the Archean mafic plains is an oceanic plateau such
as the Ontong Java or Caribbean plateaus (Storey *et al.*, 1991; Mahoney and
Coffin, 1997). Here, as in the Archean sucessions, we find thick sequences of
pillow basalt and massive sheet flows, and, in the lower parts of the Caribbean
plateau, picrites take the place of komatiites (Kerr *et al.*, 1996).

The chemical compositions of Archean tholeiitic basalts broadly resemble those of basalts in modern oceanic basins (MORB) and have been compared with modern mid-ocean ridge basalts (Ohta *et al.*, 1996). This comparison does not stand close scrutiny. Although the basalts from both settings are tholeiitic and have low K contents, their trace-element characteristics are different: rather than being depleted in the more incompatible trace elements, like normal MORB, Archean tholeiites have flat REE patterns (Figure 14.3). The concentrations of elements with intermediate compatibilities, like the

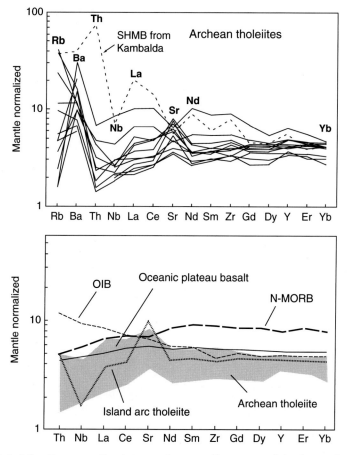

Fig. 14.3. Mantle-normalized trace-element diagrams of Archean tholeiitic basalts compared with representative patterns of basalts from modern tectonic settings. Most of the Archean tholeiites have moderately depleted incompatible elements, flat HREE and no Nb anomalies. This pattern is best matched among modern basalts by basalts from oceanic plateaus, but the overall levels of trace elements are lower. Some Archean basalts are enriched in incompatible trace elements and have negative Nb anomalies. These are either crustally contaminated or they are part of an island arc sequence (From Arndt *et al.* (1997a).)

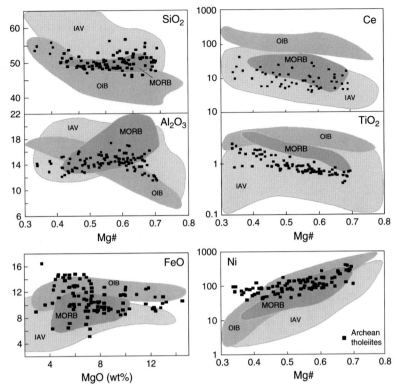

Fig. 14.4. Variation diagrams in which the major- and trace-element contents of Archean tholeiites are compared with the compositions of basalts from modern tectonic settings. The Archean basalts are characterized by relatively high FeO and Ni and relatively low Al_2O_3 and TiO_2 compared with the modern basalts. (From Arndt *et al.* (1997a).)

HREE or Ti, are much lower than in MORB, and Ni and Cr contents, for a given MgO content, are relatively high (Figure 14.4). Their Fe contents are also elevated. The closest modern analogue of these Archean tholeiites is basalts in unusually thick oceanic crust such as in the Kolbeinsey Ridge, north of Iceland or the basalts in Cretaceous oceanic plateaus (Storey *et al.*, 1991; Arndt *et al.*, 1997a).

There is, in addition, one very important difference between the mafic plain facies of an Archean greenstone belt and a Cretaceous oceanic plateau: felsic volcanic rocks, which are largely absent from Cretaceous oceanic plateaus such as Ontong Java or the Caribbean, form a minor but ubiquitous part of almost every Archean volcanic succession. Very rarely do these felsic units outcrop well, and their abundance, or their very presence, was not apparent during early mapping. More recent surveys in which fieldwork is supported by information from drill cores have proven, however, that they are almost always there. Two localities in the Abitibi belt provide examples where otherwise homogeneous mafic–ultramafic sequences include felsic horizons. Most of Munro Township in the Abitibi belt of Canada is underlain by a complex association of flows and sills with mafic or ultramafic compositions, but the Potter Mine, a volcanogenic massive sulfide deposit within an intermediate–felsic volcanic centre, lies just along strike from the classic Pyke Hill komatiite locality (Satterly, 1951; Coad, 1977; Gibson and Gamble, 2000). In the stratigraphically higher Blake River Group, which overlies the Munro sequence, a monotonous series of pillowed and massive tholeiites, several kilometres thick, is punctuated by thin, laterally persistent bands of felsic tuffs, each only a few hundred metres thick (Figure 14.2). The thin bands of sedimentary rocks that alternate with komatiites in the Kambalda sequence contain a component of felsic tuff (Bavinton, 1981). Two other examples of mafic plain facies – the lower Onverwacht Group in the Barberton belt in South Africa and the Reliance Formation of the Belingwe belt of Zimbabwe – seem, on the other hand, to be mainly free of internal felsic volcanic rocks.

Tuffs and lava flows of dacitic–rhyolitic composition are known in parts of the Cretaceous Kerguellen plateau (Ingle *et al.*, 2002; Kieffer *et al.*, 2002), and this might lead to the argument that felsic volcanic rocks are not entirely foreign to modern ocean plateaus. Drilling undertaken as part of ODP Leg182 has shown, however, that large parts of this plateau are made up of drowned continental flood volcanics and that these parts do not constitute a true oceanic plateau (Frey *et al.*, 2000). A closer modern analogy of the felsic volcanic rocks of Archean greenstone belts might be with the granophyres and rhyolites of Iceland.

The few examples of post-Archean komatiitic lavas were emplaced in oceanic settings that in some ways resemble the Archean mafic plains. Mid-Proterozoic komatiitic basalts of Gilmour Island and other parts of the Circum-Superior belt erupted into a young ocean basin that opened following rifting of the Archean craton (St-Onge and Lucas, 1990). As described in Chapter 2, the Gilmour komatiitic flows occur as long, thin, laterally persistent units which are intercalated with pillowed and massive basalts (Baragar and Scoates, 1981; Arndt, 1982). Here, direct comparison can be made with the volcanic rifted margins on both sides of the North Atlantic, where thick sequences of submarine basalts and minor picrites were deposited upon now-drowned subaerial flood basalts (Saunders *et al.*, 1997).

Despite its young age, the setting for the eruption of the Cretaceous Gorgona komatiites is poorly understood. The island is a horst elevated above surrounding sedimentary units, now 30 km off the coast of the South American continent. The age of the Gorgona volcanic rocks is the same as that of basalts of the Caribbean basin, making it very likely that both are parts of the same oceanic plateau. This plateau became strongly deformed as it migrated between the South and North American continents, and our knowledge of the position of the Gorgona rocks in the overall stratigraphy of the plateau is incomplete (Kerr, 2005).

Arcs and convergent margins

In other parts of many greenstone belts, komatiites are spatially associated with substantial amounts of intermediate–felsic volcanic rock (Jensen, 1985). In the Hunter Mine Group, located in the Quebec section of the Abitibi belt (Dostal and Mueller, 1997), and at the Kidd Creek Mine in Ontario (Wyman *et al.*, 1998; Barrie, 1999), komatiite flows have flooded onto rhyolite complexes (Figures 1.1, 14.5; and at the Perseverance Mine in the Yilgarn Craton and throughout the Boorara Domain of the Norseman–Wiluna Belt, komatiite erupted onto and intruded into a series of felsic lavas and pyroclastic rocks (Chapter 10 and Barnes *et al.* (1988), Trofimovs *et al.* (2003), Hill *et al.* (2004)). Hollings *et al.* (1999) and Wyman *et al.* (1998) described locations in other greenstone belts where komatiite forms part of volcanic sequences composed otherwise of mafic–felsic calc-alkaline volcanic rocks.

The petrological characteristics and chemical compositions of the calc-alkaline rocks demonstrate that they are related to subduction. Most telling is the range of rock types, from basalt through andesite to dacite and rhyolite. Histograms representing the abundance of different rock types often contain a gap between basalt and andesite (Goodwin, 1979), and andesite containing abundant zoned, inclusion-ridden plagioclase phenocrysts like those from modern subduction settings are rare. However, a relatively complete range of rock compositions, from mafic through intermediate to felsic, distinguishes these volcanic series from the conspicuously bimodal, (ultramafic)mafic–felsic association of the Archean mafic plains. The petrological information is complemented by major- and trace-element indices diagnostic of magmas from island arcs and convergent margins, indices that include relatively high SiO_2 and Al_2O_3 coupled with low FeO and TiO_2 contents, enrichment of incompatible trace elements, and negative anomalies of the HFSE. Throughout the Canadian Superior province (Wyman *et al.*, 2002) and even in the 3.8 Ga supracrustal sequences of the Isua region of Greenland (Polat *et al.*, 2002), rock types such as shoshonite, adakite and boninite have been identified. These geochemically distinct rock types, which are normally restricted to subduction settings, provide further evidence that these greenstone belts formed as island arcs or convergent margins.

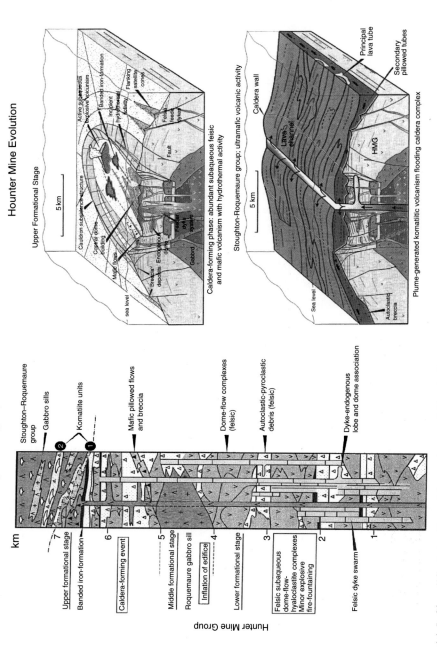

Fig. 14.5. Simplified stratigraphic section and block diagrams showing a felsic volcanic centre in the Hunter Mine Group flooded by komatiite lava flows (from Dostal and Mueller (1997)).

Pillow structures, hyaloclastites and interbedded sedimentary rocks provide clear evidence that most of the volcanic rocks erupted subaqueously, but towards the tops of some lava piles, such as those associated with Mattabi-type massive sulfide deposits of the Abitibi belt (Morton and Franklin, 1987; Hannington *et al.*, 2004), the presence of pumice, shoal-line breccias and fall deposits point to subaerial eruption. In the Boorara Domain of the Norseman–Wiluna belt, Western Australia, Trofimovs *et al.* (2003) described submarine debris flows incorporating komatiite and dacite blocks, implying eruption on the flanks of a stratovolcanic complex.

There is general agreement, therefore, that some Archean mafic–felsic calc-alkaline rocks formed in a subduction environment, either as part of an island arc or at convergent margins. Together with the abundant granitoids that constitute the major component of most granite-greenstone belts, they provide evidence that plate tectonics was operating happily through the Archean. The simple notion that Archean geodynamics was just like that of today is spoiled, however, by the association of komatiite and felsic volcanics. This association has led to considerable debate in the literature (Hollings *et al.*, 1999; Sproule *et al.*, 2002; Wyman *et al.*, 2002; Parman *et al.*, 2004), debate that centres on the question of how mantle plumes, the presumed source of the komatiites in the opinion of most authors, could have cohabitated with subduction zones. I discuss this issue later in the chapter.

Eruption in shallow water on continental basement

In several parts of northern Canada we find a curious association of komatiite with mature, quartz-rich continental arenites. In the 1970s, at a time when we were only starting to come to grips with the notion that ultramafic lavas formed an important component of many Archean volcanic successions, Mikkel Schau (1977) described a series of interlayered komatiites and quartz-ites in the 2.7 Ga Prince Albert Group in the Rae Province of the Northwest Territories of Canada. The quartzites contain rounded glassy quartz grains, muscovite, Cr-mica and rare garnets, they immediately overlie older gneissic basement, and they display prominent cross-bedding indicating a clastic origin and deposition in a high-energy environment. This combination of features led Schau to suggest that the rocks had been deposited in a shallow-water envir-onment in a tectonically stable continental platform.

The komatiites occur as 0.5–2 m thick units at the interface between base-ment gneiss and overlying sedimentary rocks, or within a sequence of quart-zites and magnetite–quartz iron formations. The rocks are now largely altered to tremolite + chlorite or actinolite-rich schists but in places platy olivine spinifex texture is preserved. Major-element analyses have shown that these rocks contain about 25–40% MgO and have chemical characteristics like those of Munro-type komatiites.

Trevor McHattie (personal communication) has recognized three strati-graphic sequences: (1) a lower, ~2730 Ma, sequence dominated by komatiite and basalt and containing minor intermediate–felsic volcanic rocks; (2) a middle, ~2710 Ma, sequence containing ultramafic–felsic volcanic rocks, iron formation and siliciclastic sediments; and (3) an upper, 2690 Ma, sequence dominated by sedimentary rocks but containing a substantial komatiite flow sequence. These rocks overlie older granitoid basement that yields detrital zircons with ages ranging from 2.8 to 3.8 Ga. The komatiites display a wide range of geochemical characteristics. Those in the lower parts of the sequence are strongly enriched in incompatible trace elements, while those higher in the sequence are depleted. McHattie explained their formation in a model of two-stage melting: the first magmas are interpreted as low-degree deep-sourced melts and the later melts are thought to have formed by the melting at shallow depths of the residue left after extraction of the first melts. Active extension at a continent margin thinned the lithosphere to the extent that high-degree melting at shallow depth was possible during the second stage.

Donaldson and de Kemp (1998), Bleeker *et al.* (2000) and Mueller and Pickett (2005) described comparable sequences in several parts of the Slave province, also in northern Canada (Figure 14.6). At these localities, a basal sequence comprising quartzites, thin rhyolite tuffs and komatiitic flows or sills is overlain by a thicker sequence of volcanic rocks. This sequence is made up mainly of pillowed tholeiites but contains komatiitic units near its base. Bleeker *et al.* (2000) mentioned the presence of subaerial komatiites but I have not been able to find a detailed description of these units. The sedimen-tological characteristics of the basal quartzites are consistent with a shallow-water environment but features of the overlying units point to deposition in deeper waters. Donaldson and de Kemp (1998) described how structures in the quartz arenites vary from the herring-bone cross-beds typical of the tidal environment to hummocky cross stratification, demonstrating deposition below the storm-wave base. Mueller and Pickett (2005) calculated that the lateral extent of quartz arenites is on the order of 5–8 km from a near-shore facies to deeper water, and proposed that both the arenites and the komatiites were deposited in narrow rift basins in a continental platform.

Much the same relationship was reported by Goutier *et al.* (1998, 1999) in the La Grande area of the Superior Province. Komatiites also erupted into a sequence of mature quartzites in the Steep Rock–Lumby Lake region in the Superior Province of Canada (Tomlinson *et al.*, 1999) and in the Ravensthorpe

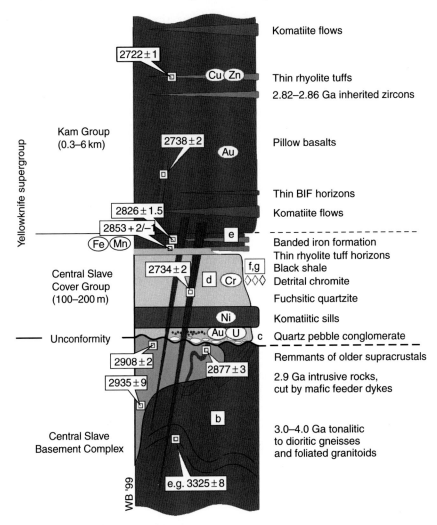

Fig. 14.6. Generalized stratigraphic column of the Central Slave Cover Group, the autochthonous cover of the Central Slave Basement Complex (Bleeker *et al.*, 2000). Komatiite flows and sills are interlayered with clastic sedimentary rocks and mafic volcanics.

belt of the Yilgarn Craton. In the correlative Forrestania and Lake Johnston belts further to the north, there is a close association of extensive sheeted komatiite units with persistent banded iron formations (Perring *et al.*, 1995), implying a continental-shelf environment. Associations of komatiite and banded iron formations (BIF) are also known in the Windarra area in the eastern Norseman–Wiluna belt (Marston, 1984).

Bleeker *et al.* (2000) and Mueller *et al.* (2005) have suggested that a mantle plume was the source of the komatiites and that its arrival at the base of the continental lithosphere was the cause of rifting that led to the opening of an ocean basin.

The sedimentary and volcanic stratigraphy of the Belingwe belt in Zimbabwe has become the focus of considerable attention following the suggestion by Kusky and Kidd (1992) and Kusky and Winsky (1995) that the belt is an allochthonous remnant of an oceanic plateau that had been thrust onto the older basement of the Zimbabwe Craton. Bickle *et al.* (1993, 1994), Nisbet *et al.* (1993b) and later Hunter *et al.* (1998) responded to this suggestion with a detailed appraisal of the geological relations in the region and a convincing rebuttal of the allochthonous emplacement model. They made the case that the sedimentary and volcanic rocks of the Belingwe belt are autochthonous and they argued that these rocks were deposited on an unconformity on an older gneissic basement (Figure 14.7). The sequence in the type area on the eastern margin of the Belingwe belt opens with a basal conglomerate containing clasts of gneissic rocks and granitoids from the basement, then continues with quartz sandstones displaying a range of shallow-water sedimentary structures and is associated with chert and iron formation. The overlying, almost completely volcanic Reliance Formation consists dominantly of komatiites and komatiitic basalts, with only rare intercalated cherts. The progression is from shallow continental platform to deeper-water volcanics, rather like that inferred for parts of the Slave and Rae Province by Bleeker *et al.* (2000). Further evidence for eruption onto a continental basement is provided by the trace-element and Nd and Pb isotope data of Chauvel *et al.* (1993), which indicate that the komatiites are contaminated with older granitic crust, and Shimizu *et al.*'s (2004) discovery of xenoliths and xenocrysts of crustal material in the volcanic rocks.

At Kambalda in the Yilgarn Craton of Australia, the evidence of a continental setting is less direct. Here the sequence is almost entirely volcanic. A basal succession of tholeiitic basalts is overlain first by komatiites (which host the Ni-sulfide ore deposits) and then komatiitic basalts (Moore *et al.*, 2000). Sedimentary rocks are rare, limited to narrow beds of chert and felsic tuffs interlayered with the volcanic rocks (Bavinton, 1981).

The arguments for a continental setting stem in part from evaluations of the overall geology of the region (Groves *et al.*, 1984) but the most convincing evidence is geochemical. In the view of Chauvel *et al.* (1985), Arndt and Jenner (1986), Barley (1986) and Lesher and Arndt (1995), the trace-element and Nd isotopic compositions of the komatiites and komatiitic basalts indicate that these rocks have been contaminated with up to 25% of older granitic crust (Chapters 5 and 6). Compston *et al.* (1986) and Claoué-Long *et al.* (1988) obtained direct evidence for this contamination by separating zircons from the volcanic rocks and showing that these minerals were xenocrysts from granites

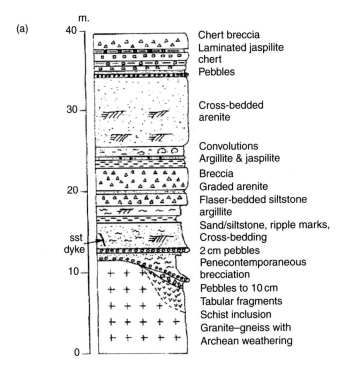

(a)

m.

40 — Chert breccia
Laminated jaspilite
chert
Pebbles

30 — Cross-bedded
arenite

Convolutions
Argillite & jaspilite
Breccia
20 — Graded arenite
Flaser-bedded siltstone
argillite
Sand/siltstone, ripple marks,
sst — Cross-bedding
dyke — 2 cm pebbles
Penecontemporaneous
10 — brecciation
Pebbles to 10 cm
Tabular fragments
Schist inclusion
Granite–gneiss with
Archean weathering

0 —

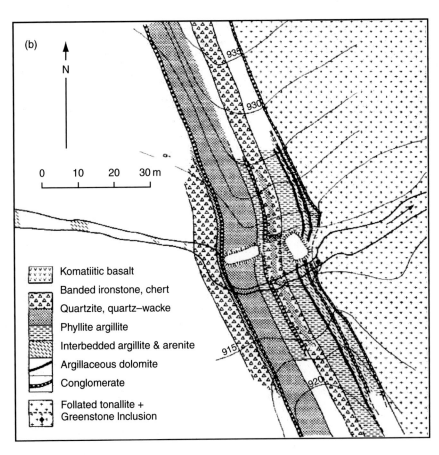

(b)

N

0   10   20   30 m

935
930
915
920

Komatiitic basalt
Banded ironstone, chert
Quartzite, quartz–wacke
Phyllite argillite
Interbedded argillite & arenite
Argillaceous dolomite
Conglomerate
Follated tonallite +
Greenstone Inclusion

with ages ranging from 2.7 Ga to 3.5 Ga. On this basis it is clear that the komatiites encountered and assimilated older granitic rocks on their way to the surface, and it seems most probable that they met these rocks as they passed through continental lithosphere. The enigma posed by the presence of cherts and pillow lavas, which indicate that the volcanic sequence is submarine and imply that the continental lithosphere was submerged at the time the komatiites erupted, was discussed by Arndt (1998). Eruption of submarine lavas onto drowned continental platforms is very rare in modern geodynamic settings. Possible solutions include more voluminous oceans in the Archean, a consequence of hotter mantle that led to more thorough degassing of subducting oceanic plates, and/or the presence in ocean basins of more voluminous ocean ridges or large oceanic plateaus during the periods when lavas erupted onto the drowned continental platforms.

14.2 The geodynamic setting of Archean greenstone belts

The nature of the basement underlying Archean greenstone belts occupied the authors of many papers on Archean tectonics in the 1970s and 1980s. Some Archean komatiites and basalts certainly erupted onto older sialic crust, as at Belingwe and Kambalda, but the eruptive setting in most greenstone belts was oceanic, either in a volcanic plateau or as part of the oceanic crust. The lower contacts of Archean greenstone belts very commonly are tectonic, or they are obliterated by the intrusion of later granitoids. Gneissic rocks older than the volcanics are indeed found in many granite-greenstone terrains (e.g. Barberton (Kröner and Todt, 1988)) but unless the volcanic rocks show evidence of interaction with the granitoid rocks, there is no guarantee that the latter were present in the region at the time the volcanic rocks were erupted: they could have formed in a quite different setting, and could have been juxtaposed against the volcanic rocks at a later time during continent accretion.

An interesting model for the formation of most granite-greenstone belts has been developed by Kimura *et al.* (1993) using evidence from the southern Superior Province of Canada. Following Hoffman (1991), they advocated an allochthonous origin for most of the greenstone sequences, drawing an

Caption for Fig. 14.7.
(a) Stratigraphic sequence through the volcano-sedimentary formations of the Belingwe greenstone belts which unconformably overlie old granitoid basement; (b) map showing the volcano-sedimentary units to the east and the granitoid basement (from Nisbet *et al.* (1993b)).

analogy with Phanerozoic tectonic processes in which the volcanism is related to both plate formation and subduction. Using a combination of geological relationships and geochemical data, they found similarities between some of the volcanic sequences with (a) thicker-than-normal modern oceanic crust, (b) Cretaceous oceanic plateaus and (c) subduction-related volcanics. In several sequences in the Larder Lake and Beardmore–Geraldton regions, they cited an absence of interdigitation of lava and terrigenous sediment as evidence that the eruption site of the lavas was remote from any deposition site of terrigenous sediments. Together with the MORB-like trace-element patterns in some of the basalts, these geological relationships led them to advocate a mid-ocean ridge setting. An alternation of lava and sediment is attributed to tectonic stacking by layer-parallel faults. In contrast, in the Malartic–Val d'Or block in another part of the province, an abundance of komatiite and a paucity of terrigenous sediment were taken to indicate that this region formed as part of an oceanic plateau. Finally, in the Noranda region, mafic–felsic calc-alkaline volcanic rocks were interpreted as parts of an island arc. According to Kimura *et al.*, 3–4 km thick slices of oceanic crust or oceanic plateau were tectonically imbricated with slices of terrigenous sedimentary rocks during accretion to a growing Archean continent, and then calc-alkaline volcanics erupted above subduction zones that had developed adjacent to the continent. Back-stepping of these subduction zones produced a series of east–west-trending belts whose ages decline from about 3 Ga in the northwest to 2.7 Ga in the south.

This model has been questioned by Ayer *et al.* (2002) who used the results of new geological mapping and precise U–Pb zircon dating to support an auto-chthonous model for the Abitibi belt rather than one involving accretion of allochthonous terranes. These authors identified seven assemblages with ages from 2745 Ma to 2698 Ma, each comprising a coherent, upward-facing, stratigraphic section. About 20% of the samples they analysed contained inherited zircons with ages similar to those found in underlying assemblages. Five of the assemblages were dominantly mafic–ultramafic and two domi-nantly calc-alkalic. The extensive intermingling of the different magma series in a number of the assemblages led Ayer *et al.* (2002) to propose complex, large-scale and long-lived interaction between mantle plumes and subduction zone magmas of a type not seen in modern terranes.

The difference between the two models may, at least in part, be one of scale. At the scale of the southern Superior Province as a whole, progressive south-ward younging of the ages of the volcanism, the shear zones, the plutonism and the sedimentation, all point to accretionary development. On the other hand, at the scale of an individual granite-greenstone belt, other observations point to allochthonous development. These observations include: the presence throughout the Superior Province (Thurston, 2002) and the Pilbara Craton (Van Kranendonk *et al.*, 2004) of volcano-sedimentary sections that become younger outward from granitoid batholiths; the presence of internal unconformities in these sections (Thurston, 2002; Van Kranendonk *et al.*, 2004; Van Kranendonk, 2006); the presence in some of the young volcanic units of zircon xenocrysts identical in age to the underlying volcanic units; and geochemical evidence of contamination of mafic and ultramafic magmas by granitoid material.

In addition, although the accretion model of Kimura *et al.* (1993) may apply to regions such as the southern Superior Province and the Barberton belt where there is little evidence for proximal, significantly older continental crust, it is inappropriate for other belts. The komatiites from the eastern Yilgarn and Belingwe belts apparently interacted extensively with far older continental crust before reaching the surface (Groves *et al.*, 1984; Chauvel *et al.*, 1985; Compston *et al.*, 1986), and the association of komatiite and shallow-water terrigenous sediments in northern Canada points to eruption of komatiite in a subaerial or shallow-water setting, again on older and apparently stable continental crust (Schau, 1977; Bleeker *et al.*, 2000; Sanborn-Barrie *et al.*, 2004). The question of how komatiite is capable of erupting in such diverse settings is the theme of the next section.

14.3 A mantle plume does not know what it will meet at the surface

Coffin and Eldholm (1993) provided a dramatic illustration of the scale of the geodynamic processes that lead to the eruption of large volumes of mafic and ultramafic lava. By calculating the volume of Phanerozoic oceanic and continental plateaus and by assuming that the magmas formed through 5–30% partial melting of the mantle, they calculated that the volume of the mantle source was enormous. For example, they represented the source of the Ontong Java oceanic plateau as a huge sphere with a diameter of at least 600 km, and possibly as much as 1400 km (Figure 14.8), and they pointed out that such a source must have extended into the lower mantle.

With these figures in mind, we can start to understand why komatiite erupts in such a wide range of tectonic settings. If we accept that komatiite forms in a mantle plume and if we suppose that the plume rose from deep in the mantle, probably from the core–mantle boundary, then we can assume that the plume had no idea what it would meet when it arrives at the surface. In the examples described above, the mafic and ultramafic magmas that formed through partial melting of the source appear to have risen rapidly to the surface, paying little attention to whether the rocks they passed through had formed at an oceanic ridge or plateau, in an island arc, or as part of a continent.

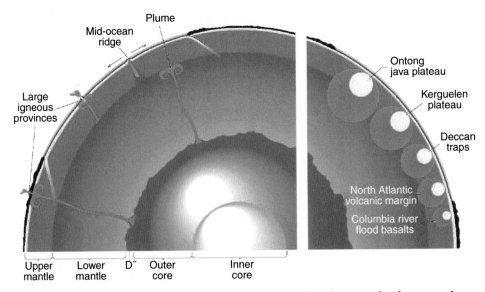

Fig. 14.8. Coffin and Eldholm's (1993) diagram showing mantle plumes and subduction zones, on the left, and spheres representing the volume of the source regions of oceanic and continental volcanic provinces, on the right.

The plume, a vast blob of abnormally hot, low-density mantle, would have bulldozed upwards, either shoving aside any tectonic plates encountered near the surface, or sliding along the base of the lithosphere and oozing into thin zones, as described by Ebinger and Sleep (1998) or Saunders *et al.* (1992, 1997). In many cases the impact would have weakened the continental lithosphere, providing a locus of subsequent rifting that preceded the opening of ocean basins (Courtillot *et al.*, 1999). A plume ascending beneath a sedimentary basin would have filled the basin with lava, and towards the basin margins, the mafic–ultramafic lava might have erupted as mature sediments were being deposited. The komatiite–quartzite associations might be explained in this way. And if the plume, our 1000 km wide bully, encountered a puny, 100 km thick slab of descending oceanic lithosphere, it would snuff out the subduction, at least along that segment of the subduction zone (compare the size of plume with that of a subduction zone in Figure 14.8). It is important to recognize that voluminous mafic–ultramafic magma only forms within 100–200 km of the surface and, if a plume is to reach this level, it must disrupt the down-going slab. Subduction-related volcanism would cease around the site of impact, to be replaced by the eruption of mafic–ultramafic lavas generated within the plume. Other parts of the slab would continue to subduct, and in these regions ultramafic magma flowing out from the site of plume impact might mingle with locally derived calc-alkaline volcanics.

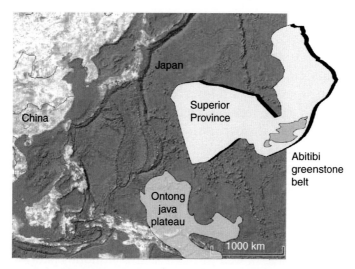

Fig. 14.9. Map of the SW Pacific ocean basin. The outlines of the Cretaceous Ontong Java oceanic plateau, the Archean Superior Province and the Abitibi greenstone belt are superimposed on the map. Note that the dimensions of the Ontong Java plateau are comparable with those of the island arcs of the western Pacific and far larger than the entire Abitibi belt, which is the biggest Archean granite-greenstone belt. The Phanerozoic oceanic plateau dwarfs even the largest of Archean volcanic provinces.

Rather than providing evidence for the cohabitation of plumes and subduction zones, the association of komatiites and calc-alkaline felsic volcanics records the conquest, by a plume, of part of a subduction zone. Where ultramafic and felsic volcanic rocks intermingle, the two magma types must have come from separate sources, the ultramafic lavas from a plume and the felsic rocks from a near-by subduction zone.

In Figure 14.9, we can compare the size of the Ontong Java plateau with the arcs of the southwest Pacific to show the scale at which the process may have happened. Evidence of long-distance transport of magmas away from mantle plumes is emerging through studies of modern large igneous provinces (Ernst and Baragar, 1992; Elliot *et al.*, 1999) and when combined with the descriptions of Lesher *et al.* (1984) and Hill *et al.* (1987, 1995) of vast komatiite flows in the Yilgarn belt of Australia, a model of wide-ranging komatiite eruptions seems entirely plausible.

14.4 Passage through the lithosphere

Melting in mantle plumes, as described in Chapter 13, produces ultramafic magmas that escape upward and migrate towards the lithosphere. What happens once they arrive there? How does komatiite pass through kilometres of cooler, reactive, rheologically and chemically heterogeneous rocks? Komatiite liquid is as dense as many solid crustal rocks and, if isolated in the crust, it ponds in magma chambers. It is a fragile magma, much hotter than the surrounding rocks and, if it loses too much heat, it crystallizes to olivine–porphyritic basalt. How does it reach the surface? The question was first raised by Nisbet (1982) and then treated in more detail by Huppert and Sparks (1985b). The conclusion reached in the two papers is that komatiite ascends through the lithosphere in magma-filled cracks, driven by the difference in density of buoyant magma and the denser surrounding mantle rocks. By using reasonable values for magma and rock density and assuming that the lithosphere was in a state of tension, Huppert and Sparks calculated that the lengths of lithosphere fracture reached tens–hundreds of kilometres, sufficient for the magma to have flowed through near-continuous pathways for much of the distance between the source and the surface. Because the upward flow rate through the fractures depends strongly on viscosity, which is very low in komatiites, ascent velocities would be large. Huppert and Sparks estimated maximum values of about $11 \, \mathrm{m \, s^{-1}}$ (about 40 km per hour) and discharge rates from 500 to $10^5 \, \mathrm{m^3 \, s^{-1}}$. These rates are far larger than those of most modern basaltic eruptions; they are similar to early estimates for continental flood volcanism but these rates have been called into question with the development of inflation models for the emplacement of these lava flows (Hon *et al.*, 1994; Self *et al.*, 1997). Very probably komatiite erupted more rapidly than any other terrestrial magma.

   Is rifting required for komatiite to erupt? In numerous papers it is stated that komatiite is emplaced in regions of rifting and extension, but close reading of these papers reveals that the main reason for making the association is the assumption, explicit or implicit, that komatiite will only reach the surface if its passage is aided by the presence of faults, in, by implication, an extensional setting. Two issues come into play here. First, does komatiite eruption require the presence of faults; and, second, does the faulting imply rifting? The first issue was treated above: the presence of fractures seems necessary for komatiite to pass up through tens of kilometres of low-density continental crust.

   With regard to the second issue, it is instructive to consider the eruption of flood basalts. While some clearly erupt at rifted margins, such those on both sides of the North Atlantic (Saunders *et al.*, 1997), in the Limbombo monocline in the eastern part of the Karoo (Cox *et al.*, 1984) or in the Proterozoic Cape Smith belt (Baragar and Scoates, 1981; St-Onge and Lucas, 1990), in other provinces a rifted setting is by no means evident. For example, during the emplacement of most of the Karoo province (Erlank, 1984) and during the eruption of the Ethiopian flood basalts (Ukstins *et al.*, 2002), the onset of magmatism preceded rifting by tens of millions of years. In these settings, as in many other continental large igneous provinces, mafic dykes, which form an integral part of the magma suite, have a radial distribution centred at the locus of eruption. The extension that was exploited (or created) by these dykes is directly associated with doming of the lithosphere above the rising mantle plume. In many cases normal rifting leading to the formation of an ocean basin may follow the doming as far-field tectonic forces tear apart the continent at sites weakened by plume interaction (Courtillot *et al.*, 1999), but even in these cases the dominant basaltic (and picritic) volcanism was pre-rift.

In Section 14.1 I argued that oceanic plateaus are the best analogue of the mafic plain facies of Archean greenstone belts. There is some evidence that the Ontong Java plateau formed at or near a mid-ocean ridge, but, given the dimensions of this province, it is evident that the rifting itself played only a secondary role in the generation of magma. Both continental and oceanic plateaus may be associated with regions where the lithosphere is not as thick as normal, and these abnormally thin regions are not necessarily caused by rifting: they may, for example, result from delamination of the lower part of the lithosphere.

What does all this mean? Komatiite erupts from volcanic centres, perhaps in the middle of broad domes created by plume upwelling, perhaps in zones of thinned, maybe rifted, crust within the periphery of the region affected by the plume. If the hypothesis that komatiite forms in the heads of enormous mantle plumes is correct, then deformation of the lithosphere, whether continental or oceanic, will be directly linked to upwelling of the mantle source. This is an active process largely independent of the thickness, composition and stress state of the lithosphere and the presence of rifting of the lithosphere at or near the locus of plume impact will be only of secondary importance.

14.5 The setting of komatiitic volcanism

Many of the peculiar aspects of the setting in which komatiites erupted – the large flat submarine plains, for example – could conceivably be related to the peculiar nature of oceanic crust in the Archean. It might be argued that komatiite is the product of melting of mantle that upwelled below Archean mid-ocean ridges. We have no direct evidence, however, that this was ever the case. The hot source of a komatiite is not normal mantle. Were it so, then a large depth interval of the upper mantle would be at temperatures well above the solidus, an unstable situation. Throughout the Archean, komatiites no doubt did erupt at mid-oceanic ridges, where they would have formed vast volcanic plateaus rather like Iceland or arguably Ontong Java. Elsewhere, they erupted in oceanic or continental intraplate settings as volcanic plateaus, or they were superimposed on subduction zones.

Once formed, the volcanic plateaus must sooner or later have accreted to the continents. In places like Barberton, where there is little direct evidence of nearby older continental crust, the plateaus may have formed the nuclei of new continents. In most other places, however, the komatiite-bearing plateaus were accreted to older continents, as happened when the mainly juvenile Abitibi belt collided with the future North American continent, which was growing from north to south. And in some places, like Kambalda and Belingwe, the volcanic plateaus formed upon older continental crust.

Subduction continued as the plateaus accreted, building the island arcs and convergent margins that formed parts of the growing continent, and producing the voluminous granitoids that invaded the supracrustal terranes (de Wit *et al.*, 1992). Ultramafic lava flowed out and away from the domed-up segments of now-defunct arc to the still-active flanking parts of the subduction zone. There they mingled with the mafic–felsic calc-alkaline products of the subduction zone. Part of the alteration that bedevils geochemical investigations of komatiites was associated with the regional metamorphism and deformation that accompanied orogenesis (Binns *et al.*, 1976; Jolly, 1982; Cloete, 1999; Moyen *et al.*, 2006). This metamorphism is unlike that in modern orogenic belts, being characterized by modest geothermal gradients in large regions of low-grade metavolcanics, and associated with multiple episodes of very heterogeneous deformation (Choukroune *et al.*, 1997). The supracrustal rocks are affected by tight upright folding and are cut by faults and shear zones, but strain is very strongly concentrated along localized shear zones. Away from these zones the rocks show remarkably little internal deformation (Chapters 3 and 4), a characteristic that has helped preserve the textures and volcanic structures that provide the basis of our knowledge of the mineralogy and volcanic morphology of komatiites.

But what are we to make of the thin persistent tuffs that are found even in the mafic plain facies of greenstone belts? Their petrographic and geochemical compositions indicate that they are not arc related. As pointed out in Section 14.1, similar tuff beds seem to be largely absent from Cretaceous oceanic plateaus. One possible explanation for their formation is that they formed through partial melting in the lower parts of the volcanic plateaus. Sleep and Windley (1982) pointed out that a consequence of higher temperatures in the Archean mantle would be the formation of oceanic crust that was much thicker than modern crust, perhaps as much as 20 km thick in the mid–late Archean. The temperature gradient across this crust, from near 0 °C at the surface to maybe 1400 °C at a depth of 20 km, was much less than that across thin modern oceanic crust. This means that the upper layer in which temperatures were below 400 °C was thicker than that of modern crust (Russell and Arndt, 2005; Zahnle *et al.*, 2007). The temperature of 400 °C corresponds to the limit between greenschist and amphibolite facies, a limit that separates strongly hydrated low-temperature rocks from less hydrated high-temperature rocks. The circulation of seawater through crust that formed at an Archean spreading centre would lead to the hydration of a relatively thick layer at the top of the crust. Furthermore, the lavas that constituted Archean oceanic crust would have formed from mantle that was hotter than the present-day mantle and these lavas would have had picritic compositions. Olivine phenocrysts in the lavas alter to water-rich secondary phases like serpentine or chlorite, further increasing the water content of the thick hydrated layer.

An oceanic plateau constructed on this crust would directly overlie the layer of hydrated basalt. Because of its low density, the hydrated layer would form a barrier to the ascent of magmas migrating upwards from the plume. As the plateau grew thicker, hot ultramafic magma would pond beneath the layer and heat brought in with the magma might then partially melt the hydrated basalts. The resultant felsic magmas would erupt along with the komatiites and other plume-derived basalts, to form the felsic tuff beds that are an enigmatic part of the mafic plains facies of many Archean greenstone belts.

14.6 Conclusions

(1) Komatiites erupted in a large variety of tectonic settings. Most were emplaced in an oceanic environment, some as parts of the oceanic crust but the majority within oceanic plateaus. Other komatiites erupted onto island arcs and still others erupted onto drowned continental platforms. The oceanic settings are comparable to modern plate tectonic environments; the other settings are not. The overall picture is one in which plate tectonics functioned much as it does at present; but the way it operated was influenced by generally higher mantle temperatures. The higher temperatures led to the formation of thicker oceanic crust, more voluminous oceans and to the generation of mantle plumes that were more abundant, larger and hotter than those of the present time.

(2) A plume model for the formation of komatiite readily explains the diversity of tectonic settings in which komatiite erupted. Plumes are generated deep in the mantle and they ascend into whatever tectonic setting is present at the surface.

Mafic–ultramafic magmas generated by melting in the plume pass through the lithosphere and erupt in ocean basins, onto island arcs or onto continental platforms, whichever the case may be.

(3) When a plume rises beneath a subduction zone, it terminates subduction at the site of impact. Magmas from the plume are generated simultaneously with magmas from flanking portions of the subducting plate. In intervening regions, komatiites from the plume and magmas from subduction zones erupt together.

References

Abraham, E. M. (1953) Geology of Sothman Township. *Ontario Department of Mines, Annual Report*, **62**, 36p.

Agee, C. B. (1998) Crystal–liquid density inversions in terrestrial and lunar magmas. *Physics of Earth and Planetary Interiors*, **107**, 63–74.

Agee, C. B., Walker, D. (1988a) Static compression and olivine flotation in ultrabasic silicate liquid. *Journal of Geophysical Research*, **93**, 3437–3449.

(1988b) Mass balance and phase density constraints on early differentiation of chondritic mantle. *Earth and Planetary Science Letters*, **90**, 144–156.

Aitken, B. G., Echeverría, L. M. (1984) Petrology, and geochemistry of komatiites and tholeiites from Gorgona Island, Colombia. *Contributions to Mineralogy and Petrology*, **86**, 94–105.

Albarède, F. (1983) Thermal limitations to the ascent of hydrated magmas. *Comptes Rendus de l'Academie de Sciences de Paris*, **296**, 1441–1444.

Allègre, C. J. (1982) Genesis of Archaean komatiites in a wet ultramafic subducted plate. In: N. T. Arndt and E. G. Nisbet (eds.) *Komatiites*, pp. 495–500. London: George Allen and Unwin.

Allègre, C. J., Dupré, B., Lewin, E. (1986) Thorium/uranium ratio of the Earth, *Chemical Geology*, **56**, 219–227.

Allsopp, H. L., Viljoen, M. J., Viljoen, R. P. (1973) Strontium isotopic studies of the mafic and felsic rocks of the Onverwacht group of the Swaziland sequence. *Geologische Rundschau*, **62**, 902–912.

Anderson, D. L. (1994) Komatiites and picrites: Evidence for a depleted "plume" source, *Earth and Planetary Science Letters*, **128**, 303–311.

Anderson, S. W., Stofan, E. R., Smrekar, S. E., Guest, J. E., Wood, B. (1999) Pulsed inflation of pahoehoe lava flows: implications for flood basalt emplacement. *Earth and Planetary Science Letters*, **168**, 7–18.

(2000) Reply to: Self *et al.* discussion of 'Pulsed inflation of pahoehoe lava flows: implications for flood basalt emplacement'. *Earth and Planetary Science Letters*, **179**, 425–428.

Anhaeusser, C. R. (1985). Archaean layered ultramafic complexes in the Barberton Mountain Land, South Africa. In: L. D. Ayres, P. C. Thurston, K. D. Card and W. Weber (eds.) *Evolution of Archaean Supracrustal Sequences*, pp. 281–301. Geological Association of Canada Special Paper 28. Ottawa: Geological Association of Canada.

415

Arndt, N. T. (1975) Ultramafic rocks of Munro Township and their volcanic setting. Unpublished Ph.D. thesis, University of Toronto.

(1976a) Melting relations of ultramafic lavas (komatiites) at 1 atm and high pressure. *Carnegie Institution Washington Yearbook*, **75**, 555–562.

(1976b) Ultramafic lavas in Munro Township: economic and tectonic implications. In: D. F. Strong (ed.) *Metallogeny and Plate Tectonics*, pp. 617–658. Geological Association of Canada Special Paper 14. Ottawa: Geological Society of Canada.

(1977a) Partitioning of nickel between olivine and ultrabasic and basic komatiite liquids. *Carnegie Institution Washington Yearbook*, **76**, 553–557.

(1977b) Thick, layered periodotite-gabbro lava flows in Munro Township, Ontario. *Canadian Journal of Earth Sciences*, **14**, 2620–2637.

(1977c) Ultrabasic magmas and high-degree melting of the mantle. *Contributions to Mineralogy and Petrology*, **64**, 205–211.

(1982) Proterozoic spinifex-textured basalts of Gilmour Island, Hudson Bay. *Geological Survey of Canada*, Paper 83–1A, 137–142.

(1986a) Differentiation of komatiite flows. *Journal of Petrology*, **27**, 279–303.

(1986b) Komatiites: A dirty window to the Archean mantle. *Terra Cognita*, **6**, 59–66.

(1986c) Spinifex and swirling olivines in a komatiite lava lake, Munro Township, Canada. *Precambrian Research*, **34**, 139–155.

(1991) High Ni in Archean tholeiites. *Tectonophysics*, **187**, 411–420.

(1994) Archean komatiites. In: K. C. Condie (ed.) *Archean Crustal Evolution*, pp. 11–44. Amsterdam: Elsevier.

(1998) Why was flood volcanism on submerged continental platforms so common in the Precambrian? *Precambrian Research*, **97**, 155–164.

(2003) Komatiites, kimberlites and boninites. *Journal of Geophysical Research*, **108**, B6, 2293, doi:10.1029/2002JB002157.

Arndt, N. T., Albarède, F., Cheadle, M., Ginibre, C., Herzberg, C., Jenner, G., Chauvel, C., Lahaye, Y. (1998a) Were komatiites wet? *Geology*, **26**, 739–742.

Arndt, N. T., Albarède, F., Nisbet, E. G. (1997a) Mafic and ultramafic magmatism. In: M. J. de Wit and L. D. Ashwal (eds.) *Greenstone Belts*, pp. 233–254. New York: Oxford University Press.

Arndt, N. T., Brooks, C. (1980) Komatiites: Penrose conference report. *Geology*, **8**, 155–156.

Arndt, N. T., Brügmann, G. E., Lehnert, K., Chauvel, C., Chappell, B. W. (1987) Geochemistry, petrogenesis and tectonic environment of Circum-Superior Belt basalts, Canada. In: T. C. Pharaoh, R. D. Beckinsale and D. Rickard (eds.) *Geochemistry and Mineralization of Proterozoic Volcanic Suites*. pp. 133–145. Geological Society Special Publication. London: Geological Society.

Arndt, N. T., Bruzak, G., Reischmann, T. (2001) The oldest continental and oceanic plateaus: geochemistry of basalts and komatiites of the Pilbara Craton, Australia. In: R. Ernst & K. L. Buchan (eds) *Mantle Plumes: Their Identification through Time*, pp. 359–388, Geological Society of America, Special Paper, 352. Boulder: Geological Society of America.

Arndt, N. T., Chauvel, C., Fedorenko, V., Czamanske, G. (1998b) Two mantle sources, two plumbing systems: tholeiitic and alkaline magmatism of the Maymecha River basin, Siberian flood volcanic province. *Contributions to Mineralogy and Petrology*, **133**, 297–313.

Arndt, N. T., Fleet, M. E. (1979) Stable and metastable pyroxene crystallization in layered komatiite flows. *American Mineralogist*, **64**, 856–864.

Arndt, N. T., Fowler, A., (2004) Textures in komatiites and variolitic basalts In: K. Eriksson, W. Altermann, D. R. Nelson, W. U. Mueller and O. Catuneau (eds). *The Precambrian Earth: Tempos and Events*, pp. 298–311. Amsterdam: Elsevier.

Arndt, N. T., Francis, D. M., Hynes, A. J. (1979) The field characteristics and petrology of Archaean and Proterozoic komatiite. *Canadian Mineralogist*, **17**, 147–163.

Arndt, N. T., Jenner, G. A. (1985) Kambalda komatiites and basalts: evidence for subduction of sediments in the Archaean mantle. *Terra Cognita*, **5**, 206.

(1986) Crustally contaminated komatiites and basalts from Kambalda, Western Australia. *Chemical Geology*, **56**, 229–255.

Arndt, N. T., Kerr, A. C., Tarney, J. (1997b) Dynamic melting in plume heads: the formation of Gorgona komatiites and basalts. *Earth and Planetary Science Letters*, **146**, 289–301.

Arndt, N. T., Lehnert, K., Vasil'ev, Y. (1995) Meimechites: highly magnesian alkaline magmas from the subcontinental lithosphere? *Lithos*, **34**, 41–59.

Arndt, N. T., Lesher, C. M. (1992) Fractionation of REE by olivine and the origin of Kambalda komatiites, Western Australia. *Geochimica et Cosmochimica Acta*, **56**, 4191–4204.

Arndt, N. T., Lesher, C. M., Czamanske, G. K. (2004a) Mantle-derived magmas and magmatic Ni–Cu–(PGE) deposits. *Economic Geology*, 100th Aniversary Volume, 5–24.

Arndt, N. T., Lesher, C. M., Houlé, M. G., Lewin, E., Lacaze, Y. (2004b) Intrusion and crystallization of a spinifex-textured komatiite sill in Dundonald Township, Ontario. *Journal of Petrology*, **45**, 2555–2571.

Arndt, N. T., Naldrett, A. J., Pyke, D. R. (1977) Komatiitic and iron-rich tholeiitic lavas of Munro Township, northeast Ontario. *Journal of Petrology*, **18**, 319–369.

Arndt, N. T., Nesbitt, R. W. (1982) Geochemistry of Munro Township basalts. In: N. T. Arndt and E. G. Nisbet (eds.) *Komatiites*, pp. 309–330. London: George Allen and Unwin.

(1984) Magma mixing in komatiitic lavas from Munro Township, Ontario. In: A. Kröner, G. N. Hanson and A. M. Goodwin (eds.) *Archaean Geochemistry*, pp. 99–114. Berlin: Springer-Verlag.

Arndt, N. T., Nisbet, E. G. (1982a) *Komatiites*. London: George Allen & Unwin.

(1982b) What is a komatiite? In: N. T. Arndt and E. G. Nisbet (eds.) *Komatiites*, pp. 19–28. London: George Allen and Unwin.

Arndt, N. T., Teixeira, N. A., White, W. M. (1989) Bizarre geochemistry of komatiites from the Crixás greenstone belt, Brazil. *Contributions to Mineralogy and Petrology*, **101**, 187–197.

Arth, J. G., Arndt, N. T., Naldrett, A. J. (1977) Genesis of Archaean komatiites from Munro Township, Ontario; trace-element evidence. *Geology*, **5**, 590–594.

Asahara, Y., Ohtani, E. (2001) Melting relations of the hydrous primitive mantle in the CMAS–H$_2$O system at high pressures and temperatures, and implications for generation of komatiites. *Physics of the Earth and Planetary Interiors*, **125**, 31–44.

Asahara, Y., Ohtani, E., Suzuki, A. (1998) Melting relations of hydrous and dry mantle compositions and the genesis of komatiites. *Geophysical Research, Letters*, **25**, 2201–2204.

Asimov, P. D., Hirschmann, M. M., Ghiorso, M. S., O'Hara, M. J., Stolper, E. M. (1995) The effect of pressure-induced solid–solid phase transitions on decompression melting of the mantle. *Geochimica et Cosmochimica Acta*, **59**, 4489–4506.

Asimow, P. D., Hirschmann, M. M., Stolper, E. M. (2001) Calculation of peridotite partial melting from thermodynamic models of minerals and melts. IV. Adiabatic decompression and the composition and mean properties of mid-ocean ridge basalts. *Journal of Petrology*, **42**, 963–998.

Ayer, J. A., Amelin, Y., Corfu, F., *et al.* (2002) Evolution of the southern Abitibi greenstone belt based on U–Pb geochronology: autochthonous volcanic construction followed by plutonism, regional deformation and sedimentation. *Precambrian Research*, **115**, 63–95.

Ayer, J. A., Thurston, P. C., Bateman, R., *et al.* (2005a) Overview of results from Greenstone Architecture Project, Discover Abitibi Initiative, *Ontario Geological Survey Open File Report 6154*, 131 pp.

(2005b) Digital compilation of maps and data from the Greenstone Architecture Project, Discover Abitibi Initiative, *Ontario Geological Survey, Miscellaneous Release – Data 155.*

Bailey, E. B., McCallien, W. J. (1953) Serpentine lavas, the Ankara mélange and the Anatolism thrust. *Transactions of the Royal Society of Edinburgh*, **62**, 403–442.

Ballhaus, C., Ryan, C. G., Mernach, T. P., Green, D. H. (1994) The partitioning of Fe, Ni, Cu, Pt and Au between sulfide, metal, and fluid phases: a pilot study. *Geochemica et Cosmochimica Acta*, **58**, 811–826.

Baragar, W. R. A., Lamontagne, C. G. (1980) The Circum-Ungava Belt in eastern Hudson Bay: the geology of the Sleeper Islands and parts of the Ottawa and Belcher Islands. *Current Research, Part A*, Paper 80–1a, 89–94. Ottawa: Geological Survey of Canada.

Baragar, W. R. A., Scoates, R. F. J. (1981) The Circum-Superior Belt: a Proterozoic plate margin? In: A. Kröner (ed.) *Precambrian Plate Tectonics*. Amsterdam: Elsevier.

Barley, M. E. (1986) Incompatible element enrichment in Archaean basalts: a consequence of contamination by older sialic crust rather than mantle heterogeneity. *Geology*, **14**, 947–950.

(1997) The Pilbara Craton. In: M. J. de Wit and L. D. Ashwal (eds.) *Greenstone Belts*, pp. 657–664. Oxford: Oxford Science Publications.

Barley, M. E., Kerrich, R., Reudavy, I., Xie, Q. (2000) Late Archaean Ti-rich, Al-depleted komatiites and komatiitic volcaniclastic rocks from the Murchison Terrane in Western Australia. *Australian Journal of Earth Sciences*, **47**, 873–883.

Barnes, R. G., Lewis, J. C., Gee, R. D. (1973) Archean ultramafic lavas from Mount Clifford. *Geological Survey of Western Australia Annual Report, 1973*, pp. 59–70.

Barnes, Sarah-Jane (1983) A comparative study of olivine and clinopyroxene spinifex flows from Alexo, Abitibi greenstone belt, Canada. *Contributions to Mineralogy and Petrology*, **83**, 293–308.

Barnes, Sarah-Jane, Often, M. (1990) Ti-rich komatiites from northern Norway. *Contributions to Mineralogy and Petrology*, **105**, 42–54.

Barnes, Sarah-Jane, Gorton, M. P., Naldrett, A. J. (1983) A comparative study of olivine and clinopyroxene spinifex flows from Alexo Abitibi greenstone belt, Ontario, Canada. *Contributions to Mineralogy and Petrology*, **83**, 292–308.

Barnes, Sarah-Jane, Lightfoot, P. C. (2005) Formation of magmatic nickel sulfide deposits and processes affecting their copper and platinum group element contents. *Economic Geology*, 100th Anniversary Volume, 179–214.

Barnes, Sarah-Jane, Melezhik, V. A., Sokolov, S. V. (2001) The composition and mode of formation of the Pechenga nickel deposits, Kola Peninsula, northwestern Russia. *Canadian Mineralogist*, **39**, 447–471.

Barnes, Sarah-Jane, Naldrett, A. J. (1986) Variations in platinum group element concentrations in the Alexo mine komatiite, Abitibi greenstone belt, northern Ontario. *Geological Magazine*, **123**, 515–524.

(1987) Fractionation of the platinum-group elements and gold in some komatiites of the Abitibi Greenstone Belt, Northern Ontario. *Economic Geology*, **82**, 165–183.

Barnes, Sarah-Jane, Naldrett, A. J., Gorton, M. P. (1985) The origin of the fractionation of the platinum-group elements in terrestrial magmas. *Chemical Geology*, **53**, 303–323.

Barnes, Sarah-Jane, Picard, C. P. (1993) The behavior of platinum-group elements during partial melting, crystal fractionation, and sulphide segregation: an example from the Cape Smith Fold Belt, Northern Quebec. *Geochimica et Cosmochimica Acta*, **57**, 79–87.

Barnes, Stephen J. (1998) Chromite in komatiites, 1. Magmatic controls on crystallization and composition. *Journal of Petrology*, **39**, 1689–1720.

(2000) Chromite in komatiites, II. Modification during greenschist to mid-amphibolite facies metamorphism. *Journal of Petrology*, **41**, 387–409.

(2004) Komatiites and nickel sulfide ores of the Black Swan area, Yilgarn Craton, Western Australia. 4. Platinum group element distribution in the ores, and genetic implications. *Mineralium Deposita*, **39**, 752–765.

(2006) Komatiite-hosted nickel sulfide deposits: geology, geochemistry, and genesis. *Society of Economic Geologists Special Publication*, **13**, 51–118.

(2007) Cotectic precipitation of olivine and sulfide liquid from komatiite magma, and the origin of komatiite-hosted disseminated nickel sulfide mineralization at Mt Keith and Yakabindie, Western Australia. *Economic Geology*, **102**, 299–304.

Barnes, Stephen, J., Barnes Sarah-Jane (1990) A new interpretation of the Katiniq nickel deposit, Ungava, Northern Quebec. *Economic Geology*, **85**, 1269–1272.

Barnes, Stephen, J., Coats, C. J. A., Naldrett, A. J. (1982) Petrogenesis of a Proterozoic nickel sulphide – komatiite association: the Katiniq Sill, Ungava, Quebec. *Economic Geology*, **77**, 413–429.

Barnes, Stephen, J., Hill, R. E. T. (1995) Poikilitic chromite in komatiitic cumulates. *Mineralogy and Petrology*, **54**, 85–92.

Barnes, Stephen, J., Hill, R. E. T. (2000) Metamorphism of komatiite-hosted nickel sulfide deposits. In P. G. Spry, B. Marshall and F. M. Vokes (eds.) *Metamorphosed and Metamorphogenic Ore Deposits*, pp. 203–216. Boulder: Society of Economic Geologists.

Barnes, Stephen, J., Hill, R. E. T., Evans, N. J. (2004a) Komatiites and nickel sulfide ores of the Black Swan area, Yilgarn Craton, Western Australia. 3. Komatiite geochemistry, and implications for ore forming processes. *Mineralium Deposita*, **39**, 729–751.

Barnes, Stephen, J., Hill, R. E. T., Gole, M. J. (1988) The Perseverance ultramafic complex, Western Australia: product of a komatiite lava river. *Journal of Petrology*, **29**, 305–331.

Barnes, Stephen, J., Hill, R. E. T., Perring, C. S., Dowling, S. E. (2004) Lithogeochemical exploration for komatiite-associated Ni-sulfide deposits: strategies and limitations. *Mineralogy and Petrology*, **82**, 259–293.

(1999) Komatiite flow fields and associated Ni-sulphide mineralisation with examples from the Yilgarn Block, Western Australia. In: R. R. Keays, C. M. Lesher, P. C. Lightfoot and C. E. G. Farrow (eds.) *Dynamic Processes in Magmatic Ore Deposits and their Application in Mineral Exploration*, pp. 159–194, Geological Association of Canada, Short Course 13. Ottawa: Geological Association of Canada.

Barnes, Stephen J., Lesher, C. M., Keays, R. R. (1995) Geochemistry of mineralised and barren komatiites from the Perseverance Nickel Deposit, Western Australia. *Lithos*, **34**, 209–234.

Barnes, Stephen, J., Lesher, C. M., Sproule, R. A. (2007) Geochemistry of komatiites in the Eastern Goldfields Superterrane, Western Australia, and the Abitibi Greenstone Belt, Canada, and implications for the distribution of associated Ni–Cu–PGE deposits. *Applied Earth Science (Transactions of Mining and Metallurgy Series B)* in press.

Barnes, Stephen, J., Roeder, P. L. (2001) The range of spinel compositions in terrestrial mafic and ultramafic rocks. *Journal of Petrology*, **42**, 2279–2302.

Barrett, F. M., Binns, R. A., Groves, D. I. (1977) Structural history and metamorphic modification of Archaean volcanic-type nickel deposits. *Economic Geology*, **72**, 1195–1223.

Barrie, T. C. (1999) Komatiite flows of the Kidd Creek footwall, Abitibi subprovince, Canada. *Economic Geology Monograph*, **10**, 143–162.

Barrière, M. (1976) Flowage differentiation: limitation of the Bagnold effect to the narrow intrusions. *Contributions to Mineralogy and Petrology*, **55**, 139–145.

Bavinton, O. A. (1981) The nature of sulfidic metasediments at Kambalda and their broad relationships with associated ultramafic rocks and nickel. *Research Economic Geology*, **76**, 1606–1628.

Beattie, P. (1993) Olivine-melt and orthopyroxene-melt equilibria. *Contributions to Mineralogy and Petrology*, **115**, 103–111.

Beattie, P., Ford, C., Russell, D. (1991) Partition coefficients for olivine melt and orthopyroxene-melt. *Contributions to Mineralogy and Petrology*, **109**, 212–224.

Beaty, D. W., Taylor, H. P., Jr. (1982) The oxygen isotope geochemistry of komatiites: Evidence for water-rock interaction. In: N. T. Arndt and E. G. Nisbet (eds.) *Komatiites*, pp. 267–280. London: George Allen and Unwin.

Bédard, J. H. (2005) Partitioning coefficients between olivine and silicate melts. *Lithos*, **83**, 394–419.

Bell, K., MacDonald, R. (1982) Geochronological calibration of the Precambrian Shield in Saskatchewan. In: *Summary of Investigations 1982*, Saskatchewan Geological Survey. pp. 17–22.

Benn, K., Mareschal, J.-C., Condie, K. C. (2006) *Archean Geodynamics and Environments*. Geophysical Monograph Series, American Geophysical Union, 164. Washington: American Geophysical Union.

Beresford, S., Cas, R., Lahaye, Y., Jane, M. (2002) Facies architecture of an Archean komatiite-hosted Ni-sulphide ore deposit, Victor, Kambalda, Western Australia: implications for komatiite lava emplacement. *Journal of Volcanology and Geothermal Research*, **118**, 57–75.

Beresford, S., Stone, W. E., Cas, R., Lahaye, Y., Jane, M. (2005) Volcanological controls on the localization of the komatiite-hosted Ni–Cu–(PGE) Coronet deposit Kambalda, Western Australia. *Economic Geology*, **100**, 1457–1467.

Beresford, S. W., Cas, R. A. F. (2001) Komatiitic invasive lava flows, Kambalda, Western Australia. *Canadian Mineralogist*, **39**, 525–535.

Beresford, S. W., Cas, R. A. F., Lambert, D. D., Stone, W. E. (2000) Vesicles in thick komatiite lava flows, Kambalda, Western Australia. *Journal of the Geological Society*, **157**, 11–14.

Berry, L. G. (1940) Geology of the Langmuir–Sheraton area. *Ontario Department of Mines, Annual Report*, 49. Toronto: Ontario Department of Mines.

Beswick, A. E. (1982) Some geochemical aspects of alteration and genetic relations in komatiitic suites. In: N. T. Arndt and E. G. Nisbet (eds.) *Komatiites*. pp. 281–308. London: George Allen and Unwin.

(1983) Primary fractionation and secondary alteration within an Archaean ultramafic lava flow. *Contributions to Mineralogy and Petrology*, **82**, 221–231.

Bickle, M. J. (1978) Heat loss from the Earth: a constraint on Archaean tectonics from the relation between geothermal gradients and the rate of heat production. *Earth and Planetary Science Letters*, **40**, 301–315.

(1982) The magnesium contents of komatiitic liquids. In: N. T. Arndt and E. G. Nisbet (eds.) *Komatiites*, pp. 479–494. London: George Allen and Unwin.

(1986) Implications of melting for stabilisation of the lithosphere and heat loss in the Archaean. *Earth and Planetary Science Letters*, **80**, 314–324.

Bickle, M. J., Arndt, N. T., Nisbet, E. G., *et al.* (1993) Geochemistry of the igneous rocks of the Belingwe greenstone belt: alteration, contamination and petrogenesis. In: M. J. Bickle and E. G. Nisbet (eds.) *The geology of the Belingwe Greenstone Belt, Zimbabwe*, pp. 175–214. Rotterdam: Balkema.

Bickle, M. J., Ford, C. E., Nisbet, E. G. (1977) The petrogenesis of peridotitic komatiites; evidence from high-pressure melting experiments. *Earth and Planetary Science Letters*, **37**, 97–106.

Bickle, M. J., Martin, A., Nisbet, E. G. (1975) Basaltic and peridotitic komatiites and stromatolites above a basal unconformity in the Belingwe greenstone belt, Rhodesia. *Earth and Planetary Science Letters*, **27**, 155–162.

Bickle, M. J., Nisbet, E. G. (1993) *The Geology of the Belingwe Greenstone Belt, Zimbabwe*, pp. 239. Rotterdam: Balkema.

Bickle, M. J., Nisbet, E. G., Martin, A. (1994) Archean greenstone belts are not oceanic crust. *Journal of Geology*, **102**, 121–138.

Binns, R. A., Groves, D. I. (1976) Iron–nickel partition in metamorphosed olivine–sulphide assemblages from Perseverance, Western Australia. *American Mineralogist*, **61**, 782–787.

Binns, R. A., Groves, D. I., Gunthorpe, R. J. (1977) Nickel sulphides in Archaean ultramafic rocks of Western Australia. In: A. V. Siderenko (ed.), *Correlation of the Precambrian*, vol. 2, pp. 369–380. Moscow: Nauka.

Binns, R. A., Gunthorpe, R. J., Groves, D. I. (1976) Metamorphic patterns and development of greenstone belts in the eastern Yilgarn Block, Western Australia. In: B. F. Windley (ed.) *The Early History of the Earth*, pp. 303–313. New York: John Wiley.

Bizzarro, M., Simonetti, A., Stevenson, R. K., David, J. (2002) Hf isotope evidence for a hidden mantle reservoir. *Geology*, **30**, 771–774.

Blais, S., Auvray, B., Jahn, B. M., Taipale, K. (1987) Processus de fractionnement dans les coulées komatiitiques archéennes: cas des laves à spinifex de la ceinture de roches vertes de Tipasjärvi (Finlande orientale). *Canadian Journal of Earth Sciences*, **24**, 953–966.

Bleacher, R., Creeley, R. (2003) Shield volcano slope distributions: an approach for characterizing Martian volcanic provinces. *Proceedings of Lunar and Planetary Science Conference XXXIV*, abstract 1794.pdf.

Bleeker, W. (1990) New structural-metamorphic constraints on Early Proterozoic oblique collision along the Thompson Nickel Belt, northern Manitoba, Canada. In: J. F. Lewry and M. R. Stauffer (eds.) *The Early Proterozoic Trans-Hudson Orogen of North America*, pp. 57–94, Geological Association of Canada Special Paper. Ottawa: Geological Association of Canada.

Bleeker, W. (1999) Structure, stratigraphy and primary setting of the Kidd Creek volcanogenic massive sulfide deposit: a semiquantitative reconstruction. In: *The Giant Kidd Creek Volcanogenic Massive Sulfide Deposit, Western Abitibi Subprovince, Canada*. pp 71–122, Economic Geology Monograph 10.

Bleeker, W., Stern, R., Sircombe, K. (2000) Why the Slave Province, Northwest Territories, got a little bigger. *Geological Survey of Canada, Current Research,* 2000-C2, 9p.

Blichert-Toft, J., Albarède, F. (1998) The Lu–Hf isotope geochemistry of chondrites and the evolution of the mantle-crust system. *Earth and Planetary Science Letters,* **148**, 243–258.

Blichert-Toft, J., Arndt, N. T. (1999) Hf isotope compositions of komatiites. *Earth and Planetary Science Letters,* **171**, 439–451.

Blichert-Toft, J., Arndt, N. T., Gruau, G. (2004) Hf isotopic measurements on Barberton komatiites: effects of incomplete sample dissolution and importance for primary and secondary magmatic signatures. *Chemical Geology,* **207**, 261–275.

Blichert-Toft, J., Boyet, M., Telouk, P., Albarede, F. (2002) $^{147}Sm/^{143}Nd$ and 176Lu/176Hf in eucrites and the differentiation of the HED parent body. *Earth and Planetary Science Letters,* **204**, 167–181.

Bolhar, R., Woodhead, J. D., Hergt, J. M. (2003) Continental setting inferred for emplacement of the 2.9–2.7 Ga Belingwe Greenstone Belt, Zimbabwe. *Geology,* **31**, 295–298.

Bottinga, Y., Weill, D. P. (1970) Densities of liquid silicate systems calculated from partial molar volumes of oxide components. *American Journal of Science,* **269**, 169–182.

(1972) The viscosity of magmatic silicate liquids. A model for calculation. *American Journal of Science,* **272**, 438–475.

Bouquain, S., Arndt, N. T., Hellebrand, E. (2008) Crystallochemistry and origin of pyroxenes in komatiites and lunar basalts. *Contributions to Mineralogy and Petrology,* in preparation.

Bowen, N. L. (1928) *Evolution of the Igneous Rocks,* p. 334. Princeton: Princeton University Press.

Boyet, M., Blichert-Toft, J., Rosing, M., Storey, M., Télouk, P., Albarède, F. (2003) ^{142}Nd evidence for early Earth differentiation. *Earth and Planetary Science Letters,* **214**, 427–442.

Boyet, M., Carlson, R. W. (2005) ^{142}Nd evidence for early (> 4.53 Ga) global differentiation of the silicate earth. *Science,* **309**, 576–581.

Brandl, G., de Wit, M. J. (1997) The Kaapvaal Craton, South Africa. In: M. J. de Wit and L. D. Ashwal (eds.) *Greenstone Belts,* pp. 581–607. Oxford: Oxford Science Publications.

Brandon, A. D., Walker, R. J., Puchtel, I. S., Becker, H., Humayun, M., Revillon, S. (2003) $^{186}Os–^{187}Os$ systematics of Gorgona Island komatiites: implications for early growth of the inner core. *Earth and Planetary Science Letters,* **206**, 411–426.

Brenan, J. M., McDonough, W. F., Dalpé, C. (2003) Experimental constraints on the partitioning of rhenium, and some platinum-group elements between olivine and silicate melt. *Earth and Planetary Science Letters,* **212**, 135–150.

Brenner, T. L., Teixeira, N. A., Olivera, J. A. L., Franke, N. O., Thompson, J. F. H. (1990) The O'Toole nickel deposit, Morro do Ferro greenstone belt, Brazilia. *Ecornomic Geology,* **85**, 904–920.

Brevart, O., Dupré, B., Allègre, C. (1986) Lead–lead age of komatiitic lavas and limitations on the structure and evolution of the Precambrian mantle. *Earth and Planetary Science Letters,* **77**, 293–302.

Brooks, C., Hart, S. R. (1974) On the significance of komatiite. *Geology,* **2**, 107–110.

Brown, M. A. N, Jolly, R. J. H., Stone, W., Coward, M. P. (1999) Nickel ore troughs in Archaean volcanic rocks, Kambalda, Western Australia: indicators of early extension. *Geological Society Special Publications,* **155**, 197–211.

Brügman, G. E., Arndt, N. T., Hofmann, A. W., Tobschall, H. J. (1987) Noble metal abundances in komatiite suites from Alexo, Ontario, and Gorgona Island, Colombia. *Geochimica et Cosmochimica Acta*, **51**, 2159–2170.

Brügmann, G. E., Naldrett, A. J., Duke, J. M. (1989) Platinum-group element distribution in the komatiitic Dumont sill, northwestern Quebec, Canada. *Bulletin of the Geological Society of Finland*, **61**, 23.

Buck, P. S., Vallance, S. A., Perring, C. S., Hill, R. E. T., Barnes, S. J. (1998) Maggie Hays nickel deposit. In: D. A. Berkman and D. H. Mackenzie (eds.) *Geology of Australian and Papua New Guinean Mineral Deposits*, pp. 357–364. Melbourne: Australian Institute of Mining and Metallurgy.

Burnham, O. M., Halden, N., Layton-Matthews, D. *et al.* (2003) *Geology, Stratigraphy, Petrogenesis and Metallogenesis of the Thompson Nickel Belt, Manitoba. Final Report for CAMIRO Project 97E–07.* Sudbury: Mineral Exploration Research Centre.

Burnham, O. M., Lesher C. M., Keays, R. R. (1999) Geochemistry of mafic–ultramafic complexes and associated basalts in the Raglan Block. In C. M. Lesher (ed.) *Komatiitic Peridotite-Hosted Fe–Ni–Cu–(PGE) Sulphide Deposits in the Raglan Area, Cape Smith Belt, New Quebec*, pp. 159–173, Guidebook Series, vol. 2. Mineral Exploration Research, Centre. Sudbury, ON: Laurentian University.

Burt, D. R. L., Sheppy, N. R. (1975) *Mount Keith Nickel Deposit.* Melbourne: Australian Institute of Mining and Metallurgy.

Butler, J. C. (1986) The role of spurious correlation in the development of a komatiite alteration model. *Journal of Geophysical Research*, **91**, E275–E280.

Byerly, G. R. (1999) Komatiites of the Mendon Formation: late-stage ultramafic volcanism in the Barberton greenstone belt. In: D. R. Lowe and G. R. Byerly (eds.) *Geological Evolution of the Barberton Greenstone Belt*, pp. 189–212, Geological Society of America, Special Paper 329. Boulder: Geological Society of America.

Byerly, G. R., Kröner, A., Lowe, D. R., Todt, W., Walsh, M. M. (1996) Prolonged magmatism and time constraints for sediment deposition in the early Archean Barberton greenstone belt: evidence from the Upper Onverwacht and Fig Tree groups. *Precambrian Geology*, **78**, 125–138.

Cameron, W. E., Nisbet, E. G. (1982) Phanerozoic analogues of komatiitic basalts. In: N. T. Arndt and E. G. Nisbet (eds.) *Komatiites*, pp. 29–50. London: George Allen and Unwin.

Campbell, I. H. (1978) Some problems with the cumulus theory. *Lithos*, **11**, 311–323.

(1987) Distribution of orthocumulate textures in the Jimberlana Intrusion. *Journal of Geology*, **95**, 35–54.

Campbell, I. H., Arndt, N. T. (1982) Pyroxene accumulation in spinifex-textured rocks. *Geological Magazine*, **119**, 605–610.

Campbell, I. H., Naldrett, A. J. (1979) The influence of silicate:sulphide ratios on the geochemistry of magmatic sulphides. *Economic Geology*, **74**, 1503–1505.

Campbell, I. H., Roeder, P. L., Dixon, J. M. (1978) Crystal buoyancy in basaltic liquids and other experiments with a centrifuge furnace. *Contributions to Mineralogy and Petrology*, **67**, 369–377.

Canil, D. (1997) Vanadium partitioning and the oxidation state of Archaean komatiite magmas. *Nature*, **389**, 842–845.

Capdevila, R., Arndt, N. T., Letendre, J., Sauvage, J. F. (1999) Diamonds in volcaniclastic komatiite from French Guiana. *Nature*, **399**, 456–458.

Card, K. D. (1990) A review of the Superior Province of the Canadian Shield, a product of Archean accretion. *Precambrian Research*, **48**, 99–156.

Caro, G., Bourdon, B., Birck, J. L., Moorbath, S. (2003) ^{146}Sm–^{142}Nd evidence from Isua metamorphosed sediments for early differentiation of the Earth's mantle. *Nature*, **423**, 428–432.

Carslaw, H. S., Jaeger, J. C. (1959) *Conduction of Heat in Solids*. New York: Oxford University Press.

Cas, R., Self, S., Beresford, S. (1999) The behaviour of the fronts of komatiite lavas in medial to distal settings. *Earth and Planetary Science Letters*, **172**, 127–139.

Cas, R. A. F., Beresford, S. W. (2001) Field characteristics and erosional processes associated with komatiitic lavas: implications for flow behavior. *Canadian Mineralogist*, **39**, 505–524.

Cattell, A. (1987) Enriched komatiitic basalts from Newton Township, Ontario: their genesis by crustal contamination of depleted komatiite magma. *Geological Magazine*, **124**, 303–309.

Cattell, A., Arndt, N. T. (1987) Low- and high-alumina komatiites from a Late Archean sequence, Newton Township, Ontario. *Contributions to Mineralogy and Petrology*, **97**, 218–227.

Cattell, A., Krogh, T. E., Arndt, N. T. (1984) Conflicting Sm–Nd and U–Pb zircon ages for Archean lavas from Newton Township, Abitibi Belt, Ontario. *Earth and Planetary Science Letters*, **70**, 280–290.

Cattell, A. C., Taylor, R. N. (1990) Archaean basic magmas. In: R. P. Hall and D. J. Hughes (eds.) *Early Precambrian Basic Magmatism*, pp. 11–39. Glasgow: Blackie.

Cawthorn, R. G., Strong, D. F. (1974) The petrogenesis of komatiites and related rocks as evidence for a layered upper mantle. *Earth and Planetary Science Letters*, **23**, 369–375.

Chauvel, C., Dupré, B., Arndt, N. T. (1993) Pb and Nd isotopic correlation in Belingwe komatiites and basalts. In: M. J. Bickle and E. G. Nisbet (eds.) *The Geology of the Belingwe Greenstone Belt, Zimbabwe*, pp. 167–174. Rotterdam: Balkema.

Chauvel, C., Dupre, B., Jenner, G. A. (1985) The Sm–Nd age of Kambalda volcanics is 500 Ma too old! *Earth and Planetary Science Letters*, **74**, 315–324.

Chavagnac, V. (2004) A geochemical and Nd isotopic study of Barberton komatiites (South Africa): implication for Archean mantle. *Lithos*, **75**, 253–281.

Choukroune, P., Ludden, J. N., Chardon, D., Calvert, A. J., Bouhallier, H. (1997) Archaean crustal growth and tectonic processes: a comparison of the Superior Province, Canada and the Dharwar Craton, India. In: J. P. Burg, M. Ford (eds.) *Orogeny Through Time*, pp. 63–98. Geological Society, Special Publication, No. 121. London: Geological Society.

Chown, E. H., Daigneault, R., Mueller, W., Mortensen, J. K. (1992) Tectonic evolution of Northern Volcanic Zone, Abitibi Belt, Quebec. *Canadian Journal of Earth Science*, **29**, 2211–2225.

Clague, D. A., Uto, K., Satake, K., Davis, A. S. (2002) Eruption style and flow emplacement in the submarine North Arch Volcanic Field, Hawaii. *Geophysical Monograph*, **128**, 65–84.

Claoué-Long, J. C., Compston, W., Cowden, A. (1988) The age of the Kambalda greenstones resolved by ion-microprobe: implications for Archaean dating methods. *Earth and Planetary Science Letters*, **89**, 239–259.

Claoué-Long, J. C., Thirlwall, M. F., Nesbitt, R. W. (1984) Revised Sm–Nd systematics of Kambalda greenstones, Western Australia. *Nature*, **307**, 697–701.

Clarke, D. B. (1970) Tertiary basalts of Baffin Bay: possible primary magma from the mantle. *Contributions to Mineralogy and Petrology*, **25**, 203–224.

Cloete, M. (1994) Aspects of the volcanism and metamorphism of the Onverwacht Group lavas in the south-western portion of the Barberton Greenstone Belt. Ph.D. thesis, University of the Witwatersrand.

(1999) Aspects of volcanism and metamorphism of the Onverwacht Group lavas in the southwestern portion of the Barberton Greenstone Belt. *Memoire of the Geological Survey of South Africa*, **84**, 1–229.

Coad, P. (1977) The Potter Mine – a komatiite-related exhalative Cu–Zn sulfide deposit. M.Sc. thesis, University of Toronto.

(1979) Nickel sulphide deposits associated with ultramafic rocks of the Abitibi Belt and economic potential of mafic-ultramafic intrusions. *Ontario Geological Survey Study*, **20**, 84 pp.

Coffin, M. F., Eldholm, O. (1993) Scratching the surface: estimating dimensions of large igneous provinces. *Geology*, **21**, 515–518.

(2000) Large igneous provinces and plate tectonics. *Geophysical Monograph*, **121**, 309–326.

Collerson, K. D., Jesseau, C. W., Bridgwater, D. (1976) Contrasting types of bladed olivine in ultramafic rocks from the Archean of Labrador. *Canadian Journal of Earth Sciences*, **13**, 442–450.

Compston, W., Williams, I. S., Campbell, I. H., Gresham, J. J. (1986) Zircon xenocrysts from the Kambalda volcanics: age constraints and direct evidence for older continental crust below the Kambalda–Norseman greenstones. *Earth and Planetary Science Letters*, **76**, 299–311.

Condie, K. C. (1984) Secular variation in the composition of basalts: an index to mantle evolution. *Journal of Petrology*, **26**, 545–563.

(1990) Geochemical characteristics of Precambrian basaltic greenstones. In: R. P. Hall and R. N. Hughes (eds.) *Early Precambrian Basic Magmatism*, pp. 40–55. Glasgow: Blackie.

(1994) Greenstones through time. In: K. C. Condie (ed.) *Archean Crustal Evolution*, pp. 85–120. Amsterdam: Elsevier.

Costa, C. N., Ferreira-Filho, C. F., Osborne, G. A., Araujo, S. M., Lopes, R. O. (1997) Geology and geochemistry of the Boa Vista Nickel Sulfide Deposit, Crixas Greenstone Belt, Central Brazil. *Revista Brasileira de Geociencias*, **27**, 365–376.

Cotterill, P. (1969) The chromite deposits of Selukwe, Rhodesia. In: *Magmatic Ore Deposits, a Symposium*, pp. 154–186. Society of Economic Geologists Monograph 4. Littleton Co.: Society of Economic Geologists.

Courtillot, V., Jaupart, C., Manighetti, I., Tapponier, P., Besse, J. (1999) On causal links between flood basalts and continental breakup. *Earth and Planetary Science Letters*, **166**, 177–195.

Cowden, A. (1988) Emplacement of komatiite lava flows and associated nickel sulphides at Kambalda, Western Australia. *Economic Geology*, **83**, 436–442.

Cowden, A., Archibald, N. J. (1987) Massive-sulfide fabrics at Kambalda and their relevance to the inferred stability of monosulfide solid-solution. *Canadian Mineralogist*, **25**, 37–50.

Cowden, A., Donaldson, M. J., Naldrett, A. J., Campbell, I. H. (1986) Platinum group elements and gold in the komatiite-hosted Fe–Ni–Cu sulphide deposits at Kambalda, Western Australia. *Economic Geology*, **81**, 1226–1235.

Cowden, A., Roberts, D. E. (1990) Komatiite-hosted nickel sulphide deposits, Kambalda. In: F. E. Hughes (ed.) *Geology of the Mineral Deposits of Australia and Papua New Guinea*, pp. 567–581. Melbourne: Australian Institute of Mining and Metallurgy.

Cox, K. G. (1980) A model for continental flood vulcanism. *Journal of Petrology*, **21**, 629–650.

(1988) The Karoo Province. In: J. D. Macdougall (ed.) *Continental Flood Basalts*, pp. 239–272. Dordrecht: Kluwer.

Cox, K. G., Duncan, A. R., Bristow, J. W., Taylor, S. R., Erlank, A. J. (1984) Petrogenesis of the basic rocks of the Lebombo. *Special Publication of the Geological Society of South Africa*, **13**, 149–169.

Creaser, R. A., Papanastassiou, D. A., Wasserburg, G. J. (1991) Isotopic analysis of Os and Re with negative thermal ion mass spectrometry and application to the age and evolution of iron meteorites. *Meteoritics*, **27**, 212.

Crocket, J. H. (2002) Platinum-group element geochemistry of mafic and ultramafic rocks. In Cabri, L. J. Cabri (ed.), *The Geology, Geochemistry, Mineralogy, and Mineral Beneficiation of the Platinum-Group Elements*, Canadian Institute of Mining, Metallurgy and Petroleum, Special Volume 54, pp. 177–210.

Crocket, J. H., Leng, D. M., Good, D. J., Stone, W. E., Stone, M. S. (2005) The spinifex layer of the Boston Creek ferropicrite, Abitibi Belt, Ontario: mineralogical and geochemical evidence for an unusual history of clinopyroxene growth and magma recharge. *Canadian Mineralogist*, **43**, 1759–1780.

Crockett, J. H., MacRae, W. E. (1986) Platinum-group element distribution in komatiitic and tholeiitic volcanic rocks from Munro Township. *Economic Geology*, **81**, 1242–1251.

Dann, J. C. (2000) The Komati Formation, Barberton Greenstone Belt, South Africa, part I: new map and magmatic architecture. *South African Journal of Earth Sciences*, **6**, 681–730.

(2001) Vesicular komatiites, 3.5-Ga Komati Formation, Barberton Greenstone Belt, South Africa: inflation of submarine lavas and origin of spinifex zones. *Bulletin of Volcanology*, **63**, 462–481.

Danyushevsky, L. V., Gee, M. A. M., Nisbet, E. G., Cheadle, M. J. (2002a) Olivine-hosted melt inclusions in Belingwe komatiites: implications for cooling history, parental magma composition and its H_2O content. *Geochemica et Cosmochimica Acta*, **15A**, A168.

Danyushevsky, L. V., Sokolov, S., Falloon, T. J. (2002b) Melt inclusions in olivine phenocrysts: using diffusive re-equilibration to determine the cooling history of a crystal, with implications for the origin of olivine–phyric volcanic rocks. *Journal of Petrology*, **43**, 1651–1671.

Davis, B. T. C., England, J. L. (1964) The melting of forsterite up to 50 kbar. *Journal of Geophysical Research*, **69**, 1113–1116.

Davis, P. C. (1997) Volcanic stratigraphy of the Late Archean Kidd–Munro assemblage in Dundonald and Munro Townships and genesis of associated nickel and copper-zinc deposits, Abitibi Greenstone Belt, Dundonald and Munro Townships, Ontario. MSc thesis. University of Alabama.

(1999) *Classic Komatiite Localities and Magmatic Fe–Ni–Cu–(PGE) Sulphide Deposits of the Abitibi Greenstone Belt, Ontario-Quebec.* Guidebook Series vol. 1, Mineral Exploration Research Centre Sudbury, ON: Laurentian University.

de Wit, M. J., Ashwal, L. D. (1997) *Greenstone Belts*. Oxford: Oxford Scientific Publishers.

de Wit, M. J., Hart, R. A., Hart, R. J. (1987) The Jamestown ophiolite complex, Barberton mountain belt: a section through 3.5 Ga oceanic crust. *Journal of African Earth Sciences*, **6**, 681–730.

de Wit, M. J., Hart, R., Pyle, D. (1983) Mg-metasomatism of the oceanic crust; Implications for the formation of ultramafic rock types. *EOS (Transactions, American Geophysical Union)*, **64**, 333.

de Wit, M. J., Roering, C., Hart, R. J., *et al.* (1992) Formation of an Archaean continent. *Nature*, **357**, 553–562.

de Wit, M. J., Stern, C. R. (1980) A 3500 Ma ophiolite complex from the Barberton Greenstone Belt, South Africa: Archaean oceanic crust and its geotectonic implications. In: *Abstract, 2nd International Archaean Symposium, Perth*, pp. 85–87. Perth: Geological Society of Australia.

De-Vitry, C., Libby, J. W., Langworthy, P. J. (1998) Rocky's Reward Nickel Deposit. In: D. A. Berkman and D. H. Mackenzie (eds.), *Geology of Australian and Papua New Guinean Mineral Deposits*, pp. 315–320. Melbourne: Australian Institute of Mining and Metallurgy.

DePaolo, D. J. (1983) The mean life of continents: estimates of continent recycling rates from Nd and Hf isotopic data and implications for mantle structure. *Geophysical Research Letters*, **10**, 705–708.

DePaolo, D. J., Wasserburg, G. J., (1979) Sm–Nd age of the Stillwater Complex and the mantle evolution curve for neodymium. *Geochimica et Cosmochica Acta*, **43**, 999–1008.

Dick, H. J. B., Bullen, T. (1984) Chromian spinel as a petrogenetic indicator in abyssal and alpine-type peridotites and spatially associated lavas. *Contributions to Mineralogy and Petrology*, **86**, 54–76.

Dimroth, E., Cousineau, P., Leduc, M., Sanschagrin, Y. (1978) Structure and organization of Archean subaqueous basalt flows, Rouyn–Noranda area, Quebec. *Canadian Journal of Earth Sciences*, **15**, 902–918.

Dimroth, E., Imreh, L., Cousineau, P., Leduc, M., Sanschagrin, Y. (1985) Paleographic analysis of mafic submarine flows and its use in the exploration for massive sulphide deposits. In: L. D. Ayres, P. C. Thurston, K. D. Card and W. Weber (eds.) *Evolution of Archean Supracrustal Sequences*, pp. 203–222, Geological Association of Canada Special Paper. Ottawa: Geological Society of Canada.

Dixon, J. E., Stolper, E. M. (1995) An experimental study of water and carbon dioxide solubilities in mid-ocean ridge basaltic liquids. Part II. Applications to degassing. *Journal of Petrology*, **36**, 1633–1646.

Donaldson, C. H. (1974) Olivine crystal types in harrisitic rocks of the Rhum pluton and Archean spinifex rocks. *Bulletin of the Geological Society of America*, **85**, 1721–1726.

(1976) An experimental study of olivine morphology. *Contributions to Mineralogy and Petrology*, **57**, 187–213.

(1979) An experimental investigation of the delay in nucleation of olivine in mafic lavas. *Contributions to Mineralogy and Petrology*, **69**, 21–32.

(1982) Spinifex-textured komatiites: a review of textures, mineral compositions, and layering. In: N. T. Arndt and E. G. Nisbet (eds.) *Komatiites*, pp. 211–244. London: George Allen and Unwin.

Donaldson, J. A., de Kemp, E. A. (1998) Archaean quartz arenites in the Canadian Shield; examples from the Superior and Churchill provinces. *Sedimentary Geology*, **120**, 153–176.

Donaldson, M. J. (1981) Redistribution of ore elements during serpentinisation and talc-carbonate alteration of some Archaean dunites, Western Australia. *Economic Geology*, **76**, 1968–1713.

Dostal, J., Mueller, W. U. (1997) Komatiite flooding of a rifted Archean rhyolitic arc complex: geochemical signature and tectonic significance of the Stoughton-Roquemaure Group, Abitibi greenstone belt, Canada. *Journal of Geology*, **105**, 545–563.

(2001) Archean hyaloclastites: fragmentation process and composition of the 2.72 Ga Stoughton–Roquemaure komatiites-komatiitic basalts. (*EOS Transactions of the American Geophysical Union*), Fall Meeting, 632.

Dowling, S. E., Barnes, Stephen J., Hill, R. E. T., Hicks, J. (2004) Komatiites and nickel sulfide ores of the Black Swan area, Yilgarn Craton, Western Australia. 2. Geology and genesis of the orebodies. *Mineralium Deposita*, **39**, 707–728.

Dowling, S. E., Hill, R. E. T. (1993) The Mount Keith ultramafic complex and the Mount Keith nickel deposit. *Australian Geological Survey Organisation Record 1993/54*, 165–170.

(1998) Exploration model – komatiite hosted nickel sulphide deposits, Australia. In the Jubilee volume 'Earth Evolution, Australian Environments and Resources'. *AGSO Journal of Australian Geology and Geophysics*, **17**, 121–127.

Draper, D. S., Xirouchakis, D., Agee, C. B. (2003) Trace element partitioning between garnet and chondritic melt from 5 to 9 GPa: implications for the onset of the majorite transition in the martian mantle. *Physics of the Earth and Planetary Interiors*, **139**, 149–169.

Drever, H. I., Johnston, R. (1957) Crystal growth of forsteritic olivine in magmas and melts. *Transactions of the Royal Society of Edinburgh*, **63**, 289–317.

Duchac, K., Hanor, J. S. (1987) Origin and timing of the metasomatic silicification of an early Archaean komatiite sequence, Barberton Mountain Land, South Africa. *Precambrian Research*, **37**, 125–146.

Duke, J. M. (1986a) Petrology and economic geology of the Dumont Sill: an Archean intrusion of komatiitic affinity in northwestern Quebec. *Geological Survey of Canada, Economic Geology Report*, No. 35.

(1986b) The Dumont nickel deposit: a genetic model for disseminated magmatic sulphide deposits of komatiitic affinity. In: M. J. Gallacher, R. A. Ixer, C. R. Neary, and H. M. Prichard (eds.) *Metallogeny of Basic and Ultrabasic Rocks*, pp. 151–160. London: The Institute of Mining and Metallurgy.

Dupré, B., Arndt, N. T. (1986) Pb isotopic compositions of Archean komatiites and sulfides. *Chemical Geology*, **85**, 35–56.

Dupré, B., Chauvel, C., Arndt, N. T. (1984) Pb and Nd isotopic study of two Archean komatiitic flows from Alexo, Ontario. *Geochimica et Cosmochimica Acta*, **48**, 1965–1972.

Dupré, B., Echeverría, L. M. (1984) Pb isotopes of Gorgona Island (Colombia): isotopic variations correlated with magma type. *Earth and Planetary Science Letters*, **67**, 186–190.

Ebel, D. S., Naldrett, A. J. (1996) Fractional crystallization of sulfide ore liquids at high temperature. *Economic Geology*, **91**, 607–621.

Ebinger, C. J., Sleep, N. H. (1998) Cenozoic magmatism throughout East Africa resulting from impact of a single plume. *Nature*, **395**, 1788–1791.

Echeverría, L. M. (1980) Tertiary or Mesozoic komatiites from Gorgona Island, Colombia; field relations and geochemistry. *Contributions to Mineralogy and Petrology*, **73**, 253–266.

(1982) Komatiites from Gorgona Island, Colombia. In: N. T. Arndt and E. G. Nisbet (eds.) *Komatiites*, pp. 199–210. London: George Allen & Unwin.

Echeverría, L. M., Aitken, B. (1980) Pyroclastic rocks: another manifestation of ultramafic volcanism of Gorgona Island, Colombia. *Contributions to Mineralogy and Petrology*, **92**, 428–436.

Eckstrand, O. R. (1972) Ultramafic flows and nickel deposits in the Abitibi orogenic belt. In: *Report of Activities. Part A: April to October, 1971*, pp. 78–81, Canadian Geological Survey Paper 72-1. Ottawa: Canadian Geological Survey.

(1975) The Dumont serpentinite: a model for control of nickeliferous opaque assemblages by alteration products in ultramafic rocks. *Economic Geology*, **70**, 83–201.

Eckstrand, O. R., Williamson, B. L. (1985) Vesicles in the Dundonald komatiites. *Program and Abstracts Geological Association of Canada–Mineralogical Association of Canada Annual Meeting*, **10**, A-16.

Elias, M. (2006) Lateritic nickel mineralization of the Yilgarn Craton. *Society of Economic Geologists Special Publication*, **13**, 195–210.

Elkins-Tanton, L. T., Draper, D., Agee, C., Jewell, J., Thorpe, A., Hess, P. (2006) The last lavas erupted during the main phase of the Siberian flood basalts: Results from experimental petrology. *Contributions to Mineralogy and Petrology* doi:10.1007/s00410-006-0140-1.

Elliot, D. H., Fleming, T. H., Kyle, P. R., Foland, K. A. (1999) Long-distance transport of magmas in the Jurassic Ferrar Large Igneous Province, Antarctica. *Earth and Planetary Science Letters*, **167**, 89–104.

Erlank, A. J. (1984) *Petrogenesis of the Volcanic Rocks of the Karoo Province*. Geological Society of South Africa Special Publication 13. Pretoria: Geological Society of South Africa.

Ernst, R. E., Baragar, W. R. A. (1992) Evidence from magnetic fabric for the flow pattern of magma in the Mackenzie giant radiating dyke swarm. *Nature*, **356**, 511–513.

Evans, D. M., Cowden, A., Barrett, R. M. (1989) Deformation and thermal erosion at the Foster nickel deposit, Kambalda St. Ives, Western Australia. In: M. D. Prendergast and M. J. Jones (eds.) *Magmatic Sulfides – The Zimbabwe Volume*, pp. 215–227. London: Institute of Mining and Metallurgy.

Fan, J., Kerrich, R. (1997) Geochemical characteristics of alumina depleted and undepleted komatiites and HREE-enriched low-Ti tholeiites, Western Abitibi greenstone belt: a heterogeneous mantle plume-convergent margin environment. *Geochemica et Cosmochimica Acta*, **61**, 4723–4744.

Faure, F., Arndt, N. T., Libourel, G. (2006) Formation of spinifex texture in komatiites: an experimental study. *Journal of Petrology*, **47**, 1591–1610.

Faure, F., Trolliard, G., Nicollet, C., Montel, J.-M. (2003) A developmental model of olivine morphology as a function of the cooling rate and the degree of undercooling. *Contributions to Mineralogy and Petrology*, **145**, 251–263.

Fedorenko, V., Czamanske, G. (1997) Results of new field and geochemical studies of the volcanic and intrusive rocks of the Maymecha-Kotuy area, Siberian Flood-Basalt Province, Russia. *International Geological Review*, **39**, 479–531.

Fiorentini, M. L., Rosengren, N., Beresford, S. W., Grguric, B., Barley, M. E. (2007) Controls on the emplacement and genesis of the MKD5 and Sarah's Find Ni–Cu–PGE deposits, Mount Keith, Agnew–Wiluna Greenstone Belt, Western Australia. *Mineralium Deposita*, **126**, 847–877.

Fitton, J. G., Mahoney, J. J., Wallace, P. J., Saunders, A. D. (2004) Origin and evolution of the Ontong Java Plateau. *Geological Society Special Publication*, **229**, 384pp.

Fleet, M. E., MacRae, N. D. (1975) A spinifex rock from Munro Township, Ontario. *Canadian Journal of Earth Sciences*, **12**, 928–939.

Fleet, M. E., Tronnes, R. G., Stone, W. E. (1996) Partitioning of platinum-group elements (Os, Ir, Ru, Pt, Pd) and gold between sulfide liquid and basalt melt. *Geochemica et Cosmochimica Acta*, **60**, 2397–2412.

Foster, J. G., Lambert, D. D., Frick, L. R., Maas, R. (1996) Re–Os isotopic evidence for genesis of Archaean nickel ores from uncontaminated komatiites. *Nature*, **382**, 703–705.

Fowler, A. D., Berger, B., Shore, M., Jones, M. I., Ropchan, J., (2002) Supercooled rocks: development and significance of varioles, spherulites, dendrites and spinifex in Archaean volcanic rocks, Abitibi Greenstone belt, Canada. *Precambrian Research*, **115**, 311–328.

Francis, D. (1995) The implications of picritic lavas for the mantle sources of terrestrial magmatism. *Lithos*, **34**, 89–106.

Francis, D., Ludden, J., Johnstone, R., Davis, W. (1999) Picrite evidence for more Fe in Archean mantle reservoirs. *Earth and Planetary Science Letters*, **167**, 197–213.

Francis, D. M. (1985) The Baffin Bay lavas and the value of picrites as analogues of primary magmas. *Contributions to Mineralogy and Petrology*, **89**, 144–154.

Francis, D. M., Hynes, A. J. (1979) Komatiite-derived tholeiites in the Proterozoic of New Quebec. *Earth and Planetary Science Letters*, **44**, 473–481.

Francis, D. M., Hynes, A. J., Ludden, J. N., Bédard, J. H. (1981) Crystal fractionation and partial melting in the petrogenesis of a Proterozoic high-MgO volcanic suite, Ungava, Quebec. *Contributions to Mineralogy and Petrology*, **78**, 27–36.

Frei, R., Jensen, B. K. (2003) Re–Os, Sm–Nd isotope- and REE systematics on ultramafic rocks and pillow basalts from the Earth's oldest oceanic crustal fragments (Isua Supracrustal Belt and Ujaragssuit Nunât area, W Greenland). *Chemical Geology*, **196**, 163–191.

Frei, R., Rosing, M., Waight, T. E., *et al.* (2002) Hydrothermal-metasomatic and tecto-metamorphic processes in the Isua supracrustal belt (West Greenland), a multi-isotopic investigation of their effects on the Earth's oldest oceanic crustal sequence. *Geochemica et Cosmochimica Acta*, **66**, 467–486.

Frey, F. A., Coffin, M. F., Wallace, P. J., *et al.* (2000) Origin and evolution of a submarine large igneous province: the Kerguelen Plateau and Broken Ridge, southern Indian Ocean. *Earth and Planetary Science Letters*, **176**, 73–89.

Frost, K. M., Groves, D. I. (1989a) Magmatic contacts between immiscible sulfide and komatiite melts; implications for genesis of Kambalda sulfide. *Economic Geology*, **84**, 1697–1704.

(1989b) Ocellar units at Kambalda: evidence for sediment assimilation by komatiite lavas. In: M. D. Prendergast and M. J. Jones (eds.) *Magmatic Sulphides – the Zimbabwe Volume*, pp. 207–214. London: Institution of Mining and Metallurgy.

Fujii, T., Kushiro, I. (1977) Density, viscosity and compressibility of basaltic liquid at high pressure. *Carnergie Institution of Washington Year Book*, **76**, 419–423.

Gaetani, G. A., Grove, T. L. (1997) Partitioning of moderately siderophile elements among olivine, silicate melt, and sulfide melt: constraints on core formation in the Earth and Mars. *Geochimica et Cosmochimica Acta*, **61**, 1829–1846.

Gangopadhyay, A., Sproule, R. A., Walker, R. J., Lesher, C. M. (2005) Re–Os systematics of komatiites and komatiitic basalts at Dundonald Beach, Ontario, Canada: evidence for a complex alteration history and implications of a late-Archean chondriitic mantle source. *Geochemica et Cosmochimica Acta*, **69**, 5087–5098.

Gansser, A. (1950). Geological and petrological notes on Gorgona Island in relation to North–West S. America. *Schweizerische Mineralogische und Petrographische M. Heilungen*, **30**, 219–237.

Gansser, A., Dietrich, V. J., Cameron, W. E. (1979) Palaeogene komatiites from Gorgona Island. *Nature*, **278**, 546.

Gélinas, L., Brooks, C., Trzcienski, W. E. (1977a) Archean variolites quenched immiscible liquids. *Canadian Journal of Earth Sciences*, **13**, 210–230.

Gélinas, L., Lajoie, J., Brooks, C. (1977b) The origin and significance of Archaean ultramafic volcaniclastics from Spinifex Ridge, Lamotte Township, Quebec. *Geological Association Of Canada Special Paper* **16**, 297–309.

Gibson, H. L., Gamble, A. P. D. (2000) A reconstruction of the volcanic environment hosting Archean seafloor and subseafloor VMS mineralization at the Potter Mine, Munro Township, Ontario, Canada. Volcanic environments and massive sulfide deposits, Special Publication 3: Hobart, Tasmania, Centre for Ore Deposit Research, University of Tasmania, pp. 65–66.

Gillies, S. L. (1993) Physical volcanology of the Katinniq Peridotite Complex and associated Fe– Ni–Cu–(PGE) Mineralization, Cape Smith Belt, Northern Quebec. Msc. thesis. Tuscaloosa, University of Alabama.

Ginibre, C., Arndt, N. T., Hallot, E., Lesher, C. E., Cashman, K. V. (1997) An experimental study of spinifex textures in komatiites from Gorgona, Colombia. *Terra Nova*, **9**, 203.

Giovenazzo, D., Picard, C., Guha, J. (1989) Tectonic setting of Ni–Cu–PGE deposits in the central part of the Cape Smith Belt. *Geoscience Canada*, **16**, 134–136.

Gole, M. J., Barnes, Stephen J., Hill, R. E. T. (1987) The role of fluids in the metamorphism of komatiites, Agnew nickel deposit, Western Australia. *Contributions to Mineralogy and Petrology*, **96**, 151–162.

(1989) The Geology of the Agnew Nickel Deposit, Western Australia. Ottawa: *Bulletin of Canadian Institute of Mining and Metallurgy*, **82**, 46–56.

(1990) Partial melting and recrystallization of Archaean komatiites by residual heat from rapidly accumulated flows. *Contributions to Mineralogy and Petrology*, **105**, 704–714.

Goodwin, A. M. (1979) Archean volcanic studies in the Timmins–Kirkland Lake–Noranda region of Ontario and Quebec. *Geological Survey of Canada Bulletin*, 278.

Goutier, J., Dion, C., Lafrance, I., David, J., Parent, M., Dion, D.-J. (1998) *Géologie de la région des Lacs Langelier et Threefold (33F/03 et 33F/04)*. Géologie Québec, RG 98-18. Québec: Ministère de Ressources naturelles.

Goutier, J., Dion, C., Ouellet, M.-C., David, J., Parent, M. (1999) *Géologie de la région des Lacs Guillaumat et Sakami (33F/002 et 33F/07)*. Géologie Québec, RG 99-15. Québec: Ministère de Ressources naturelles.

Greeley, R., Fagents, S. A., Harris, R. S., Kadel, S. D., Williams, D. A., Guest, J. E. (1998) Erosion by flowing lava – field evidence. *Journal of Geophysical Research*, **103**, 27325–27345.

Green, A. H., Melezhik, V. (1999), Ferropicrites, subvolcanic intrusions, and associated Ni–Cu–PGE sulphide mineralization with emphasis on the Pechenga camp. In R. R. Keays, C. M. Lesher, P. C. Lightfoot, and C. E. G. Farrow (eds.), *Dynamic*

Processes in Magmatic Ore Deposits and Their Application in Mineral Exploration, Geological Association of Canada, Short Course Notes, vol. 13, pp. 287–328. Sudbury: Geological Society of Canada.

Green, A. H., Naldrett, A. J. (1981) The Langmuir volcanic peridotite-associated nickel sulphide deposits: Canadian equivalents of the Western Australian occurrences. *Economic Geology*, **76**, 1503–1523.

Green, D. H. (1975) Genesis of Archean peridotitic magmas and constraints on Archean geothermal gradients and tectonics. *Geology*, **3**, 15–18.

(1981) Petrogenesis of Archean ultramafic magmas and implications for Archaean tectonics. In: A. Kröner (ed.) *Precambrian Plate Tectonics*, pp. 469–490. Amsterdam: Elsevier.

Green, D. H., Nicholls, I. A., Viljoen, M. J., Viljoen, R. P. (1975) Experimental demonstration of the existence of peridotitic liquids in earliest Archaean magmatism. *Geology*, **3**, 11–14.

Gresham, J. J., Loftus-Hills, G. D. (1981) The geology of the Kambalda nickel field, Western Australia. *Economic Geology*, **76**, 1373–1416.

Grguric, B. A., Rosengren, N. M., Fletcher, C. M., Hronsky, J. M. A. (2006) Type 2 deposits: geology, mineralogy and processing of the Mount Keith and Yakabindie orebodies, Western Australia. *Society of Economic Geologists Special Publications*, **13**, 119–138.

Grove, T. L., Bence, A. E. (1979) Crystallization kinetics in a multiply saturated basalt magma: An experimental study of Luna 24 ferrobasalt. *Proceedings of the 10th Lunar and Planetary Science Conference*, 439–478.

Grove, T. L., de Wit, M. J., Dann, J. (1997) Komatiites from the Komati type section, Barberton, South Africa. In: M. J. de Wit and L. D. Ashwal (eds.) *Greenstone Belts*, pp. 422–437. Oxford: Oxford Science Publications.

Grove, T. L., Gaetani, G. A., de Wit, M. J. (1994) Spinifex textures in 3.49 Ga Barberton Mountain Belt komatiites: evidence for crystallization of water-bearing, cool magmas in the Archean. *EOS (Transactions, American Geophysical Union)*, **75**, 354.

Grove, T. L., Gaetani, G. A., Parman, S., Dann, J., de Wit, M. J. (1996) Origin of spinifex textures in 3.49 Ga komatiite magmas from the Barberton Mountainland South Africa. *EOS (Transactions, American Geophysical Union)*, **77**, 281.

Grove, T. L., Parman, S. (2004) Thermal evolution of the Earth as recorded by komatiites. *Earth and Planetary Science Letters*, **219**, 173–187.

Grove, T. L., Parman, S. W., Dann, J. C. (1999) Conditions of magma generation for Archean komatiites from the Barberton Mountainland, South Africa. In: Y. Fei, C. M. Bertka and B. O. Mysen (eds.) *Mantle Petrology: Field Observations and High-Pressure Experimentation*, pp. 155–167. Houston: The Geochemical Society.

Groves, D. I., Barrett, F. M., Brotherton, R. H. (1981) Exploration significance of the chrome-spinels in mineralised ultramafic rocks and Ni–Cu cores. In J. R. R. De Villiers and G. A. Cawthorn (eds.) *ICAM 81 – The Proceedings of the First International Conference on Applied Mineralogy*: Geological Survey of South Africa Special Report 7, pp. 21–30. Johannebury: Geological Survey of South Africa.

Groves, D. I., Keays, R. R. (1979) Mobilization of ore forming elements during alteration of dunites, Mt Keith-Betheno, Western Australia. *Canadian Mineralogist*, **17**, 373–389.

Groves, D. I., Korkiakoski, E. A., McNaughton, N. J., Lesher, C. M., Cowden, A. (1986) Thermal erosion by komatiites at Kambalda, Western Australia and the genesis of nickel ores, *Nature*, **319**, 136–139.

Groves, D. I., Lesher, C. M., Gee, R. D. (1984) Tectonic setting of the sulphide nickel deposits of the Western Australian Shield. In D. L. Buchanan and M. J. Jones (eds.) *Sulphide Deposits in Mafic and Ultramafic Rocks.* pp. 1–13. London: Institution of Mining and Metallurgy.

Gruau, G., Chauvel, C., Arndt, N. T., Cornichet, J. (1990a) Aluminium depletion in komatiites and garnet fractionation in the early Archean mantle: hafnium isotopic constraints. *Geochimica et Cosmochimica Acta*, **54**, 3095–3101.

Gruau, G., Chauvel, C., Jahn, B. M. (1990b) Anomalous Sm–Nd ages for the early Archean Onverwacht Group volcanics: Significance and petrogenetic implications. *Contributions to Mineralogy and Petrology*, **104**, 27–34.

Gruau, G., Jahn, B. M., Glikson, A. Y., Davy, R., Hickman, A. H., Chauvel, C. (1987) Age of the Talga-Talga Subgroup, Pilbara Block, Western Australia, and early evolution of the mantle: new Sm–Nd isotopic evidence. *Earth and Planetary Science Letters*, **85**, 105–116.

Gruau, G., Martin, H., Leveque, B., Capdevila, R., Marot, A. (1985) Rb–Sr and Sm–Nd geochronology of Lower Proterozoic granite greenstone terrains in French Guiana, South America. *Precambrian Research*, **30**, 63–80.

Gruau, G., Rosing, M., Bridgwater, D., Gill, R. C. O. (1996) Resetting of the Sm–Nd systematics during metamorphism of >3.7Ga rocks, implications for isotopic models of early Earth differentiation. *Chemical Geology*, **133**, 225–240.

Gruau, G., Tourpin, S., Fourcade, S., Blais, S. (1992) Loss of isotopic (Nd, O) and chemical (REE) memory during metamorphism of komatiites: new evidence from eastern Finland. *Contributions to Mineralogy and Petrology*, **112**, 66–82.

Gudfinnsson, G. H., Presnall, D. C. (2000) Melting behavior of model lherzolite in the system $CaO–MgO–Al_2O_3–SiO_2–FeO$ at 0.7 to 2.8 GPa. *Journal of Petrology*, **41**, 1241–1269.

Gurenko, A. A., Chaussidon, M. (1995) Enriched and depleted primitive melts included in olivine from Icelandic tholeiites: origin by continuous melting of a single mantle column. *Geochimica et Cosmochimica Acta*, **58**, 2905–2917.

Hall, A. L. (1918) The geology of the Barberton gold mining district. *Geological Survey of South Africa Memoire*, **9**, 1–324.

Hallberg, J. A. (1972) Geochemistry of the Archean volcanic rocks in the Eastern Goldfields region of Western Australia. *Journal of Petrology*, **13**, 45–56.

Hallberg, J. A., Williams, D. A. C. (1972). Archaean mafic and ultramafic rock associations in the Eastern Goldfields region, Western Australia. *Earth and Planetary Science Letters*, **15**, No. 2, 191–200.

Hamilton, P. J., Evensen, N. M., O'Nions, R. K., Smith, H. S., Erlank, A. J. (1979a) Sm–Nd dating on the Onverwacht Group volcanics, Southern Africa. *Nature*, **279**, 298–300.

Hamilton, P. J., O'Nions, R. K., Evensen, N. M. (1979b) Sm–Nd dating basic and ultrabasic volcanics. *Earth and Planetary Science Letters*, **36**, 263–268.

Hannington, M. D., Santaguida, F., Kjarsgaard, I. M., Cathles, L. M. (2004) Regional-scale hydrothermal alteration in the Central Blake River Group, western Abitibi subprovince, Canada: implications for VMS prospectivity. *Mineralium Deposita*, **38**, 393–422.

Hanski, E. J. (1980) Komatiitic and tholeiitic metavolcanics of the Siivikkovaara area in the Archean Kuhmo greenstone belt, eastern Finland. *Bulletin of the Geological Survey of Finland*, **52**, 67–100.

(1992) Petrology, of the Pechenga ferropicrites and cogenetic, Ni-bearing gabbro–wehrlite intrusions, Kola Peninsula, Russia. *Bulletin of the Geological Survey of Finland*, **367**, 192p.

(1993) Globular ferropicritic rocks at Pechenga, Kola Peninsula (Russia): liquid immiscibility versus alteration. *Lithos*, **29**, 197–216.

Hanski, E., Huhma, H., Rastas, P., Kamenetsky, V. S. (2001) The Palaeoproterozoic komatiite–picrite association of Finnish Lapland. *Journal of Petrology*, **42**, 855–876.

Hanski, E. Smolkin, V. F. (1989) Pechenga ferropicrites and other early Proterozoic picrites in the eastern part of the Baltic Shield. *Precambrian Research*, **45**, 63–82.

Hanson, G. N., Langmuir, C. H. (1978) Modelling of major elements in mantle-melt systems using a trace element approach. *Geochimica et Cosmochimica Acta*, **42**, 725–741.

Hargreaves, R., Ayres, L. D. (1979) Morphology of Archean metabasalt flows, Utik Lake, Manitoba. *Canadian Journal of Earth Sciences*, **16**, 1452–1466.

Harker, A. (1908) The geology of the small isles of Inverness-shire. *Memoir of the Geological Survey of Scotland*.

Hart, S. R., Brooks, C. (1977) Geochemistry and evolution of early Precambrian mantle. *Contributions to Mineralogy and Petrology*, **61**, 109–128.

Hart, S. R., Davis, K. E. (1978) Ni partitioning between olivine and silicate melt. *Earth and Planetary Science Letters*, **40**, 203–219.

Hauff, F., Hoernle, K., Tilton, G., Graham, D. W., Kerr, A. C. (2000) Large volume recycling of oceanic lithosphere over short time scale: geochemical constraints from the Caribbean Large Igneous Province. *Earth and Planetary Science Letters*, **174**, 247–263.

Hawkesworth, C. J., Moorbath, S., O'Nions, R. K., Wilson, J. F. (1975) Age relationships between greenstone belts and 'granites' in the Rhodesian Archaean craton. *Earth and Planetary Science Letters*, **25**, 251–262.

Hawksworth, C. J., O'Nions, R. K., Pankhurst, R. J., Hamilton, P. J., Evensen, N. M. (1977). Sm–Nd isotopic investigations of Isua supercrustals and implications for mantle evolution. *Earth and Planetary Science Letters*, **36**, 253.

Hazen, R. M. (1977) Effects of temperature and pressure on the crystal structure of ferromagnesian olivine. *American Mineralogist*, **62**, 286–295.

Heath, C., Lahaye, Y., Stone, W. E., Lambert, D. D. (2001) Origin of variations in nickel tenor along the strike of the Edwards lode nickel sulfide orebody, Kambalda, Western Australia. *Canadian Mineralogist*, **39**, 655–671.

Hegner, E., Kröner, A., Hofmann, A. W. (1984) Age and isotope geochemistry of the Archean Pongola and Usushwana suites in Swaziland, southern Africa: a case for crustal contamination of a mantle-derived magma. *Earth and Planetary Science Letters*, **70**, 267–279.

Heliker, C. C., Mangan, M. T., Maltox, T. N., Kauahikaua, J. P., Helz, T. N. (1998) The character of long-term eruptions: inferences from episodes 50–53 of the Pu'u O'o-Kupaianaha eruption of Kilauea volcano. *Bulletin of Volcanology*, **59**, 381–393.

Helz, R. T. (1982) Determination of the $P–T$ dependence of the first appearance of Fe–S liquid in natural basalts to 20 kb. *EOS (Transactions, American Geophysical Union)* **58**, 523.

Hémond, C., Arndt, N. T., Lichtenstein, U., Hofmann, A. W., Oskarsson, N., Steinthorsson, S. (1993) Iceland hotspot: Nd–Sr–O Isotopes and trace element constraints. *Journal of Geophysical Research*, **98**, 15833–15850.

Herzberg, C. (1992) Depth and degree of melting of komatiite. *Journal of Geophysical Research*, **97**, 4521–4540.

(1995) Generation of plume magmas through time: an experimental perspective. *Chemical Geology*, **126**, 1–17.

(1999) Phase equilibrium constraints on the formation of cratonic mantle. In: Y. Fei, C. M. Bertka and B. O. Mysen (eds.) *Mantle Petrology: Field Observations and High-Pressure Experimentation*, pp. 13–46. Houston: The Geochemical Society.

Herzberg, C., O'Hara, M. J. (1998) Phase equilibrium constraints on the origin of basalts, picrites, and komatiites. *Earth-Science Reviews*, **44**, 39–79.

(2002) Plume-associated ultramafic magmas of Phanerozoic age. *Journal of Petrology*, **43**, 1857–1883.

Herzberg, C., Ohtani, E. (1988) Origin of komatiite at high pressure. *Earth and Planetary Science Letters*, **88**, 321–329.

Herzberg, C., Zhang, J. (1996) Melting experiments on anhydrous peridotite KLB-1: compositions of magmas in the upper mantle and transition zone. *Journal of Geophysical Research*, **101**, 8271–8295.

Herzberg, C. T. (1983) Solidus and liquidus temperatures and mineralogies for anhydrous garnet-lherzolite to 15 GPa. *Physics of Earth and Planetary Interiors*, **32**, 193–202.

Herzberg, C. T., O'Hara, M. J. (1985) Origin of mantle peridotite and komatiite by partial melting. *Geophysical Research Letters*, **12**, 541–544.

Hill, R. E. T. (2001) Komatiite volcanology, volcanological setting and primary geochemical properties of komatiite-associated nickel deposits. *Geochemistry: Exploration, Environment, Analysis*, **1**, 365–381.

Hill, R. E. T., Barnes, Stephen J., Dowling, S. E., Thordarson, T. (2004) Komatiites and nickel sulfide orebodies of the Black Swan area, Yilgarn Block, Western Australia. 1. Petrology, and volcanology of host rocks. *Mineralium Deposita*, **39**, 684–706.

Hill, R. E. T., Barnes, Stephen J., Gole, M. J., Dowling, S. E. (1995) The physical volcanology of komatiites as deduced from field relationships in the Norseman–Wiluna greenstone belt, Western Australia. *Lithos*, **34**, 159–188.

Hill, R. E. T., Gole, M. J. (1990) *Nickel Sulphide Deposits of the Yilgarn Block*. In: F. E. Hughes (ed.) Melbourne: Australian Institute of Mining and Metallurgy, pp. 557–559.

Hill, R. E. T., Gole, M. J., Barnes, Stephen J. (1987) *Physical Volcanology of Komatiites*. Excursion Guide No. 1. Perth: Geological Society of Australia.

(1989) Olivine adcumulates in the Norseman–Wiluna greenstone belt, Western Australia: implications for the volcanology of komatiites. In M. D. Prendergast and M. J. Jones (eds.) *Magmatic Sulphides – The Zimbabwe Volume*, pp. 189–206. London: Institute of Mining and Metallurgy.

Hill, R. E. T., Perring, C. S. (1996) *The Evolution of Archaean Komatiite Flow Fields – Are They Inflationary Sheet Flows*. Townsville, Queensland: James Cook University.

Hirose, K. (1997) Melting experiments on lherzolite KLB-1 under hydrous conditions and generation of high-magnesian andesitic melts. *Geology*, **25**, 42–44.

Hirose, K., Kawamoto, T. (1995) Hydrous partial melting of lherzolite at 1 GPa: the effect of H_2O on the genesis of basaltic magmas. *Earth and Planetary Science Letters*, **133**, 463–473.

Hirschmann, M. M., Ghiorso, M. S., Wasylenki, L. E., Asimow, P. D., Stolper, E. M. (1998) Calculation of peridotite partial melting from thermodynamic models of minerals and melts. I. Review of methods and comparison to experiments. *Journal of Petrology*, **39**, 1091–1115.

Hoatson, D. M., Jaireth, S., Jaques, A. L. (2006) Nickel sulfide deposits in Australia: Characteristics, resources and potential. *Ore Geology Reviews*, **29**, 177–241.

Hoatson, D. M., Sun, S. S., Duggan, M. B., Davies, M. B., Daly, S. J., Purvis, A. C. (2005) Late Archean Lake Harris komatiite, Central Gawler Craton, South Australia; Geologic setting and geochemistry. *Economic Geology*, **100**, 349–374.

Hoefs, J., Binns, R. A. (1978) Oxygen isotope compositions in Archaean rocks from Western Australia, with special reference to komatiites. In: R. E. Zartman (ed.) *Fourth International Conference, Geochronology, Cosmochronology, Isotope Geology*, pp. 180–182. Boulder: United States Geological Survey.

Hoffman, P. F. (1991) On accretion of granite–greenstone terranes. In: F. Robert (ed.) *Nuna Conference on Greenstone Gold and Crustal Evolution*, pp. 32–45. Ottawa: Geological Society of Canada.

Hoffman, S. E., Wilson, M., Stakes, D. S. (1986) Inferred oxygen isotope profile of Archaean oceanic crust, Onverwacht Group, South Africa. *Nature*, **321**, 55–58.

Hofmann, A. (2005) The geochemistry of sedimentary rocks from the Fig Tree Group, Barberton greenstone belt: Implications for tectonic, hydrothermal and surface processes during mid-Archaean times. *Precambrian Research*, **143**, 23–49.

Hofmann, A. W. (1988) Chemical differentiation of the Earth: the relationship between mantle, continental crust, and oceanic crust. *Earth and Planetary Science Letters*, **90**, 297–314.

Hollings, P., Wyman, D., Kerrich, R. (1999) Komatiite–basalt–rhyolite volcanic associations in Northern Superior Province greenstone belts: significance of plume-arc interaction in the generation of the proto continental Superior Province Province. *Lithos*, **46**, 137–161.

Hon, K., Kauahikaua, J., Denlinger, R., Mackay, K. (1994) Emplacement and inflation of pahoehoe sheet flows: observations and measurements of active lava flows on Kilauea Volcano, Hawaii. *Geological Society of America Bulletin*, **106**, 351–370.

Hon, K., Self, S., Thondarson, T., Keszthelyi, L. (1995) The emplacement of flood basalts, *IUGG XXI General Assembly*, abstracts, 436.

Hopf, S., Head, D. L., (1998) Mount Keith nickel deposit. In D. A. Berkman and D. H. Mackenzie (eds), *Geology of Australian and Papua New Guinean Mineral Deposits*, Monograph 22, pp. 307–314. Melbourne: Australian Institute of Mining and Metallurgy.

Houlé, M. G., Davis, P. C., Lesher, C. M., Arndt, N. T. (2002) Extrusive and intrusive komatiites, komatiitic basalt and peperite and ore genesis at the Dundonald Ni–Cu–(PGE) Deposit, Abitibi Greenstone Belt, Canada. *9th International Platinum Symposium Abstract Volume, Billings, MT*, pp. 181–184. Durham NC: Duke University.

Houlé, M. G., Gibson, H. L., Lesher, C. M. *et al.* (2008a) Komatiitic basalt sills and multi-generational peperite at Dundonald Beach, Abitibi Greenstone Belt, Ontario: implications for the stratigraphic architecture of komatiite volcanoes and distribution of subseafloor and seafloor komatiite-associated Ni–Cu–(PGE) deposits, *Economic Geology*, in press.

Houlé, M. G., Lesher, C. M., Gibson, H. L., Fowler, A. D., Sproule, R. A. (2001). Physical volcanology of komatiites in the Abitibi greenstone belt, Superior Province, Canada (Project Unit 99–021). Summary of field work and other activities. *Ontario Geological Survey Open File Report*, **6070**, p. 13.1–13.16.

Houlé, M. G., Lesher, C. M., Gibson, H. L., Sproule, R. A. (2002). Recent advances in komatiite volcanology in the Abitibi greenstone belt, Ontario. Summary of fieldwork and other activities. *Ontario Geological Survey Open File Report*, **6100–7**, 1–19.

Houlé, M. G., Préfontaine, S., Fowler, A. D., Gibson, H. L. (2008b). Endogenous growth in komatiite lava flows: evidence from spinifex-textured sills at Pyke Hill and Serpentine Mountain, Western Abitibi Greenstone Belt, northeastern Ontario, Canada. *Bulletin of Volcanology* (submitted).

Hronsky, J. M. A., Schodde, R. C. (2006) Nickel exploration history of the Yilgarn Craton: from the Nickel Boom to today. *Society of Economic Geologists Special Publication*, **13**, 1–11.

Hulbert, L. J., Hamilton, M. A., Horan, M. F., Scoates, R. F. J. (2005) U–Pb zircon and Re–Os isotope geochronology of mineralized ultramafic intrusions and associated nickel ores from the Thompson nickel belt, Manitoba, Canada. *Economic Geology*, **100**, 29–41.

Hunter, M. A., Bickle, M. J., Nisbet, E. G., Martin, A., Chapman, H. J. (1998) Continental extensional setting for the Archean Belingwe greenstone belt, Zimbabwe. *Geology*, **26**, 883–886.

Huppert, H. E., Sparks, R. S. J. (1985a) Cooling and contamination of mafic and ultramafic magmas during ascent through continental crust. *Earth and Planetary Science Letters*, **74**, 371–386.

(1985b) Komatiites I: eruption and flow. *Journal of Petrology*, **26**, 694–725.

(1989) Chilled margins in igneous rocks. *Earth and Planetary Science Letters*, **92**, 397–405.

Huppert, H. E., Sparks, R. S. J., Turner, J. S., Arndt, N. T. (1984) Emplacement and cooling of komatiite lavas. *Nature*, **309**, 19–22.

Imreh, L. (1978) *Album photographique de coulées meta-ultramafiques sous-marines archéennes dans le sillon de La Motte-Vassan* [*Photographic Album of Submarine Archean Meta-Ultramafic Flows in the LaMotte–Vassan Belt*]. Quebec: Ministere des Richesses Naturelles.

Ingle, S., Weis, D., Frey, F. A. (2002) Indian continental crust recovered from Elan Bank, Kerguelen Plateau (ODP Leg 183, Site 1137). *Journal of Petrology*, **43**, 1241–1257.

Inoue, T. (1994) Effect of water on melting phase relations and melt composition in the system Mg_2SiO_4–$MgSiO_3$–H_2O up to 15 GPa. *Physics of Earth and Planetary Interiors*, **85**, 237–263.

Inoue, T., Rapp, R. P., Zhang, J., Gasparik, T., Weidner, D. J., Irifune, T. (2000) Garnet fractionation in a hydrous magma ocean and the origin of Al-depleted komatiites: melting experiments of hydrous pyrolite with REEs at high pressure. *Earth and Planetary Science Letters*, **177**, 81–87.

Inoue, T., Sawamoto, H. (1992) High pressure melting of pyrolite under hydrous condition and its geophysical implications. In: Y. Syono, M. H. Manghnani (eds.) *High-Pressure Research: Application to Earth and Planetary Sciences*, pp. 323–331. Washington: Terra Publishing Company.

Irvine, T. N. (1982) Terminology for layered intrusions. *Journal of Petrology*, **23**, 127–162.

Jahn, B.-M., Auvray, B., Blais, S., *et al.* (1980) Trace element geochemistry and petrogenesis of Finnish Greenstone Belts. *Journal of Petrology*, **21**, 201–244.

Jahn, B.-M., Condie, K. C. (1976) On the age of the Rhodesian greenstone belts. *Contributions to Mineralogy and Petrology*, **57**, 317–330.

Jahn, B.-M., Gruau, G., Glickson, A. Y. (1982) Komatiites of the Onverwacht Group, South Africa: REE chemistry, Sm–Nd age and mantle evolution. *Contributions to Mineralogy and Petrology*, **80**, 25–40.

Jahn, B.-M., Shih, C.-Y. (1974) Trace element geochemistry of Archaean volcanic rocks. *Geochimica et Cosmochimica Acta*, **38**, 611–627.

Jarvis, R. A. (1995) On the cross-sectional geometry of thermal erosion channels formed by turbulent lava flows. *Journal of Geophysical Research*, **100**, 10127–10140.

Jensen, L. S. (1985) Stratigraphy and petrogenesis of Archean metavolcanic sequences, southwestern Abitibi Subprovince, Ontario. In: L. D. Ayres, P. C. Thurston, K. D. Card and W. Weber (eds.) *Evolution of Archean Subcrustal Sequences*, pp. 65–87, Geological Society of Canada Special Publication 28. Toronto: Geological Society of Canada.

Jensen, L. S., Langford, F. F. (1985) Geology and petrogenesis of the Archean Abitibi Belt in the Kirkland Lake area, Ontario. Ontario Geological Survey Miscellaneous Paper 123. Toronto: Ontario Geological Survey.

Jerram, D. A., Cheadle, M. J. (2000) On the cluster analysis of grains and crystals in rocks. *American Mineralogist*, **85**, 47–67.

Jerram, D. A., Cheadle, M. J., Philpotts, A. R. (2003) Quantifying the building blocks of igneous rocks; are clustered crystal frameworks the foundation? *Journal of Petrology*, **44**, 2033–2051.

Jochum, K. P., Arndt, N. T., Hofmann, A. W. (1990) Nb–Th–La in komatiites and basalts: constraints on komatiite petrogenesis and mantle evolution. *Earth and Planetary Science Letters*, **107**, 272–289.

Jolly, W. T. (1982) Progressive metamorphism of komatiites and related Archaean lavas of the Abitibi area, Canada. In: N. T. Arndt and E. G. Nisbet (eds.) *Komatiites*, pp. 245–266. London: George Allen and Unwin.

Kamenetsky, V., Sobolev, A., McDonough, W. (2003) Melt inclusion evidence for a volatile-enriched (H_2O, Cl, B) component in parental magmas of Gorgona Island komatiites. *Geophysical Research Abstracts*, **5**, 14774.

Kamo, S. L. (1994) Reassessment of Archean crustal development in the Barberton Mountain Land, from U–Pb dating. *Tectonics*, **13**, 167–192.

Kareem, K. M., Byerly G. R. (2003) Petrology, and geochemistry of 3.3 Ga komatiites – Weltevreden Formation, Barberton Greenstone Belt. *Lunar and Planetary Science* **XXXIV**, abstract 20.

Kato, T., Ringwood, A. E., Irifune, T. (1988) Experimental determination of element partitioning between silicate perovskites, garnets and liquids: constraints on early differentiation of the mantle. *Earth and Planetary Science Letters*, **89**, 123–145.

Kato, T., Ringwood, A. E., Irifune, T. (1996) Constraints on element partition coefficients between $MgSiO_3$ perovskite and liquid determined by direct measurements. *Earth and Planetary Science Letters*, **90**, 65–68.

Kauahikaua, J., Cashman, K. V., Mattox, T. N. *et al.* (1998) Observations on basaltic lava streams in tubes from Kilauea Volcano, Island of Hawaii. *Journal of Geophysical Research*, **103**, 27303–27323.

Kawamoto, T., Hervig, R. L., Holloway, J. R. (1996) Experimental evidence for a hydrous transition zone in the early Earth's mantle. *Earth and Planetary Science Letters*, **142**, 587–592.

Kawamoto, T., Holloway, J. R. (1997) Melting temperature and partial melt chemistry of H_2O-saturated mantle peridotite to 11 GPa: implications for the stability of H_2O fluid in the Earth's mantle and kimberlite generation. *Science*, **276**, 240–243.

Keays, R. R. (1982) Palladium and iridium in komatiites and associated rocks: application to petrogenetic problems. In: N. T. Arndt and E. G. Nisbet (eds.) *Komatiites*, pp. 435–458. London: George Allen and Unwin.

(1995) The role of komatiite and picritic magmatism and S-saturation in the formation of ore deposits. *Lithos*, **34**, 1–18.

Keays, R. R., Ross, J. R., Woolrich, P. (1981) Precious metals in volcanic peridotite-associated nickel sulfide deposits in Western Australia, II. Distribution within ores and host rocks at Kambalda. *Economic Geology*, **76**, 1645–1674.

Keele, R. A., Nickel, E. H. (1974) The geology of a primary millerite-bearing sulfide assemblage and supergene alteration at the Otter Shoot, Kambalda, Western Australia. *Economic Geology*, **69**, 1102–1117.

Keep, F. E. (1929) The geology of the Shabani Mineral Belt, Belingwe District. *Bulletin of the Geological Survey of Southern Rhodesia*, **12**.

Kerr, A., Marriner, G., Arndt, N. T., *et al.* (1995a) The petrogenesis of Gorgona komatiites, picrites and basalts: new field, petrographic and geochemical constraints. *Lithos*, **37**, 245–260.

Kerr, A., Tarney, J., Marriner, G., Klaver, G. T., Saunders, A. D., Thirlwall, M. F. (1996) The geochemistry and petrogenesis of the late-Cretaceous picrites and basalts of Curacao, Netherlands Antilles: a remnant of an oceanic plateau. *Contributions to Mineralogy and Petrology*, **37**, 245–260.

Kerr, A., Tarney, J., Marriner, G., Nivia, A., Saunders, A. (1997) The Caribbean–Colombian Cretaceous Igneous Province: the internal anatomy of an oceanic plateau. In: J. J. Mahoney and M. F. Coffin (eds.) *Large Igneous Provinces: Continental, Oceanic and Planetary Flood Volcanism*, pp. 123–144. Washington: American Geophysical Union.

Kerr, A. C. (2005) La Isla de Gorgona, Colombia: a petrological enigma? *Lithos*, **84**, 77–101.

Kerr, A. C., Arndt, N. T. (2001) A note on the IUGS reclassification of the high-Mg and picritic volcanic rocks. *Journal of Petrology*, **42**, 2169–2171.

Kerr, A. C., Saunders, A. D., Tarney, J., Berry, N. H., Hards, V. L. (1995b) Depleted mantle plume geochemical signatures: no paradox for plume theories. *Geology*, **23**, 843–846.

Kerr, R. C. (2001) Thermal erosion by laminar lava flows. *Journal of Geophysical Research – Solid Earth*, **106**, 26453–26465.

Kerr, R. C., Tait, S. R. (1985) Convective exchange between pore fluid and an overlying reservoir of dense fluid: a postcumulus process in layered intrusions. *Earth and Planetary Science Letters*, **75**, 147–156.

(1986) Crystallisation and compositional convection in a porous medium with application to layered igneous intrusions. *Journal of Geophysical Research*, **91**, 3591–3608.

Keszthelyi, L., Self, S., Thordarson, T. (1999) Applications of recent studies on the emplacement of basaltic lava flows to the Deccan Traps. *Memoir of Geological Society of India*, **43**, 485–520.

(2006) Flood lavas on Earth, Io and Mars. *Journal of the Geological Society*, **163**, 253–264.

Kieffer, B., Arndt, N. T., Weis, D. (2002) Petrology, and geochemistry of a bimodal alkalic shield volcano, Site 1139, Kerguelen plateau. *Journal of Petrology*, **43**, 1259–1286.

Kimura, G., Ludden, J. N., Desrochers, J.-P., Hori, R. (1993) A model of ocean-crust accretion for the Superior Province, Canada. *Lithos*, **30**, 337–355.

Kinzler, R., Grove, T. L., Recca, S. I. (1990) An experimental study on the effect of temperature and melt composition on the partitioning of nickel between olivine and silicate melt. *Geochimica et Cosmochimica Acta*, **54**, 1255–1265.

Kinzler, R. J., Grove, T. L. (1985) Crystallization and differentiation of Archean komatiite lavas from northeast Ontario: phase equilibrium and kinetic studies. *American Mineralogist*, **70**, 40–51.

Kirkpatrick, R., Reck, B. H., Pelly, I. Z., Kuo, L.-C. (1983) Programmed cooling experiments in the system MgO–SiO$_2$: kinetics of a peritectic reaction. *American Mineralogist*, **68**, 1095–1101.

Kröner, A., Hegner, E., Wendt, J. I., Byerly, G. R. (1996) The oldest part of the Barberton granitoid-greenstone terrain, South Africa: evidence for crust formation between 3.5 and 3.7 Ga. *Precambrian Research*, **78**, 105–124.

Kröner, A., Todt, W. (1988) Single zircon dating constraining the maximum age of the Barberton greenstone belt, southern Africa. *Journal of Geophysical Research*, **93**, 15329–15337.

Kushiro, I. (1986) Viscosity of partial melts in the upper mantle. *Journal of Geophysical Research*, **91**, 9343–9350.

Kusky, T. M., Kidd, W. S. F. (1992) Remnants of an Archaean oceanic plateau, Belingwe greenstone belt. *Geology*, **20**, 43–46.

Kusky, T. M., Winsky, P. A. (1995) Structural relationships between a shallow water platform and an oceanic plateau, Zimbabwe. *Tectonics*, **14**, 448–471.

Laflèche, M. R., Dupuy, C., Bougault, H. (1992) Geochemistry and petrogenesis of Archean mafic volcanic rocks of the southern Abitibi belt, Québec. *Precambrian Research*, **57**, 207–241.

Lahaye, Y., Arndt, N. T. (1996) Alteration of a komatiitic flow: Alexo, Ontario, Canada. *Journal of Petrology*, **37**, 1261–1284.

Lahaye, Y., Arndt, N. T., Byerly, G., Gruau, G., Fourcade, S., Chauvel, C. (1995) The influence of alteration on the trace-element and Nd isotope compositions of komatiites. *Chemical Geology*, **126**, 43–64.

Lahaye, Y., Barnes, S. J., Frick, L. R., Lambert, D. D. (2001) Re–Os isotopic study of komatiitic volcanism and magmatic sulfide formation in the southern Abitibi greenstone belt, Ontario, Canada. *Canadian Mineralogist*, **39**, 473–490.

Lajoie, J., Gelinas, L. (1978) Emplacement of Archaean peridotitic komatiites in Lamotte Township, Quebec. *Canadian Journal of Earth Sciences*, **15**, 672–677.

Lambert, D. D., Foster, J. G., Frick, L. R., Ripley, E. M., Zientek, M. L. (1998) Geodynamics of magmatic Cu–Ni–PGE sulfide deposits – new insights from the Re–Os isotope system. *Economic Geology*, **93**, 121–136.

Langmuir, C. H., Klein, E. M., Plank, T. (1992) Petrological systematics of mid-ocean ridge basalts: constraints on melt generation beneath ocean ridges. In: J. Phipps-Morgan, D. K. Blackman and J. Sinton (eds.) *Mantle Flow and Melt Generation at Mid-Ocean Ridges*, pp. 183–280, AGU Geophysical Monograph 71, Washington: American Geophysical Union.

Layton-Matthews, D., Burnham, O. M., Lesher, C. M. (2003) Trace element geochemistry of ultramafic intrusions in the Thompson nickel belt: relative roles of contamination and metasomatism. *The Gangue (GAC Mineral Deposits Division)*, **76**, 1–10.

Layton-Matthews, D., Lesher, C. M., Burnham, O. M., *et al.* (2007) Magmatic Ni–Cu–platinum-group-element deposits in the Thompson Nickel Belt. In: W. D. Goodfellow (ed.) *Mineral Deposits in Canada: A Synthesis of Major Deposit Types, District Metallogeny, the Evolution of Geological Provinces and Exploration Methods*. Geological Association of Canada Special Publication 5. Ottawa: Geological Association of Canada.

Le Bas, M. J. (2000) IUGS reclassification of the high-Mg and picritic volcanic rocks. *Journal of Petrology*, **41**, 1467–1470.

Le Maitre, R. W., Bateman, P., Dudek, A. *et al.* (eds.) (1989) *A classification of igneous rocks and glossary of terms*. Oxford: Blackwell.

Lécuyer, C., Gruau, G., Anhaeusser, C. R., Fourcade, S. (1994) The origin of fluids and the effects of metamorphism on the primary chemical compositions of Barberton komatiites: new evidence from geochemical (REE) and isotopic (Nd, O, H, ^{39}Ar/^{40}Ar) data. *Geochimica et Cosmochimica Acta*, **58**, 969–984.

Lesher, C. M. (1983) Localization and genesis of komatiite-associated Fe–Ni–Cu sulphide mineralization at Kambalda, Western Australia. Ph.D. thesis, University of Western Australia.

(1989) Komatiite-associated nickel sulfide deposits. In: J. A. Whitney and A. J. Naldrett (eds.) *Ore Deposition Associated with Magmas*. pp. 45–102. Dordrecht: Society of Economic Geologists.

(1999) *Komatiite Peridotite-Hosted Fe–Ni–Cu–(PGE) Sulphide Deposits in the Raglan Area, Cape Smith Belt, New Quebec,* Guidebook Series, vol. 2. Sudbury: Mineral Exploration Centre.

(2007) Deposits in the Raglan Area, Cape Smith Belt, New Quebec. In: W. D. Goodfellow (ed.) *Mineral Deposits of Canada: A Synthesis of Major Deposit Types, District Metallogeny, the Evolution of Geological Provinces, and Exploration Methods*, pp. 351–386. Geological Association of Canada Special Publication 5. Ottawa: Geological Association of Canada, Mineral Deposits Division.

Lesher, C. M., and Arndt, N. T. (1990) Geochemistry of komatiites at Kambalda, Western Australia: assimilation, crystal fractionation and lava replenishment. In: J. E. Glover and S. E. Ho (eds) *Third International Archaean Symposium, Perth, 1990, Extended Abstracts*, pp. 149–152. Perth: Geoconferences (W.A.).

(1995) REE and Nd isotope geochemistry, petrogenesis and volcanic evolution of contaminated komatiites at Kambalda, Western Australia. *Lithos*, **34**, 127–158.

Lesher, C. M., Arndt, N. T., Groves, D. I. (1984) Genesis of komatiite-associated nickel suphide deposits at Kambalda, Western Australia: a distal volcanic model. In: D. L. Buchanan and M. J. Jones (eds.) *Sulphide Deposits in Mafic and Ultramafic Rocks*. pp. 70–80. London: Institute of Mining and Metallurgy.

Lesher, C. M., Burnham, O. M. (2001) Multicomponent elemental and isotopic mixing in Ni–Cu–(PGE) ores at Kambalda, Western Australia. *Canadian Mineralogist*, **39**, 421–446.

Lesher, C. M., Burnham, O. M., Keays, R. R., Barnes, Stephen J., Hulbert L. (1999) Geochemical discrimination of barren and mineralized komatiites in dynamic ore-forming magmatic systems. In: R. R. Keays, C. M. Lesher, P. C. Lightfoot and C. E. G. Farrow (eds.) *Dynamic Processes in Magmatic Ore Deposits and Their Application to Mineral Exploration*. pp. 451–477. Geological Association of Canada, Short Course 13. Ottawa: Geological Association of Canada.

(2001) Geochemical discrimination of barren and mineralized komatiites associated with magmatic Ni–Cu–(PGE) sulphide deposits. *Canadian Mineralogist*, **39**, 673–696.

Lesher, C. M., Campbell, I. H. (1993) Geochemical and fluid dynamic modelling of compositional variations in Archean komatiite-hosted nickel sulfide ores in Western Australia. *Economic Geology*, **88**, 804–816.

Lesher, C. M., Groves, D. I. (1986) Controls on the formation of komatiite-associated nickel-copper sulfide deposits. In: G. H. Friedrich (ed.) *Geology and Metallogeny of Copper Deposits*. Berlin: Springer Verlag.

Lesher, C. M., Keays, R. R. (2002) Komatiite-associated Ni–Cu–(PGE) deposits: Mineralogy, geochemistry, and genesis. In: L. J. Cabri (ed.) *The Geology Geochemistry, Mineralogy, and Mineral Beneficiation of the Platinum-Group Elements*, pp. 579–617, Special Volume 54. Montreal: Canadian Institute of Mining, Metallurgy and Petroleum.

Lesher, C. M., Lee, R. F., Groves, D. I., Bickle, M. J., Donaldson, M. J. (1981) Geochemistry of komatiites from Kambalda, Western Australia: I. Chalcophile element depletion – a consequence of sulfide liquid separation from komatiite magmas. *Economic Geology*, **76**, 1714–1728.

Lesher, C. M., Stone W. E. (1996) Exploration geochemistry of komatiites. In: D. A. Wyman (ed.) *Igneous Trace Element Geochemistry Applications for Massive Sulphide Exploration*, pp. 153–204, Geological Association of Canada Short Course Notes 12. Ottawa: Geological Association of Canada.

Lévesque, M., and Lesher, C. M. (2002) Invasive Features of Mafic-Ultramafic Rocks at the Zone 3, Zone 2, and Katinniq Ni–Cu–(PGE) Deposits, Raglan Formation, Cape Smith Belt, Nouveau-Québec, 9th International Platinum Symposium, Billings, Montana, pp. 257–260.

Lewis, J. D. (1971) "Spinifex texture" in a slag, as evidence for its origin in rocks. *Annual Report of the Geological Survey of Western Australia* **1971**, 45–49.

Lewis, J. D., Williams, J. R. (1973) The petrology of an ultramafic lava near Murphy Well, Eastern Goldfields, Western Australia. *Annual Report of the Geological Survey of Western Australia* **1972**, 60–68.

Li, C. S., Ripley, E. M. (2005) Empirical equations to predict the sulfur content of mafic magmas at sulfide saturation and applications to magmatic sulfide deposits. *Mineralium Deposita*, **40**, 218–230.

Li, Z.-X. A., Lee, C.-T. A. (2004) The constancy of upper mantle fO_2 through time inferred from V/Sc ratios in basalts. *Earth and Planetary Science Letters*, **228**, 483–493.

Libby, J. W., Stockman, P. R., Cervoj, K. M., Muir, M. R. K., Whittle, M., Langworthy, P. J. (1998) *Perseverance Nickel Deposit*. Carlton: Australian Institute of Mining and Metallurgy.

Liebske, C., Schmickler, B., Terasaki, H., *et al.* (2005) Viscosity of peridotite liquid up to 13 GPa: implications for magma ocean viscosities. *Earth and Planetary Science Letters*, **240**, 589–604.

Lightfoot, P. C., Naldrett, A. J., Gorbachev, N. S., Fedorenko, V. A. (1990) Geochemistry of the Siberian trap of the Noril'sk area, USSR, with implications for the relative contributions of crust and mantle to flood basalt magmatism. *Contributions to Mineralogy and Petrology*, **104**, 631–644.

Litasov, K., Ohtani, E. (2002) Phase relations and melt compositions in CMAS–pyrolite–system up to 25 GPa. *Physics of Earth and Planetary Interiors*, **134**, 105–127.

Lofgren, G. E. (1980) Experimental studies on the dynamic crystallization of silicate melts. In: R. B. Hargraves (ed.) *Physics of Magmatic Processes*, pp. 487–552. Princeton: Princeton University Press.

Lofgren, G. E., Donaldson, C. H. (1975) Curved branching crystals and differentiation in comb-layered rocks. *Contributions to Mineralogy and Petrology*, **274**, 243–273.

Lofgren, G. E., Donaldson, C. H., Williams, R. J., Mullins, O., Usselman, T. M. (1974) Experimentally reproduced textures and mineral chemistry of Apollo 15 quartz-normative basalts. *Proceedings of the Lunar Science Conference*, **6**, 79–99.

Lowe, D. R., Byerly, G. R. (1999a) (eds.) *Geologic Evolution of the Barberton Greenstone Belt, South Africa*. Geological Society of America Special Paper 329. Boulder: Geological Society of America.

(1999b) Stratigraphy of the west-central part of the Barberton Greenstone Belt, South Africa. In: D. R. Lowe, and G. R. Byerly, (eds.) *Geologic evolution of the*

Barberton Greenstone Belt, South Africa, pp. 1–36, Geological Society of America Special Paper 329, Boulder: Geological Society of America.

Luck, J.-M., Allégre, C. (1984) [187]Re–[187]Os investigation in sulfide from Cape Smith komatiite. *Earth and Planetary Science Letters*, **64**, 205–208.

Luck, J. M., Arndt, N. T. (1986) Re/Os isochron from Archean komatiite from Alexo, Ontario. *Terra Cognita*, **5**, 323.

Ludden, J., Arndt, N. T. (1996) Mafic magmatism through time. *Chemical Geology*, **126**.

Machado, N., Brooks, C., Hart, S. R. (1986) Determination of initial [87]Sr/[86]Sr and [143]Nd/[144]Nd in primary minerals from mafic and ultramafic rocks: experimental procedure and implications for the isotopic characteristics of the Archean mantle under the Abitibi greenstone belt. *Geochimica et Cosmochimica Acta*, **50**, 2335–2348.

Mahoney, J. J., Coffin, M. F. (1997) *Large Igneous Provinces: Continental, Oceanic and Planetary Flood Volcanism*. Washington: American Geophysical Union.

Mareschal, J.-C., Jaupart, C. (2006) In: K. Benn, J.-C. Mareschal, K.C. Condie (eds.) *Archean Geodynamics and Environments*, pp. 61–73. Geophysical Monograph Series, 164. Washington: American Geophysical Union.

Marsh, B. D. (1981) On the crystallinity, probability of occurrence, and rheology of lava and magma. *Contributions to Mineralogy and Petrology*, **78**, 85–98.

Marston, R. J. (1984) Nickel mineralization in Western Australia. *Geological Survey of Western Australia Mineral Resources Bulletin*, **14**, 271.

Marston, R. J., Groves, D. I., Hudson, D. R., Ross, J. R. (1981) Nickel sulfide deposits in Western Australia: a review. *Economic Geology*, **76**, 1330–1363.

Marston, R. J., Kay, B. D. (1980) The distribution, petrology, and genesis of nickel ores at the Juan complex, Kambalda, Western Australia. *Economic Geology*, **75**, 546–565.

Martin, D., Nokes, R. (1988) Crystal settling in a vigorously convecting magma chamber. *Nature*, **332**, 534–536.

Matsumoto, T., Seta, A., Matsuda, J., Takebe, M., Chen, Y., Arai, S. (2002) Helium in Archean komatiites revisited: significantly high [3]He/[4]He ratios revealed by fractional crushing gas extraction. *Earth and Planetary Science Letters*, **196**, 213–225.

Mavrogenes, J. A., O'Neill, H. S. C. (1999) The relative effects of pressure, temperature and oxygen fugacity on the solubility of sulfide in mafic magmas. *Geochimica et Cosmochimica Acta*, **63**, 1173–1180.

McCall, G. J. H., Leishman, J. (1971) Clues to the origin of Archean eugeosynclinal peridotites and the nature of serpentinization. *Geological Society of Australia Special Publication*, **3**, 281–299.

McCulloch, M. T., Bennett, V. C. (1994) Progressive growth of the Earth's continental crust and depleted mantle: geochemical constraints. *Geochemica et Cosmochimica Acta*, **58**, 4717–4738.

McCulloch, M. T., Compston, W. (1981) Sm–Nd age of Kambalda and Konowna greenstones and heterogeneity in the Archaean mantle. *Nature*, **294**, 322–326.

McDonough, W. F. (1990) Constraints on the composition of the continental lithospheric mantle. *Earth and Planetary Science Letters*, **101**, 1–18.

McDonough, W. F., Danyushevsky, L. V. (1995) Water and sulfur contents of melt inclusions from Archean komatiites. *EOS (Transactions, American Geophysical Union)*, **76**, 266.

McDonough, W. F., Ireland, T. R. (1993) Intraplate origin of komatiites inferred from trace elements in glass inclusions. *Nature*, **365**, 432–434.

McDonough, W. F., Sun, S. S. (1995) The composition of the Earth. Chemical Geology, **120**, 223–253.

McKenzie, D. (1984) The generation and compaction of partially molten rock. *Journal of Petrology*, **25**, 713–765.

McKenzie, D., Bickle, M. J. (1988) The volume and composition of melt generated by extension of the lithosphere. *Journal of Petrology*, **29**, 625–679.

McNaughton, N. J., Frost, K. M., Groves, D. I. (1988) Ground melting and ocellar komatiites: a lead isotopic study at Kambalda, Western Australia. *Geological Magazine*, **125**, 285–295.

McQueen, K. G. (1987) Deformation and remobilization in some Western Australian nickel ores, *Ore Geology Reviews*, **2**, 269–286.

Miller, G. H., Stolper, E. M., Ahrens, T. J. (1991) The equation of state of molten komatiite, 2: Application to komatiite petrogenesis and the Hadean mantle. *Journal of Geophysical Research*, **96**, 11849–11864.

Mitchell, R. H. (1995) *Kimberlites, Orangites, and Related Rocks*. New York: Plenum Press.

Mo, X., Carmichael, I. S. E., Rivers, M., Stebbins, J. (1982) The partial molar volume of Fe_2O_3 in multicomponent silicate liquids and the pressure dependence of oxygen fugacity in magmas. *Mineralogical Magazine*, **45**, 237–245.

Moore, A. G., Cas, R. A. F., Beresford, S. W., Stone, M. (2000) Geology of an Archaean metakomatiite succession, Tramways, Kambalda Ni province, Western Australia: assessing the extent to which volcanic facies architecture and flow emplacement mechanisms can be reconstructed. *Australian Journal of Earth Sciences*, **47**, 659–673.

Moore, J. G. (1975) Mechanism of formation of pillow lava. *American Science*, **63**, 269–277.

Morton, J. L., Franklin, J. M. (1987) Two-fold classification of Archean volcanic-associated massive sulfide deposits. *Economic Geology*, **82**, 1057–1063.

Moyen, J.-F. O., Stevens, G., Kisters, A. (2006) Record of mid-Archaean subduction from metamorphism in the Barberton terrain, South Africa. *Nature*, **442**, 559–562.

Mueller, W., Pickett, C. (2005) Relative sea level change along the Slave craton coastline: characteristics of Archean continental rifting. *Sedimentary Geology*, **176**, 97–119.

Mueller, W. U. (2005) Physical volcanology of komatiites. In: P. G. Eriksson, W. Altermann, D. R. Nelson, W. U. Mueller, O. Catuneanu (eds.) *The Precambrian Earth: Tempos and Events*, pp. 277–290. Amsterdam: Elsevier.

Mueller, W. U., Corcoran, P. L., Pickett, C. (2005) Mesoarchean continental breakup: evolution and inferences from the >2.8 Ga Slave craton-cover succession, Canada. *Journal of Geology*, **113**, 23–45.

Muir, J. E., Comba, C. D. A. (1979) The Dundonald deposit: an example of volcanic-type nickel sulfide ore in ultramafic lavas. *Canadian Mineralogist*, **17**, 351–360.

Murck, B. W., Campbell, I. H. (1986) The effects of temperature, oxygen fugacity and melt composition of the behaviour of chromium in basic and ultrabasic melts. *Geochimica et Cosmochimica Acta*, **50**, 1871–1887.

Myers, J. S., Swager, C. (1997) The Yilgarn Craton, Australia. In: M. de Wit and L. D. Ashwal (eds.) *Greenstone Belts*, pp. 640–656. Oxford: Oxford University Press.

Naldrett, A. J. (1964a) Talc-carbonate alteration of some serpentinized ultramafic rocks south of Timmins, Ontario. *Journal of Petrology*, **7**, 489–499.

(1964b) Ultrabasic rocks of the Porcupine and related nickel deposits. Ph.D. thesis. Queen's University, Kingston, Ontario.

(1966) The role of sulphurization in the genesis of iron–nickel sulphide deposits of the Porcupine district, Ontario. *Canadian Institute of Mining and Metallurgy Transactions*, **69**, 147–155.

(1969) A portion of the system Fe–S–O and its application to sulfide ore magmas. *Journal of Petrology*, **10**, 171–202.

(1973) Nickel sulphide deposits: their classification and genesis, with special emphasis on deposits of volcanic association. *Canadian Institute of Mining and Metallurgy Bulletin*, **66**, 45–63.

(2004) *Magmatic Sulfide Deposits: Geology, Geochemistry and Exploration.* Heidelberg: Springer Verlag.

Naldrett, A. J., Asif, M., Schandl, E., *et al.* (1999) Platinum-group elements in the Sudbury ores: significance with respect to the origin of different ore zones and to the exploration for footwall orebodies. *Economic Geology*, **94**, 185–210.

Naldrett, A. J., Barnes, Sarah-Jane (1986) The behaviour of platinum group elements during fractional crystallization and partial melting with special reference to the composition of magmatic sulfide. *Research Fortschritte der Mineralogie*, **64**, 113–134.

Naldrett, A. J., Campbell, I. H. (1982) Physical and chemical constraints on genetic models for komatiite-related Ni-sulphide deposits. In: N. T. Arndt and E. G. Nisbet (eds.) *Komatiites*, pp. 423–434. London: George Allen and Unwin.

Naldrett, A. J., Gasparrini, E. L. (1971) Archaean nickel sulphide deposits in Canada; their classification, geological setting and genesis with some suggestions as to exploration. Special Publication 3, pp. 201–225. Perth: Geological Society of Australia.

Naldrett, A. J., Hoffman, E. L., Green, A. H., Chou, C.-L., Naldrett, S. R., Alcock, R. A. (1979) The composition of Ni-sulfide ores, with particular reference to their content of PGE and Au. *Canadian Mineralogist*, **17**, 403–416.

Naldrett, A. J., Mason, G. D. (1968) Contrasting Archaean ultramafic igneous bodies in Dundonald and Clerque Townships, Ontario. *Canadian Journal of Earth Sciences*, **5**, 111–143.

Naldrett, A. J., Turner, A. R. (1977) The geology and petrogenesis of a greenstone belt and related nickel sulfide mineralization at Yakabindie, Western Australia. *Precambrian Research*, **5**, 43–103.

Nelson, D. R. (1997) Evolution of the Archaean granite-greenstone terranes of the Eastern Goldfields, Western Australia: SHRIMP U–Pb zircon constraints. *Precambrian Research*, **83**, 57–81.

(1998) Granite-greenstone crust formation on the Archaean Earth – a consequence of two superimposed processes. *Earth and Planetary Science Letters*, **158**, 109–119.

Nesbitt, R. W. (1971) Skeletal crystal forms in the ultramafic rocks of the Yilgarn Block, Western Australia: evidence for an Archaean ultramafic liquid. *Geological Society of Australia Special Publication*, **3**, 331–347.

Nesbitt, R. W., Jahn, B. M., Purvis, A. C. (1982) Komatiites: an early Precambrian phenomenon. *Journal of Volcanology and Geothermal Research*, **14**, 31–45.

Nesbitt, R. W., Sun, S.-S. (1976) Geochemistry of Archaean spinifex-textured peridotites and magnesian and low-magnesian tholeiites. *Earth and Planetary Science Letters*, **31**, 433–453.

Nesbitt, R. W., Sun, S. S., Purvis, A. C. (1979) Komatiites: geochemistry and genesis. *Canadian Mineralogist*, **17**, 165–186.

Nisbet, E. G. (1982) The tectonic setting and petrogenesis of komatiites. In: N. T. Arndt and E. G. Nisbet (eds.) *Komatiites*. pp. 501–520. London: George Allen and Unwin.

Nisbet, E. G., Arndt, N. T., Bickle, M. J., *et al.* (1987) Uniquely fresh 2.7 Ga komatiites from the Belingwe greenstone belt, Zimbabwe. *Geology*, **15**, 1147–1150.

Nisbet, E. G., Bickle, M. J., Martin, A. (1977) The mafic and ultramafic lavas of the Belingwe greenstone belt, Rhodesia. *Journal of Petrology*, **18**, 521–566.

Nisbet, E. G., Cheadle, M. J., Arndt, N. T., Bickle, M. J. (1993a) Constraining the potential temperature of the Archaean mantle: a review of the evidence from komatiites. *Lithos*, **30**, 291–307.

Nisbet, E. G., Chinner, G. A. (1981) Controls on the eruption of mafic and ultramafic lavas: Ruth Well Cu–Ni prospect, western Pilbara. *Economic Geology*, **76**, 1729–1735.

Nisbet, E. G., Martin, A., Bickle, M. J., Orpen, J. L. (1993b) The Ngezi Group: komatiites, basalts and stromatolites on continental crust. In: M. J. Bickle and E. G. Nisbet (eds.) *The Geology of the Belingwe Greenstone Belt, Zimbabwe*, pp. 121–166. Rotterdam: Balkema.

Nowell, G. M., Pearson, D. G., Bell, D. R., *et al.* (2004) Hf isotope systematics of kimberlites and their megacrysts: new constraints on their source regions. *Journal of Petrology*, **45**, 1583–1612.

O'Driscoll, B., Donaldson, C. H., Troll, V. R., Jerram, D. A., Emeleus, C. H. (2007) An origin for harrisitic and granular olivine in the Rum Layered Suite, NW Scotland: a crystal size distribution study. *Journal of Petrology*, **48**, 253–270.

O'Hanley, D. S., Kyser, T. K. (1991) Ages of serpentinization, stable isotope compositions and evidence for oceanic serpentinization in ophiolites. *Geological Society of America Abstracts with program*, **23**, A147.

O'Hara, M. J. (1968) The bearing of phase equilibrium studies on the origin and evolution of basic and ultrabasic rocks. *Earth Science Reviews*, **4**, 69–133.

(1980) Nonlinear nature, of the unavoidable long-lived isotopic, trace and major element contamination of a developing magma chamber. *Philosophical Transactions of the Royal Society London*, **A297**, 215–227.

O'Neill, H. S. C., Dingwell, D. B., Borisov, A., Spettel, B., Palme, H. (1995) Experimental petrochemistry of some highly siderophile elements at high temperatures, and some implications for core formation and the mantle's early history. *Chemical Geology*, **120**, 255–273.

Ohta, H., Maruyama, S., Takahashi, E., Watanabe, Y., Kato, Y. (1996) Field occurrence, geochemistry and petrogenesis of the Archean mid-ocean ridge basalts (AMORBs) of the Cleaverville area, Pilbara Craton, Western Australia. *Lithos*, **37**, 199–222.

Ohtani, E. (1984) Generation of komatiite magma and gravitational differentiation in the deep upper mantle. *Earth and Planetary Science Letters*, **67**, 261–272.

Ohtani, E., Asahara, Y., Suzuki, A., Moro, N. (1997) A link between cratonic peridotite and komatiite: experimental evidence for melting of wet Archean mantle. *EOS (Transactions, American Geophysical Union)*, **78**, 750.

Ohtani, E., Kato, T., Sawamoto, H. (1986) Melting of a model chondritic mantle to 20 GPa. *Nature*, **322**, 352–353.

Ohtani, E., Kawabe, I., Moriyama, J., Nagata, Y. (1989) Partitioning of elements between majorite garnet and melt and implications for petrogenesis of komatiite. *Contributions to Mineralogy and Petrology*, **103**, 263–269.

Ohtani, E., Kumazawa, M. (1981) Melting of forsterite up to 15 GPa. *Physics of Earth and Planetary Interiors*, **27**, 32–38.

Ohtani, E., Mibe, K., Kato, T. (1996) Origin of cratonic peridotite and komatiite: evidence for melting in the wet Archean mantle. *Proceedings Japan Academy*, **72**, 113–117.

Ohtani, E., Moriyama, J., Kawabe, I. (1988) Majorite garnet stability and its implication for genesis of komatiite magmas. *Chemical Geology*, **70**, 147–156.

Ohtani, E., Nagata, Y., Suzuki, A., Kato, T. (1995) Melting relations of peridotite and the density crossover in planetary mantles. *Chemical Geology*, **120**, 207–221.

Ohtani, E., Suzuki, A., Kato, T. (1998) Flotation of olivine and diamond in mantle melt at high pressure: implications for fractionation in the deep mantle and ultradeep origin of diamond. In: M. H. Manghnani and T. Yagi (eds) *Properties of Earth and Planetary Materials*, pp. 227–239. Washington: American Geophysical Union.

Olinger, B. (1977) Compression studies of forsterite (Mg_2SiO_4) and enstatite ($MgOSiO_3$). In: M. H. Manghnani and S. I. Akimoto (eds.) *High-Pressure Research*, pp. 325–335. New York: Academic Press.

Oliver, R. L., Nesbitt, R. W., Hansen, D. M., Franzen, N. (1972) Metamorphic olivine in ultramafic rocks from Western Australia. *Contributions to Mineralogy and Petrology*, **36**, 335–342.

Page, M. L., Schmulian, M. L. (1981) The proximal volcanic environment of the Scotia nickel deposits. *Economic Geology*, **76**, 1469–1479.

Parman, S., Dann, J., Grove, T. L., de Wit, M. J. (1997) Emplacement conditions of komatiite magmas from the 3.49 Ga Komati Formation, Barberton Greenstone Belt, South Africa. *Earth and Planetary Science Letters*, **150**, 303–323.

Parman, S., Grove, T. L., Dann, J. (2001) The production of Barberton komatiites in an Archean subduction zone. *Geophysical Research Letters*, **28**, 2513–2516.

Parman, S., Grove, T. L., Dann, J., de Wit, M. J. (1996) Pyroxene compositions in 3.49 Ga komatiite: evidence of variable H_2O content. *EOS (Transactions of the American Geophysical Union)*, **77**, 280.

(2004) A subduction origin for komatiites and cratonic lithospheric mantle. *South African Journal of Geology*, **107**, 107–118.

Parman, S. W., Shimizu, N., Grove, T. L. (2003) Constraints on the pre-metamorphic trace element composition of Barberton komatiites from ion probe analyses of preserved clinopyroxene. *Contributions to Mineralogy and Petrology*, **144**, 383–396.

Patchett, P. J., Chauvel, C. (1984) The mean life of the continents is not constrained by Nd and Hf isotopes. *Geophysical Research Letters*, **11**, 151–153.

Patchett, P. J., Tatsumoto, M. (1980) A routine high-precision method for Lu–Hf isotope geochemistry and geochronology. *Contributions to Mineralogy and Petrology*, **78**, 263–267.

Peach, C. L., Mathez, E. A., Keays, R. R., Reeves, S. J. (1994) Experimentally-determined sulfide melt-silicate melt partition coefficients for iridium and palladium. *Chemical Geology*, **117**, 361–377.

Pearce, T. H. (1968) A contribution to the theory of variation diagrams. *Contributions to Mineralogy and Petrology*, **19**, 142–157.

(1987) The identification and assessment of spurious trends in Pearce-type ratio variation diagrams: a discussion of some statistical arguments. *Contributions to Mineralogy and Petrology*, **97**, 529–534.

Peck, D. L., Hamilton, M. S., Shaw, H. R. (1977) Numercial analysis of lava lake cooling models. Part II: application to Alae lava lake, Hawaii. *American Journal of Science*, **277**, 415–437.

Pedersen, A. K. (1985) Reaction between picritic magma and continental crust: early Tertiary silicic basalts and magnesian andesites from Disko, West Greenland. *Bulletin of the Geological Society of Greenland*, **152**, 126.

Perring, C. S., Barnes, Stephen J., Hill, R. E. T. (1995) The physical volcanology of Archean komatiite sequences from Forestania, Southern Cross Province, Western Australia. *Lithos*, **34**, 189–207.

(1996) Geochemistry of Archaean komatiites from the Forrestania Greenstone Belt, Western Australia: evidence for supracrustal contamination. *Lithos*, **37**, 181–197.

Polat, A., Hofmann, A. W., Rosing, M. (2002) Boninite-like volcanic rocks in the 3.7–3.8 Ga Isua greenstone belt, West Greenland: geochemical evidence for intra-oceanic subduction zone processes in the early Earth. *Chemical Geology*, **184**, 231–254.

Polat, A., Kerrich, R., Wyman, D. (1999) Geochemical diversity in oceanic komatiites and basalts from the late Archean Wawa greenstone belt, Superior Province, Canada: trace element and Nd isotope evidence for a heterogeneous mantle. *Precambrian Research*, **94**, 139–173.

Polat, A., Li, J., Fryer, B., Kusky, T., Gagnon, J., Zhang, S. (2005) Geochemical characteristics of the Neoarchean (2800–2700 Ma) Taishan greenstone belt, North China Craton: evidence for plume-craton interaction. *Chemical Geology*, **230**, 60–87.

Prendergast, M. D. (2001) Komatiite-hosted Hunters Road nickel deposit, central Zimbabwe: physical volcanology and sulfide genesis. *Australian Journal of Earth Sciences*, **48**, 681–694.

(2003) The nickeliferous Late Archean Reliance komatiitic event in the Zimbabwe craton-magmatic architecture, physical volcanology, and ore genesis. *Economic Geology*, **98**, 865–891.

Prest, V. K. (1950) Geology of the Keith–Muskego townships area. Ontario Department of Mines, Annual Report, **59**, 44p.

Puchtel, I. S., Brandon, A. D., Humayun, M. (2004a) Precise Pt–Re–Os isotope systematics of the mantle from 2.7-Ga komatiites. *Earth and Planetary Science Letters* **224**, 157–174.

Puchtel, I. S., Brügmann, G. E., Hofmann, A. W. (1999) Precise Re–Os mineral isochron and Pb–Nd–Os isotope systematics of a mafic-ultramafic sill in the 2.0 Ga Onega plateau (Baltic Shield). *Earth and Planetary Science Letters*, **170**, 447–461.

(2001a) [187]Os-enriched domain in an Archean mantle plume: evidence from 2.8 Ga komatiites of the Kostomuksha greenstone belt, NW Baltic Shield. *Earth and Planetary Science Letters*, **186**, 513–526.

Puchtel, I. S., Brügmann, G. E., Hofmann, A. W., Kulikov, V. S., Kulikova, V. V. (2001b) Os-isotope systematics of komatiitic basalts from the Vetreny belt, Baltic Shield: evidence for a chondritic source of the 2.45 Ga plume. *Contribution to Mineralogy and Petrology*, **140**, 588–599.

Puchtel, I. S., Haase, K. M., Hofmann, A. W., Chauvel, C., Kulikov, V. S. (1997) Petrology and geochemistry of crustally contaminated komatiitic basalts from the Vetreny belt, southeastern Baltic Shield: evidence for an early Proterozoic mantle plume beneath the rifted Archean continental lithosphere. *Geochimica et Cosmochimica Acta*, **61**, 1205–1222.

Puchtel, I. S., Hofmann, A. W., Jochum, K. P., Mezger, K., Shchipansky, A. A., Samsonov, A. V. (1997) The Kostomuksha Greenstone Belt, NW Baltic Shield-remnant of a late Archaean oceanic plateau. *Terra Nova*, **9**, 87–90.

Puchtel, I. S., Humayun, M. (2000). Platinum group elements in Kostamuksha komatiites and basalts: implications for oceanic crust recycling and core–mantle interaction. *Geochimica et Cosmochimica Acta*, **64**, 4227–4242.

(2001) Platinum group element fractionation in a komatiitic basalt lava lake. *Geochimica et Cosmochimica Acta*, **65**, 2979–2993.

Puchtel, I. S., Humayun, M., Campbell, A. J., Sproule, R. A., Lesher, C. M. (2004b) Platinum group element geochemistry of komatiites from the Alexo and Pyke Hill areas, Ontario, Canada. *Geochimica et Cosmochimica Acta*, **68**, 1361–1383.

Puchtel, I. S., Humayun, M., Walker, R. J. (2007) Os–Pb–Nd isotope and highly siderophile and lithophile trace element systematics of komatiitic rocks from the Volotsk suite, Baltic Shield, *Precambrian Research*, in press.

Puchtel, I. S., Zhuralev, D. Z. (1989) Petrology and geochemistry of early and late Archean komatiites from Olekma granite greenstone terrain. *28th International Geological Congress, Abstracts*.

Puchtel, I. S., Zhuralev, D. Z., Kulikova, V. V., Samsonov, A. V., Simon, A. K. (1991). Komatiite from the Vodla Block, Baltic Shield: a window to the early Archean mantle. *Doklady Akademia Nauk SSSR*, **317**, 197–203.

Puchtel, I. S., Zhuravlev, D. Z., Samsonov, A. V., Arndt, N. T. (1993) Petrology and geochemistry of metamorphosed komatiites and basalts from the Tungurcha greenstone belt, Aldan Shield. *Precambrian Research*, **62**, 399–418.

Pyke, D. R. (1970) *Geology of the Langmuir and Blackstock Townships, District of Timiskaming: Ontario, Canada*. Toronto: Ontario Department of Mines.

(1975) *On the Relationship of Gold Mineralization and Ultramafic Volcanic Rocks in the Timmins Area: Ontario, Canada*. Ontario Division of Mines Miscellaneous Paper 63. Toronto: Ontario Division of Mines.

(1982) *Geology of the Timmins area*. Ontario Geological Survey Report, 219. Toronto: Ontario Geological Survey.

Pyke, D. R., Naldrett, A. J., Eckstrand, O. R. (1973) Archean ultramafic flows in Munro Township, Ontario. *Bulletin of the Geological Society of America*, **84**, 955–978.

Ransom, B., Byerly, G. R., Lowe, D. R. (1999) Subaqueous to subaerial Archean ultramafic phreatomagmatic volcanism, Kromberg Formation, Barberton Greenstone Belt, South Africa. In: D. R. Lowe and G. R. Byerly (eds.) *Geologic Evolution of the Barberton Greenstone Belt, South Africa*, pp. 151–166, Geological Society of America Special Paper 329. Boulder: Geological Society of America.

Räsänen, J. (1996) Palaeovolcanology of the Sattasvaara komatiites, northern Finland: evidence for emplacement in a shallow water environmen. In: *IGCP Project 336 Symposium in Rovaniemi, Finland, 21–23 August 1996, Program and Abstracts*. pp.69–70. Publications of the Geology Department, University of Turku vol. 38. Turku: University of Turku.

Räsänen, J., Hanski, E., Lehtonen, M. I. (1989) Komatiites, low-Ti basalts and andesites in the Moykkelma area, Central Finnish Lapland. *Geological Survey of Finland, Report of Investigation*, **88**, 1–41.

Ray, G. L., Shimizu, N., Hart, S. R. (1983) An ion microprobe study of the partitioning of trace elements between clinopyroxene and liquid in the system diopside-albite-anorthite. *Geochimica et Cosmochimica Acta*, **47**, 2131–2140.

Redman, B. A., Keays, R. R. (1985) Archaean basic volcanism in the Eastern Goldfields Province, Yilgarn Block, Western Australia. *Precambrian Research*, **30**, 113–152.

Rehkämper, M., Halliday, A. N., Fitton, J. G., Lee, D.-C., Wieneke, M., Arndt, N. T. (1999) Ir, Ru, Pt, and Pd in basalts and komatiites: new constraints for the geochemical behavior of the platinum-group elements in the mantle. *Geochimica et Cosmochimica Acta*, **63**, 3915–3934.

Renner, R. (1989) Cooling and crystallization of komatiite flows from Zimbabwe. Ph.D. thesis, University of Cambridge.

Renner, R., Nisbet, E. G., Cheadle, M. J., Arndt, N. T., Bickle, M. J., Cameron, W. E. (1994) Komatiite flows from the Reliance Formation, Belingwe Belt, Zimbabwe: I. Petrography and mineralogy. *Journal of Petrology*, **35**, 361–400.

Révillon, S., Arndt, N. T., Chauvel, C., Hallot, E. (2000) Geochemical study of ultramafic volcanic and plutonic rocks from Gorgona Island, Colombia: plumbing system of an oceanic plateau. *Journal of Petrology*, **41**, 1127–1153.

Révillon, S., Chauvel, C., Arndt, N. T., Pik, R., Martineau, F., Fourcade, S., Marty, B. (2002) Heterogeneity of the Caribbean plateau mantle source: new constraints from Sr, O and He isotope compositions of olivine and clinopyroxene. *Earth and Planetary Science Letters*, **205**, 91–106.

Rice, A., Moore, J. M. (2001) Physical modeling of the formation of komatiite-hosted nickel deposits and a review of the thermal erosion paradigm. *Canadian Mineralogist*, **39**, 491–503.

Richard, D., Marty, B., Chaussidon, M., Arndt, N. T. (1996) Helium isotope evidence of a lower mantle component in depleted Archean komatiite. *Science*, **273**, 93–95.

Richter, F. M. (1988) A major change in the thermal state of the Earth at the Archaean-Proterozoic boundary: consequences for the nature and preservation of continental lithosphere. *Journal of Petrology*, Special Lithosphere Issue, 39–52.

Rigden, S. B., Ahrens, T. J., Stolper, E. M. (1984) Densities of liquid silicates at high pressure. *Science*, **226**, 1071–1074.

(1988) Shock compression of molten silicate: results for a model basaltic composition. *Journal of Geophysical Research*, **93**, 367–382.

Ripley, E. M. (1986) *Applications of Stable Isotope Studies to Problems of Magmatic Sulfide Ore Genesis with Special Reference to the Duluth Complex, Minnesota.* Heidelberg: Springer-Verlag.

Roddick, J. C. (1984) Emplacement and metamorphism of Archean mafic volcanics of Kambalda, Western Australia–geochemical and isotopic constraints. *Geochimica et Cosmochimica Acta*, **38**, 1305–1318.

Roeder, P. L., Emslie, R. F. (1970) Olivine–liquid equilibrium. *Contributions to Mineralogy and Petrology*, **29**, 275–282.

Rollinson, H. R., Roberts, C. R. (1986) Ratio correlations and major element mobility in altered basalts and komatiites. *Contributions to Mineralogy and Petrology*, **93**, 89–97.

Rosengren, N. M., Beresford, S. W., Grguric, B. A., Cas, R. A. F. (2005) An intrusive origin for the komatiitic dunite-hosted Mt Keith disseminated nickel sulfide deposit, Western Australia. *Economic Geology*, **100**, 149–156.

Rosengren, N. M., Grguric, B. A., Beresford, S. W., Fiorentini, M. L., Cas, R. A. F. (2007) Internal stratigraphic architecture of the komatiitic dunite-hosted MKD5 disseminated nickel sulphide deposit, Mount Keith Domain, Agnew–Wiluna Greenstone Belt, Western Australia. *Mineralium Deposita*, **126**, 821–845.

Ross, J. R., Hopkins, G. M. F. (1975) The nickel sulphide deposits of Kambalda, Western Australia. In: C. Knight (ed.) *Economic Geology of Australia and Papua-New Guinea* Vol 1, Metals, pp. 100–121, Monograph 5. Melbourne: Australian Institute of Mining and Metallurgy.

Ross, J. R., Keays, R. R. (1979) Precious metals in volcanic-type nickel sulphide deposits in Western Australia. I. Relationship with the composition of the ores and their host rocks. *Canadian Mineralogist*, **17**, 417–435.

Russell, M. J., Arndt, N. T. (2005) Geodynamic and metabolic cycles in the Hadean. *Biogeosciences*, **2**, 1–15.

Ryerson, F. J., Weed, H. C., Piwinskii, A. J. (1988) Rheology of subliquidus magmas. *Journal of Geophysical Research*, **93**, 3421–3436.

Saboia, L. A., Teixeira, N. A. (1980) Ultramafic flows of the Crixas greenstone belt, Goias, Brazil. *Precambrian Research*, **22**, 23–40.

Sanborn-Barrie, M., Rogers, N., Skulski, T., Parker, J., McNicoll, V., Devaney, J. R. (2004) Geology and tectonostratigraphic assemblages, East Uchi subprovince, Red lake and Birch-Uchi belts, Ontario. Open File 4256, Geological Survey of Canada.

Sasseville, C., Tomlinson, K. Y. (2000) Tectonostratigraphy, structure, and geochemistry of the Mesoarchean Wallace Lake greenstone belt, southeastern Manitoba, *Geological Survey of Canada, Current Research*, 2000-C14, 9pp.

Satterly, J. (1951a) Geology of Harker Township. In: *Ontario Department of Mines Annual Report, 1951*. Toronto: Ontario Department of Mines.

(1951b) Geology of Munro Township. In: *Ontario Department of Mines Annual Report, 1951*. Toronto: Ontario Department of Mines.

Saunders, A. D., Fitton, J. G., Kerr, A. C., Norry, M. J., Kent, R. W. (1997) The North Atlantic Igneous Province. In: J. J. Mahoney and M. F. Coffin (eds.) *Large Igneous Provinces: Continental, Oceanic and Planetary Flood Volcanism*, pp. 95–122. Washington: American Geophysical Union.

Saunders, A. D., Storey, M., Kent, R. W., Norry, M. J. (1992) Consequences of plume-lithosphere interactions. In: B. C. Storey, T. Alabaster and R. J. Pankhurst (eds.) *Magmatism and the Causes of Continental Break-Up*, pp. 41–60. London: The Geological Society.

Saverikko, M. (1985) The pyroclastic komatiite complex at Sattasvaara in northern Finland. *Bulletin of the Geological Society of Finland*, **57**, 55–87.

(1987) The Lapland greenstone belt: stratigraphic and depositional features in northern Finland. *Bulletin of the Geological Society of Finland*, **59**, 129–154.

Scarfe, C. (1987) Pressure dependence of the viscosity of silica melts. In: B. O. Mysen (ed.). Magmatic Processes: Physicochemical Principles. pp. 59–68. Special Publication No 1. The Geochemical Society.

Schaefer, S., Morton, P. (1991) Two komatiitic pyroclastic units, Superior Province, NW Ontario: their geology, petrography and correlation. *Canadian Journal of Earth Sciences*, **28**, 1455–1470.

Schau, M. (1977) Komatiites and quartzites in the Archaean Price Albert Group. In: W. R. A. Baragar, L. C. Coleman and J. M. Hall (eds.) *Volcanic Regimes in Canada*, pp. 341–354. Ottawa: Geological Society of Canada.

Schwindinger, K. R. (1999) Particle dynamics and aggregation of crystals in a magma chamber with application to Kilauea Iki olivines. *Journal of Volcanology and Geothermal Research*, **88**, 209–238.

Seat, Z., Stone, W. E., Mapleson, D. B., Daddow, B. C. (2004) Tenor variation within komatiite-associated nickel sulphide deposits: insights from the Wannaway Deposit, Widgiemooltha Dome, Western Australia. *Mineralogy and Petrology*, **82**, 317–339.

Self, S., Keszthelyi, L. P., Thordarson, T. (2000) Discussion of: 'Pulsed inflation of pahoehoe lava flows: implications for flood basalt emplacement', by S. W. Anderson, E. R. Stofan, E. R. Smrekar, J. E. Guest and B. Wood [*Earth and Planetary Science Letters*, **168** (1999) 7–18]. *Earth and Planetary Science Letters*, **179**, 421–423.

Self, S., Thordarson, T., Keszthelyi, L., *et al.* (1996) A new model for the emplacement of Columbia River basalts as large, inflated pahoehoe lava flow fields. *Geophysical Research Letters*, **23**, 2689–2692.

(1997) Emplacement of continental flood basalt lava flows. In: J. J. Mahoney and M. F. Coffin (eds.) *Large Igneous Provinces.* pp. 381–410. Washington: American Geophysical Union.

Shaw, H. R. (1969) Rheology of basalt in the melting range. *Journal of Petrology,* **10**, 510–535.

(1972) Viscosities of magmatic liquids: an empirical method of calculation. *American Journal of Science,* **272**, 870–893.

Shima, H., Naldrett, A. J. (1975) Solubility of sulfur in an ultramafic melt and the relevance of the system Fe–S–O. *Economic Geology,* **70**, 960–967.

Shimizu, K., Nakamura, E., Maruyama, S. (2005) The geochemistry of ultramafic to mafic volcanics from the Belingwe Greenstone Belt, Zimbabwe: magmatism in an Archean continental large igneous province. *Journal of Petrology,* **46**, 2367–2394.

Shimizu, K., Nakamura, E., Maruyama, S., Kobayashi, K. (2004) Discovery of Archean continental and mantle fragments inferred from xenocrysts in komatiites, the Belingwe greenstone belts, Zimbabwe. *Geology,* **32**, 285–288.

Shimizu, K. T., Komiya, S., Maruyama, S., Hirose, K. (2001) Water content of melt inclusion in Cr-spinel of 2.7 Ga komatiite from Belingwe Greenstone Belt, Zimbabwe. *Earth and Planetary Science Letters,* **78**, 750.

Shirey, S. B., Barnes, Sarah-Jane (1994) Re–OS and Sm–Nd isotopic constraints on basaltic komatiitic volcanism and magmatic sulphide formation in the Cape Smith Foldbelt, Quebec. *Mineralogical Magazine,* **58A**, 835–836.

(1995) Os–Nd isotope systematics of ultramafic–mafic magmatism: Cape Smith foldbelt and midcontinental rift system. *IGCP Project 336, Proceedings,* Duluth. pp. 175–176.

Shirey, S. B., Hanson, G. N. (1986) Mantle heterogeneity and crustal recycling in Archean granite-greenstone belts: evidence from Nd isotopes and trace elements in the Rainy Lake area, Superior Province, Ontario, Canada. *Geochimica et Cosmochimica Acta,* **50**, 2631–2651.

Shirey, S. B., Walker, R. J. (1995) Carius-tube digestion for low-blank rhenium-osmium analysis. *Analytical Chemistry,* **34**, 2136–2141.

Shore, M. (1996) Cooling and crystallization of komatiite flows. Ph.D. thesis, University of Ottawa.

Shore, M., Fowler, A. D. (1999) The origin of spinifex texture in komatiites. *Nature,* **397**, 691–694.

Silva, K. E., Cheadle, M. J., Nisbet, E. G. (1997) The origin of B1 zones in komatiites. *Journal of Petrology,* **11**, 1565–1584.

Skufin, P. K., Theart, H. F. J. (2005) Geochemical and tectono-magmatic evolution of the volcano-sedimantary rocks of Pechenga and other greenstone fragments within the Kola Greenstone Belt, Russia. *Precambrian Research,* **141**, 1–48.

Sleep, N. H., Windley, B. F. (1982) Archaean plate tectonics: constraints and inferences. *Journal of Geology,* **90**, 363–379.

Smith, A. D., Ludden, J. N. (1989) Nd isotopic evolution of the Precambrian mantle. *Earth and Planetary Science Letters,* **93**, 14–22.

Smith, C. B., Gurney, J. J., Skinner, E. M. W., Clement, C. R., Ebrahim, N. (1985) Geochemical character of southern African kimberlites: a new approach based on isotopic constraints. *Transactions of the Geological Society of South Africa,* **88**, 267–280.

Smith, H. S., Erlank, A. J. (1982) Geochemistry and petrogenesis of komatiites from the Barberton greenstone belt, South Africa. In: N. T. Arndt and E. G. Nisbet (eds.) *Komatiites,* pp. 347–398. London: George Allen & Unwin.

Smith, H. S., Erlank, A. J., Duncan, A. R. (1980) Geochemistry of some ultramafic komatiite lava flows from the Barberton Mountain Land, South Africa. *Precambrian Research*, **11**, 399–415.

Smith, H. S., O'Neil, J. R., Erlank, A. J. (1984) Oxygen isotope compositions of minerals and rocks and chemical alteration patterns in pillow lavas from the Barberton greenstone belt, South Africa. In: A. Kröner, G. N. Hanson and A. M. Goodwin (eds.) *Archaean Geochemistry*, pp. 115–138. Berlin: Springer-Verlag.

Sobolev, A. V., Hofmann, A. W., Sobolev, S. V., Nikogosian, I. K. (2005) An olivine free mantle source of Hawaiian shield basalts. *Nature*, **434**, 590–593.

Sobolev, A. V., Hofmann, A. W., Yaxley, G., *et al.* (2007) Estimating the amount of recycled crust in sources of mantle-derived melts. *Science*, **316**, 412–417.

Sobolev, A. V., Kamenetskiy, V. S., Kononkova, N. N. (1991) New data on Siberian meymechite petrology. *Geokhimiya*, **8**, 1084–1095.

Sobolev, V. S., Kostyuk, V. P. (1975) Magmatic crystallization based on the study of melt inclusions, partial translation in 'Fluid Inclusion Research'. *Proceedings COFFI* 9, 182–253.

Sorjonen-Ward, P., Nironen, M., Luukkonen, E. (1997) Greenstone associations in Finland. In: M. J. de Wit and L. D. Ashwal (eds.) *Greenstone Belts*. pp. 677–698. Oxford: Oxford University Press.

Sproule, R. A., Lesher, C. M., Ayer, J. A., Thurston, P. C. (2002) Secular variations in the geochemistry of komatiitic rocks from the Abitibi Greenstone Belt, Canada. *Precambrian Research*, **115**, 153–186.

Sproule, R. A., Lesher, C. M., Houlé, M., Keays, R. R., Thurston, P. C., Ayer, J. A. (2005) Chalcophile element geochemistry and metallogenesis of komatiitic rocks in the Abitibi Greenstone Belt, Canada. *Economic Geology*, **100**, 1169–1190.

St-Onge, M. R., Lucas, S. B. (1990) Evolution of the Cape Smith Belt: early Proterozoic continental underthrusting, ophiolite obduction, and thick-skinned folding. In: J. F. Lewry and M. R. Stauffer (eds.) *The Early Proterozoic Trans-Hudson Orogen of North America*, pp. 313–352. Ottawa: Geological Association Canada.

(1993a) Controls on the regional distribution of iron-nickel-copper-platinum-group element mineralization in the eastern Cape Smith Belt, Quebec. *Canadian Journal of Earth Sciences*, **31**, 206–218.

(1993b) Geology of the Eastern Cape Smith Belt: Parts of the Kangiqsujuaq, cratère du Nouveau-Québec and Lacs Nuvilik Map Areas, Quebec. *Geological Survey of Canada, Memoir*, **438**, 1–110.

St. Onge, M. R., Lucas, S. B., Scott, D. J. (1997) The Ungava Orogen and the Cape Smith Thrust Belt. In: M. J. De Wit and L. D. Ashwal (eds.), *Greenstone Belts*, pp. 772–780. Oxford: Oxford University Press.

Stamatelopoulou-Seymour, K., Francis, D. M. (1980) An Archean ultramafic turbidite from Lac Guyer, James Bay area, Quebec. *Canadian Journal of Earth Sciences*, **17**, 1576–1582.

Stolper, E. M., Walker, D., Hager, B. H., Hays, J. F. (1981) Melt segregation from partially molten source regions: the importance of melt density and source region size. *Journal of Geophysical Research*, **86**, 6261–6271.

Stolz, G. W., Nesbitt, R. W. (1981) The komatiite nickel sulfide association at Scotia: a petrochemical investigation of the ore environment. *Economic Geology*, **76**, 1480–1502.

Stone, M. S., Stone, W. E. (2000) A crustally contaminated komatiitic dyke–sill–lava complex, Abitibi greenstone belt, Ontario. *Precambrian Research*, **102**, 21–46.

Stone, W. E., Archibald, N. J. (2004) Structural controls on nickel sulphide ore shoots in Archaean komatiite, Kambalda, WA: the volcanic trough controversy revisited. *Journal of Structural Geology*, **26**, 1173–1194.

Stone, W. E., Crocket, J. H., Fleet, M. E. (1993) Sulfide-poor platinum-group mineralization in komatiitic systems: Boston Creek Flow, layered basaltic komatiite, Abitibi Belt, Ontario. *Economic Geology*, **88**, 817–836.

Stone, W. E., Crocket, J. H., Fleet, M. E. (1995) Differentiation processes in an unusual iron-rich alumina-poor Archean ultramafic/mafic igneous body. *Contributions to Mineralogy and Petrology*, **119**, 287–300.

Stone, W. E., Deloule, E., Beresford, S., Fiorentini, M. (2005) Anomalously high δD values in an Archean ferropicritic melt: implications for magma degassing. *Canadian Mineralogist*, **43**, 1745–1758.

Stone, W. E., Deloule, E., Larson, M. S., Lesher, C. M. (1997) Evidence for hydrous high-MgO melts in the Precambrian. *Geology*, **25**, 143–146.

Stone, W. E., Heydari, M., Seat, Z. (2004) Nickel tenor variations between Archaean komatiite-associated nickel sulphide deposits, Kambalda ore field, Western Australia: the metamorphic modification model revisited. *Mineralogy and Petrology*, **82**, 295–316.

Stone, W. E., Jensen, L. S., Church, W. R. (1987) Petrography and geochemistry of an unusual Fe-rich basaltic komatiite from Boston Township, northeastern Ontario. *Canadian Journal of Earth Sciences*, **24**, 2537–2550.

Stone, W. E., Masterman, E. E. (1998) Kambalda nickel deposits. In: D. A. Berkman and D. H. Mackenzie (eds.) *Economic Geology of Australia and Papua New Guinea*, pp. 347–356. Melbourne: Australasian Institute of Mining and Metallurgy.

Stone, W. R., Deloule, E., Stone, M. S. (2003) Hydromagmatic amphibole in komatiitic, tholeiitic and ferropicritic units, Abitibi greenstone belt, Ontario and Quebec: evidence for Archaean wet basic and ultrabasic melt. *Mineralogy and Petrology*, **77**, 39–65.

Storey, M., Mahoney, J. J., Kroenke, L. W., Saunders, A. D. (1991) Are oceanic plateaus sites of komatiite formation? *Geology*, **19**, 376–379.

Sun, S.-S. (1984) Geochemical characteristics of Archean ultramafic and mafic volcanic rocks: implications for mantle composition and evolution. In: A. Kröner, G. N. Hanson and A. M. Goodwin (eds.) *Archaean Geochemistry*, pp. 25–47. Berlin: Springer-Verlag.

Sun, S.-S., Nesbitt, R. W., McCulloch, M. T. (1989) Geochemistry and petrogenesis of Archaean and early Proterozoic siliceous high magnesian basalts. In: A. J. Crawford (ed.) *Boninites and Related Rocks*, pp. 148–173. London: Allen and Unwin.

Sun, S. S., Nesbitt, R. W. (1978) Petrogenesis of Archean ultrabasic and basic volcanics: evidence from rare earth elements. *Contributions to Mineralogy and Petrology*, **65**, 301–325.

Swager, C. P., Goleby, B. R., Drummond, B. J., Rattenbury, M. S., Williams, P. R. (1997) Crustal structure of granite-greenstone terranes in the Eastern Goldfields, Yilgarn craton as revealed by seismic reflection profiling. *Precambrian Research*, **83**, 43–56.

Swager, C. P., Griffin, T. J., Witt, W. K., *et al.* (1990) Geology of the Archaean Kalgoorlie terrane – an explanatory note. *Western Australian Geological Survey Record*, 1990/12.

Swanson, D. A. (1973) Pahoehoe flows from the 1969–1971 Mauna Ulu eruption, Kilauea Volcano, Hawaii. *Geological Society of America Bulletin*, **84**, 615–626.

Swanson, D. A., Wright, T. L., Helz, R. T. (1975) Linear vent systems and estimated rates of magma production and eruption for the Yakina Basalt on the Columbia Plateau. *American Journal of Science*, **275**, 877–905.

Sylvester, P. J., Harper, G. D., Byerly, G. R., Thurston, P. C. (1997) Volcanic aspects. In: M. J. de Wit and L. D. Ashwal (eds.) *Greenstone Belts*, pp. 55–90. Oxford: Oxford University Press.

Sylvester, P. J., Kamenetsky, V., McDonough, W. F. (2000) Melt inclusion evidence for komatiite genesis in the Gorgona plume. *Journal of Conference Abstracts*, **5**, 975.

Takahashi, E. (1986) Melting of a dry peridotite KLB1 up to 14 GPa: implications on the origin of peridotite upper mantle. *Journal of Geophysical Research*, **91**, 9367–9382.

Takahashi, E., Kushiro, I. (1983) Melting of a dry peridotite at high pressures and basalt magma genesis. *American Mineralogist*, **68**, 859–879.

Takahashi, E., Scarfe, C. M. (1985) Melting of peridotite to 14 GPa and the genesis of komatiite. *Nature*, 315.

Takahashi, E., Shimazaki, T., Tsuzaki, Y., Yoshida, H. (1993) Melting study of a peridotite KLB-1 to 6.5 GPa, and the origin of basaltic magmas. *Philosophical Transactions of the Royal Society of London*, A **342**, 105–120.

Thompson, M. E., Lowe, D. R., Byerly, G. R. (2005) Komatiitic tuffs of the 3.5–3.2 Ga Onverwacht Group of the southern Barberton greenstone belt, South Africa. *Geological Society of America Abstracts with Programs*, **37**, 292.

Thompson, P. M. E., Kempton, P. D., White, R. V., et al. (2004) Hf–Nd isotope constraints on the origin of the Cretaceous Caribbean plateau and its relationship to the Galapagos plume. *Earth and Planetary Science Letters*, **217**, 59–75.

Thompson, R. N (1983) Book Review: Komatiites. *Journal of Petrology*, **24**, 319–320.

Thompson, R. N., Gibson, S. A., Dickin, A. P., Smith, P. M. (2001) Early Cretaceous basalt and picrite dykes of the southern Etendeka region, NW Namibia: windows into the role of the Tristan mantle plume in Parana–Etendeka Magmatism. *Journal of Petrology*, **42**, 2049–2081.

Thomson, B. (1989) Petrology and stratigraphy of some texturally well preserved thin komatiites from Kambalda, Western Australia. *Geological Magazine*, **126**, 249–261.

Thordarson, T., Self, S. (1998) The Roza Member, Columbia River Basalt Group – a gigantic pahoehoe lava flow field formed by endogenous processes. *Journal of Geophysical Research*, **103**, 27411–27445.

Thurston, P. C. (2002) Autochthonous development of Superior Province greenstone belts? *Precambrian Research*, **115**, 11–36.

Tomlinson, K. Y., Hughes, D. J., Thurston, P. C., Hall, R. P. (1999) Plume magmatism and crustal growth at 2.9 and 3.0 Ga in the Steep Rock and Lumby Lake area, western Superior Province. *Lithos*, **46**, 103–136.

Toplis, M. J. (2005) The thermodynamics of iron and magnesium partitioning between olivine and liquid: criteria for assessing and predicting equilibrium in natural and experimental systems. *Contributions to Mineralogy and Petrology*, **149**, 22–39.

Tourpin, S., Gruau, G., Blais, S., Fourcade, S. (1991) Resetting of REE and Nd and Sr isotopes during carbonatization of a komatiite flow from Finland. *Chemical Geology*, **90**, 15–29.

Trofimovs, J., Davis, B. K., Cas, R. A. F. (2004) Contemporaneous ultramafic and felsic intrusive and extrusive magmatism in the Archaean Boorara Domain, Eastern Goldfields Superterrane, Western Australia, and its implications. *Precambrian Research*, **131** (3–4), 283–304.

Trofimovs, J., Tait, M. A., Cas, R. A. F., McArthur, A., Beresford, S. W. (2003) Can the role of thermal erosion in strongly deformed komatiite–Ni–Cu–(PGE) deposits be determined? Perseverance, Agnew-Wiluna Belt, Western Australia. *Australian Journal of Earth Sciences*, **50**, 199–214.

Tsomondo, J. M. (1995) *The Shurugwi Chromite Deposits; New Perspectives and Exploration Outlook*, pp. 25–37. Special Publication 3. Geological Society of Zimbabwe.

Turek, A., Compston, W. (1971) Rubidium–strontium geochronology in the Kalgoorlie region. *Geological Society Australia, Special Publication*, **3**, 72–74.

Turner, J. S., Huppert, H. E., Sparks, R. S. J. (1986) Komatiites II: experimental and theoretical investigations of post-emplacement cooling and crystallization. *Journal of Petrology*, **27**, 397–437.

Ukstins, I. A., Renne, P. R., Wolenden, E., Baker, J., Ayalew, D., Menzies, M. (2002) Matching conjugate volcanic rifted margins: $^{40}Ar/^{39}Ar$ chrono-stratigraphy of pre-and syn-rift bimodal flood volcanism in Ethiopia and Yemen. *Earth and Planetary Science Letters*, **198**, 289–306.

Urbain, G., Bottinga, Y., Richet, P. (1982) Viscosity of liquid silica, silicates and alumino-silicates. *Geochemica et Cosmochimica Acta*, **46**, 1061–1072.

Usselman, T. M., Hodge, D. S., Naldrett, A. J., Campbell, I. H. (1979) Physical constraints on the characteristics of nickel-sulfide ore in ultramafic lavas. *Canadian Mineralogy*, **17**, 361–372.

Van Kranendonk, M. J. (2006) Volcanic degassing, hydrothermal circulation and the flourishing of early life on Earth: A review of the evidence from c. 3490–3240 Ma rocks of the Pilbara Supergroup, Pilbara Craton, Western Australia. *Earth-Science Reviews*, **74**, 197–240.

Van Kranendonk, M. J., Collins, W. J., Hickman, A. H., Pawley, M. J. (2004) Critical tests of vertical vs. horizontal tectonic models for the Archaean East Pilbara Granite-Greenstone Terrane, Pilbara Craton, Western Australia. *Precambrian Research*, **131**, 173–211.

Vasil'yev, Y. R., Zolotukhin, V. V. (1975) *Petrologiya ul'trabazitov severa Siberskoy platformy i nekotoryye problemy ikh genezisa [Petrology of the Ultrabasites in the North Siberian Platform and Some Problems of Their Origin]* (in Russian). Novosibirsk: Nauka.

Vervoort, J., Blichert-Toft, J. (1999) Evolution of the depleted mantle: Hf isotope evidence from juvenile rocks through time. *Geochimica et Cosmochimica Acta*, **63**, 533–556.

Vervoort, J. D., Patchett, P. J., Gehrels, G. E., Nutman, A. P. (1996) Constraints on early Earth differentiation from hafnium and neodymium isotopes. *Nature*, **379**, 624–627.

Viljoen, M. J., Viljoen, R. P. (1969a) Archaean vulcanity and continental evolution in the Barberton region, Transvaal. In: T. N. Clifford and I. Gass (eds.) *African Magmatism and Tectonics*, pp. 27–39. Edinburgh: Oliver and Boyd.

(1969b) Evidence for the existence of a mobile extrusive peridotitic magma from the Komati Formation of the Onverwacht Group. *Geological Society of South Africa, Special Publication*, **21**, 87–112.

(1969c) The geology and geochemistry of the lower ultramafic unit of the Onverwacht Group and a proposed new class of igneous rocks. *Geological Society of South Africa, Special Publication*, **21**, 55–85.

Viljoen, M. J., Viljoen, R. P., Smith, H. S., Erlank, A. J. (1983) Geological, textural, and geochemical features of komatiitic flows from the Komati formation. *Geological Society of South Africa, Special Publication*, **9**, 1–20.

Viljoen, R. P., Viljoen M. J. (1982) Komatiites – an historical review. In: N. T. Arndt and E. G. Nisbet (eds.) *Komatiites*. pp. 5–18. London: George Allen and Unwin.

Visser, D. J. L. (1956) The geology of the Barberton Area. *Geological Survey of South Africa, Special Publication,* **15**, 1–253.

Vlaar, N. J. (1986) Archaean global dynamics. *Geologie en Mijnbouw,* **65**, 91–101.

Walker, D., Kirkpatrick, R. J., Longhi, J., Hays, J. F. (1976) Crystallization history of lunar picritic basalt sample 12002: phase equilibria and cooling rate studies. *Bulletin of the Geological Society of America,* **87**, 646–656.

Walker, G. P. L. (1970) Compound and simple lava flows and flood basalts. *Bulletin of Volcanology,* **35**, 579–590.

Walker, R., Shirey, S. B., Stecher, O. (1988) Comparative Re–Os, Sm–Nd and Rb–Sr isotope and trace element systematics from Archean komatiite flows from Munro Township, Abitibi Belt, Ontario. *Earth and Planetary Science Letters,* **87**, 1–12.

Walker, R. J., Echeverría, L. M., Shirey, S. B., Horan, M. F. (1991) Re–Os isotopic constraints on the origin of volcanic rocks, Gorgona Island, Colombia: Os isotopic evidence for ancient heterogeneities in the mantle. *Contributions to Mineralogy and Petrology,* **107**, 150–162.

Walker, R. J., Morgan, J. W., Horan, M. F. (1995) [187]Os enrichment in some mantle plume sources: evidence for core-mantle interaction? *Science,* **269**, 819–822.

Walker, R. J., Nisbet, E. G. (2002) [187]Os isotopic constraints on Archean mantle dynamics. *Geochimica et Cosmochimica Acta,* **66**, 3317–3325.

Walker, R. J., Storey, M., Kerr, A. C., Tarney, J., Arndt, N. T. (1999) Implications of [187]Os isotopic heterogeneities in a mantle plume: evidence from Gorgona Island and Curaçao. *Geochimica et Cosmochimica Acta,* **63**, 713–728.

Walter, M. J. (1998) Melting of garnet peridotite and the origin of komatiites and depleted lithosphere. *Journal of Petrology,* **39**, 29–60.

Wei, J. F., Tronnes, R. G., Scarfe, C. M. (1990) Phase relations of alumina-undepleted and alumina-depleted komatiites at pressures of 4–12 GPa. *Journal of Geophysical Research,* **95**, 15817–15828.

Wendlandt, R. F. (1982) Sulfide saturation of basalt and andesite melts at high pressures and temperature. *American Mineralogist,* **67**, 877–885.

Whitford, D. J., Arndt, N. T. (1977) Rare-earth element abundances in a thick, layered komatiite lava flow from Ontario, Canada. *Earth and Planetary Science Letters,* **41**, 188–196.

Wiles, J. W. (1957) The geology of the eastern portion of the Hartley gold belt. *Bulletin of the Geological Survey of Southern Rhodesia,* 44.

Williams, D. A., Kerr, R. C., Lesher, C. M. (1998) Emplacement and erosion by Archean komatiite lava flows at Kambalda: revisited. *Journal of Geophysical Research – Solid Earth,* **103**, 27533–27549.

(1999) Thermal and fluid dynamics of komatiitic lavas associated with magmatic Ni–Cu–(PGE) sulphide deposits. In: R. R. Keays, C. M. Lesher, P. C. Lightfoot and C. E. G Farrow (eds.) *Dynamic Processes in Magmatic Ore Deposits and Their Application to Mineral Exploration,* pp. 367–412, Geological Association of Canada Short Course, vol. 13. Ottawa: Geological Association of Canada.

Williams, D. A., Kerr, R. C., Lesher, C. M., Barnes, S. J. (2002) Analytical/numerical modeling of komatiite lava emplacement and thermal erosion at Perseverance, Western Australia. *Journal of Volcanology and Geothermal Research,* **110**, 27–55.

Williams, D. A. C. (1972) Archaean ultramafic, mafic and associated rocks, Mt Monger, Western Australia. *Journal of the Geological Society of Australia,* **19**, 163–188.

Williams, D. A. C., Furnell, R. G. (1979) Reassessment of part of the Barberton type area. *Precambrian Research*, **9**, 325–347.

Wilson, A. H. (1992) The geology of the Great Dyke, Zimbabwe: crystallization, layering and cumulate formation in the P1 pyroxenite of Cyclic Unit 1 of the Darwendale Subchamber. *Journal of Petrology*, **33**, 611–663.

Wilson, A. H. (2003) A new class of silica enriched, highly depleted komatiites in the southern Kaapvaal Craton, South Africa. *Precambrian Research*, **127**, 125–141.

Wilson, A. H., Carlson, R. W. (1989) A Sm–Nd and Pb isotopic study of Archaean greenstone belts in the southern Kaapvaal Craton, South Africa. *Earth and Planetary Science Letters*, **96**, 89–105.

Wilson, A. H., Shirey, S. B., Carlson, R. W. (2003) Archaean ultra-depleted komatiites formed by hydrous melting of cratonic mantle. *Nature*, **423**, 858–861.

Wilson, A. H., Versfeld, J. A., Hunter, D. R. (1989) Emplacement, crystallization and alteration of spinifex-textured komatiitic basalt flows in the Archaean Nondweni greenstone belt, southern Kaapvaal Craton, South Africa. *Contributions to Mineralogy and Petrology*, **101**, 301–317.

Woodall, R., Travis, G. A. (1970) *The Kambalda Nickel Deposits. Ninth Commonwealth Mining and Metallurgy Congress, 1969, London*, **2**, 517–533.

Wooden, J. L., Czamanske, G. K., Fedorenko, V. A., *et al.* (1993) Isotopic and trace-element constraints on mantle and crustal contributions to Siberian continental flood basalts, Noril'sk area, Siberia. *Geochimica et Cosmochimica Acta*, **57**, 3677–3704.

Woodhead, J., Kent, A. J., Hergt, J., Bolhar, R., Rowe, M. C. (2005) Volatile contents of komatiite magmas and the Archaean mantle: insights from melt inclusions in komatiites. *American Geophysical Union, Fall Meeting 2005*, abstract V52A-05.

Wyman, D., Kerrich, R., Polat, A. (2002) Assembly of Archean cratonic mantle lithosphere and crust: plume-arc interaction in the Abitibi–Wawa subduction–accretion complex. *Precambrian Research*, **115**, 37–62.

Wyman, D. A. (1999) A 2.7 Ga depleted tholeiite suite: evidence of plume-arc interaction in the Abitibi Greenstone Belt, Canada. *Precambrian Research*, **97**, 27–42.

Wyman, D. A., Bleeker, W., Kerrich, R. (1998) A 2.7 Ga komatiite low/Ti tholeiite arc tholeiite transition and inferred proto-arc geodynamic setting of the Kidd Creek deposit: evidence from precise ICP MS trace element data. *Economic Geology, Monograph*, **10**, 511–528.

Xie, Q., Kerrich, R. (1994) Silicate-perovskite and majorite signature komatiites from the Archean Abitibi greenstone belt: Implications for early mantle differentiation and stratification. *Journal of Geophysical Research*, **99**, 15799–15812.

Xie, Q., Kerrich, R., Fan, J. (1993) HFSE/REE fractionations recorded in three komatiite-basalt sequences, Archean Abitibi greenstone belt: implications for multiple plume sources and depths. *Geochimica et Cosmochimica Acta*, **57**, 4111–4118.

Zahnle, K., Arndt, N. T., Cockell, C., *et al.* (2007) Emergence of a habitable planet. In: K. Fishbaugh, D. des Marais, O. Korablev, P. Lognonne, F. Raulin (eds.) *Geology, and Habitability of Terrestrial Planets*, pp. 35–78. New York: Springer.

Zhu, W. G., Xie, M. S., Xu, J., *et al.* (1999) Experimental studies on silicate structures of basaltic glasses quenched at 1650°C and 1–3.5 GPa. *Chinese Science Bulletin*, **44**, 461–466.

Zindler, A. (1982) Nd and Sr isotopic studies of komatiites and related rocks. In: N. T. Arndt and E. G. Nisbet (eds.) *Komatiites*, pp. 399–420. London: George Allen & Unwin.

General index

Localities